Random Signal
Processing

Random Signal Processing

DWIGHT F. MIX
UNIVERSITY OF ARKANSAS

PRENTICE HALL, Englewood Cliffs, New Jersey 07632

Library of Congress Cataloging-in-Publication

Mix, Dwight F.
    Random signal processing / Dwight F. Mix.
        p.   cm.
    Includes index.
    ISBN 0-02-381852-2
    1. Signal processing—Statistical methods.   2. Stochastic
processes.   I. Title.
    TK5102.9.M59   1995
    621.382'2—dc20   94-15146
                                                            CIP

Editor: Linda Ratts
Production Supervisor: Margaret Comaskey
Production Manager: Francesca Drago
Text Designer: Robert Freese
Cover Designer: Brian Deep

© 1995 by Prentice-Hall, Inc.
A Simon & Schuster Company
Englewood Cliffs, New Jersey 07632

The author and publisher of this book have used their best efforts in
preparing this book. These efforts include the development, research,
and testing of the theories and programs to determine their effectiveness.
The author and publisher shall not be liable in any event for incidental
or consequential damages in connection with, or arising out of, the
furnishing, performance, or use of these programs.

Printed in the United States of America

10 9 8 7 6 5 4 3 2 1

ISBN 0-02-381852-2

Prentice-Hall International (UK) Limited, *London*
Prentice-Hall of Australia Pty. Limited, *Sydney*
Prentice-Hall Canada Inc., *Toronto*
Prentice-Hall Hispañoamericana, S.A., *Mexico*
Prentice-Hall of India Private Limited, *New Delhi*
Prentice-Hall of Japan, Inc., *Tokyo*
Simon & Schuster Asia Pte. Ltd., *Singapore*
Editora Prentice-Hall do Brasil, Ltda., *Rio de Janeiro*

Dedicated to

Kelly C. Overman

whose influence appears
throughout this text

Contents _____

Preface _____

This text was written for use in a course on random signal analysis taught at the senior-graduate level at the University of Arkansas. Prerequisites are a beginning statistics course and some background in Fourier, Laplace, and z transforms. Although our students have had one course in statistics, we still find it necessary to review probability and random variables, so Chapters 1 through 4 are devoted to these topics. Chapter 5 reviews the special topics in signal analysis that we need for an understanding of Chapters 6 through 9, which concentrate on random signals.

In using preliminary editions of this text in the classroom I have found there to be more material here than can be covered in one semester. This should give some flexibility to the instructor based on individual requirements. For example, if there is no prerequisite of statistics for the course, more time can be spent on probability and random variables in the first four chapters. For others, it may be possible to review the first part of the text rapidly and then spend more time on applications.

My intent has been to write this text for the students, with the hope that the less formal style, historical insights, and the order and method of presentation will be helpful in assimilating the many concepts contained herein.

For the Student

Ralph Waldo Emerson said, "Thinking is the hardest work in the world. That's why so few of us do it." Of course he meant original thinking or learning, not idle thinking or day-dreaming. There are three parts to learning: familiarity, classification, and association. We first become familiar with a new concept in the sense that we learn to recognize the name or the activity. We then classify the concept into a familiar category, and then we put it in its proper place (association) as we become more familiar with it through use. When new concepts have many aspects to them, this learning can take a great deal of effort. Such is the case for many concepts in random signal analysis.

The curriculum subjects that are most difficult are calculus, electromagnetics, and random signal analysis. Calculus is difficult because it is the first exposure to continuous math. Although we live in an analogue world, the math we study in elementary school is discrete. Not until high school or college are we exposed to calculus. Electromagnetics is difficult because the math is four-dimensional—three spatial dimensions plus time. Then, to add insult to injury, the field of numbers is complex, which in effect doubles the number of dimensions. Random signal analysis is difficult for the same reasons calculus is. There are a lot of new concepts, and most of them are interrelated. It seems that there is a new definition on each page, certainly in each section of the book. And many terms have names you have never heard before. Have you ever heard the word "ergodic"?

My best advice to you is to keep a log of new terms and concepts. Write down these new things in a special notebook and review them briefly from time to time. Otherwise there is real danger of becoming lost. But don't let me scare you. There is really nothing difficult in all this (as there is in electromagnetics); there is just a wealth of new terms.

Dwight F. Mix

CHAPTER 1 _____

Introduction

1.1 What Is Random Signal Processing? _____

Random signal processing is the study of systems with random input signals. We will concentrate on electrical systems (computer circuits, RLC circuits, etc.) with random input signals, but random signal processing applies to vibration analysis of tall buildings, safety monitoring of nuclear power plants, and many other applications where randomness is present.

We will study two types of systems. One type modifies the input signal to produce an output signal. A radio modifies the input signal and produces rock-and-roll music. A long-distance telephone circuit tries to duplicate the input voice signal in the ear of the listener. In each of these examples, an input signal produces an output signal.

A second type of system makes a decision about the input signal. The output is not a function of time, but a decision or declaration about the input signal. The input to a sonar system in a submarine is the sea noise plus the "ping" from an echo of the transmitted signal. You are all familiar with the vintage movie where the sonar operator says to Clark Gable, "I think I have something, Captain." Gable turns to the chief petty officer and asks him to "have a listen." All are silent as the Chief puts on the earphones, listens intently, and announces it is only a big fish. But we know from the music that next time will be different. The point is that the system makes a decision about the presence or absence of an enemy ship. Just because a person makes the decision does not change the nature of the system. Since the days of Clark Gable we have tried to replace the person with a black box for making decisions, but progress has been slow. People are good at making decisions if they are alert and if they are provided with good information.

An example of a self-contained system is a police officer's radar gun. The Doppler signal is the input, and the number representing the speed of

1

the approaching car is the output. A human operator makes no decision here: The radar gun determines the speed of the car.

We will include deterministic input signals in our study because many of the techniques for random signals apply to deterministic signals. Equations describe deterministic signals, and beginning courses depend on this description. Since we cannot write an equation for a random signal, we must abandon those procedures that require such a formula. But many techniques of system analysis apply to both random and deterministic signals. It is these techniques on which we will concentrate.

After a bit more introductory material in Sections 1.2 and 1.3, we will embark on a study of probability (Chapter 2), random variables (Chapters 3, 4, and 5), and stochastic processes (Chapter 6). Then we will consider applications to estimation (Chapters 7 and 8) and decision-making systems (Chapter 9). But let us first consider just what a random signal is and give some examples in the next two sections.

1.2   Random Signals

Most things can be classified in more than one way. A car can be classified as either large or small, as a sports car or a sedan, and as expensive or cheap. In this spirit, signals can be classified as energy or power signals, as discrete-time or continuous-time signals, and as deterministic or random signals. For now, we are concerned with whether a signal is deterministic or random. There seems to be no consensus on the definition of random signals, so we distinguish deterministic from random signals in the following way. A deterministic signal is one whose future values can be predicted, while a random signal is one whose future values cannot be predicted with complete accuracy. We can usually write an equation for a deterministic signal, such as $x(t) = \cos(\omega t)$, or $x(t) = \exp(at)$, while this is impossible for most random signals. Since an equation determines the value of a signal at any time, we can predict its value at any future time. There may be situations where we have no formula for the signal, but because it is deterministic we can predict its future values. For example, if we observe a periodic signal for 10 periods, we can predict that the next period will be just like the previous ones. If this signal is deterministic, then we will probably be correct.

As we have said, future values of a random signal cannot be predicted with complete accuracy. Now the phrase ''with complete accuracy'' brings up an interesting point. There are signals whose future values can be predicted, but with less than complete accuracy. Think of a World War II fighter pilot flying over enemy territory. His flight path is erratic to avoid flak. Antiaircraft gunners on the ground aim at his predicted position, hoping the plane and the projectile will arrive at the same point in space a short time in the future. Thus the gunners predict future values of the signal. Today an automatic-fire control system not only makes this prediction, but controls the firing of the gun. The signal that represents the flight path is said to be *correlated,* meaning that values of the signal at two different times are statistically related to each other. This signal is not deterministic, but it has some degree of regularity and is not completely random.

Some signals have no correlation between values at different times and are therefore more random than correlated signals. This means that we cannot base our prediction of its future values on past values of the signal. One signal of this type is the static noise in an AM radio when you are trying to listen to a distant station. This is called by several names, including Johnson noise after the scientist who first measured it, and Brownian motion after the famous Scottish botanist Robert Brown, who first observed another form of this same phenomenon.

The interesting situation is the middle one, correlated signals, because we can use their statistics to design systems. We do this by the familiar methods used for frequency-selective filter design, and by methods based on the correlation function of the signal. Most of this text is devoted in one way or another to this problem.

Two ways to describe random signals are by their entropy and by their correlation function. The most natural should be entropy, since randomness is another name for entropy. But correlation is more useful in signal processing because we can analyze and design systems directly in terms of correlation functions, while entropy is more useful in measuring system performance. The following examples show the meaning of correlation, and how it is related to the frequency content of a signal. The concept of entropy will be covered in a later section.

Let us demonstrate how to generate two random signals by tossing a coin. The first signal is uncorrelated, while the second is highly correlated. We will give a precise meaning to the term "correlated" later; for now it simply means "associated." A correlated signal is one whose value at one time is associated with its value at another time. We begin by tossing a coin, and then show how to use a sequence of coin tosses to generate a random signal.

Suppose we toss a coin 12 times and count the number of heads. Figure 1.2.1 shows three trials of this experiment. On the first trial, 4 heads occurred in the 12 tosses, at the first, fourth, ninth, and twelfth tosses. There were 4 heads and 8 tails in the second trial, and 8 heads and 4 tails in the third trial. A head is a pulse above the center line, and tails is below the line.

Let $x_0(n)$ be a digital signal whose value is the number of heads on the nth trial. Then $x_0(1)$ is the number of heads in the first 12 tosses, $x_0(2)$ is the number of heads in the second 12 tosses, etc. Thus the first three values of x_0 are given by

$$x_0(1) = 4$$
$$x_0(2) = 4$$

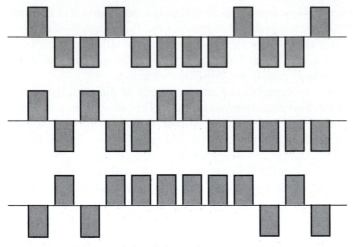

Fig. 1.2.1.   Three trials of the coin-toss experiment.

$$x_0(3) = 8$$

etc.

Figure 1.2.2 shows $x_0(n)$ for 100 values of n. The average or expected value is 6, which means that when we toss a fair coin 12 times we ''expect'' to obtain 6 heads and 6 tails. Notice that $x_0(n)$ appears to have a value of 6 more often than any other value. Notice also that signal values near 6 are more common than values near 0 or 12.

A system that generates $x_0(n)$ is shown in Fig. 1.2.3. The input signal is a binary pulse train, meaning the input pulse is either 1 or 0 with equal probability. The output is the sum of 12 input pulses. The clock circuit

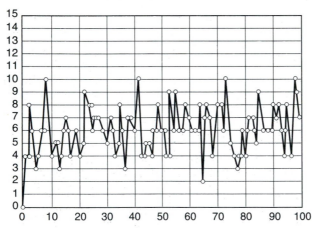

Fig. 1.2.2.   The signal $x_0(n)$.

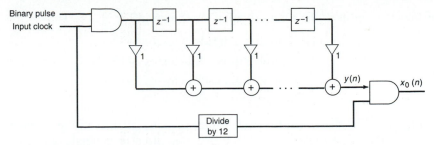

Fig. 1.2.3.   A system to generate $x_0(n)$.

produces one output pulse for every 12 input pulses. The z^{-1} blocks in the diagram denote unit delay; i.e., if the input is $x(n)$, then the output of a delay block is $x(n-1)$.

Any two values of the signal are uncorrelated, meaning that knowledge of one value helps us not a bit to guess the value at some other time. Our best guess for $x_0(n)$ is 6, independent of the value of $x_0(n)$ at any other time. This is not necessarily true for correlated signals, as we will now show.

With a slight change in our method, we can generate a highly correlated signal that we will call $x_{11}(n)$. What we'll do is use the number of heads in tosses 1 through 12 for $x_{11}(1)$, use the heads in tosses 2 through 13 for $x_{11}(2)$, the heads in tosses 3 through 14 for $x_{11}(3)$, etc. We use subscripts 0 and 11 for these two signals because x_0 has no coin tosses in common between signal values, while x_{11} has 11 coin tosses common between successive signal values. Figure 1.2.4 shows the first 18 tosses of the coin from Fig. 1.2.1. The signal $x_{11}(n)$ is again the number of heads in 12 successive tosses

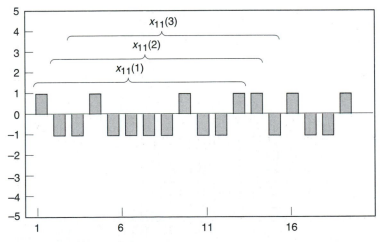

Fig. 1.2.4.   Generating $x_{11}(n)$.

of the coin, but now the tosses overlap, as shown in Fig. 1.2.4. Thus

$$x_{11}(1) = \text{number of heads in tosses } 1{-}12$$
$$x_{11}(2) = \text{number of heads in tosses } 2{-}13$$
$$x_{11}(3) = \text{number of heads in tosses } 3{-}14$$
$$\text{etc.}$$

Figure 1.2.5a shows the first 100 values of this signal. It was generated by using the same sequence of heads and tails used to generate $x_0(n)$, so you can see that every twelfth value of $x_{11}(n)$ is the same as successive values of $x_0(n)$. We can make several other comparisons between the two signals. The most obvious difference is the rapid fluctuations in $x_0(n)$ when compared with the slow meandering of $x_{11}(n)$. Signals with low correlation

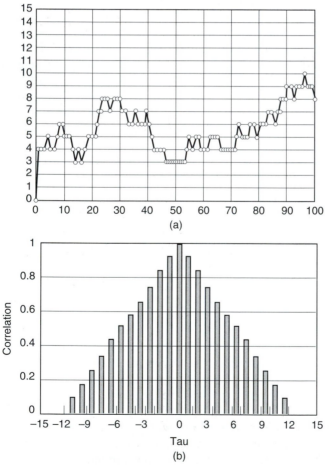

Fig. 1.2.5.   The signal $x_{11}(n)$.

between neighboring values change rapidly with time, and they contain high frequencies. Those with high correlation between neighboring values change slowly from one value to another, which means that their frequency content is limited to low frequencies. Therefore $x_0(n)$ contains higher frequencies than does $x_{11}(n)$.

Notice that $y(n)$ in Fig. 1.2.3 is the signal $x_{11}(n)$. $y(n)$ is the signal at the input of the last AND gate.

Successive values of $x_{11}(n)$ are highly correlated because they have 11 out of 12 tosses of the coin in common. We say that the correlation coefficient is 11/12. Likewise, the correlation coefficient between values of the signal separated by two time units is 10/12, by three time units it is 9/12, continuing in this fashion until the correlation coefficient between signal values separated by 12 time units is 0. These correlation coefficients are tabulated in the following list and are plotted in Fig. 1.2.5b.

Separation, τ	Correlation coefficient, $\rho(\tau)$
0	1
1	11/12
2	10/12
3	9/12
4	8/12
5	7/12
6	6/12
7	5/12
8	4/12
9	3/12
10	2/12
11	1/12
12	0

The correlation coefficients $\rho(\tau)$ are an even function of $\tau = n - k$ in Fig. 1.2.6, where n and k are the times at which $x_{11}(n)$ and $x_{11}(k)$ occur. This means two things. First, $\rho(\tau)$ is a function of the separation $n - k$, regardless of the particular value of n (or k). Second, it does not matter whether we measure from n to k, or from k to n, in measuring the separation of signal values. The correlation coefficient is the same either way.

If we plot the correlation coefficients for $x_0(n)$ they will be 0 for all τ, except for $\tau = 0$ where the value is 1. The correlation function is a delta function, meaning a function that is 0 everywhere except at the origin. Therefore we may summarize the important difference between the two signals by comparing their correlation coefficient functions. The signal with no correlation between any two values has a delta function for its correlation function, and it has the wildest fluctuation with time. This means that it contains much higher frequencies. From this you should begin to see that there is a relationship between the correlation function and the frequency content of a signal.

To emphasize this point, let us look at two additional examples of correlated signals, one with correlation coefficient 0.5 between samples, and one with correlation coefficient 0.75 between adjacent samples. We will call these x_6 and x_9, because x_6 has half of the coin tosses in common between successive samples (that's 6 common tosses), and x_9 has three-fourths of the coin tosses (9 tosses) common between adjacent samples. Figure 1.2.6a shows the signal $x_6(n)$; Fig. 1.2.6b shows its correlation coefficients. Figure 1.2.7a shows the signal $x_9(n)$; Fig. 1.2.7b shows its correlation coefficients. Notice that x_6 has more fluctuations than x_9, and the rate of fluctuation decreases as we progress from x_0 to x_6 to x_9 to x_{11}. The plot of correlation coefficients becomes wider, and the corresponding frequency content decreases, as we progress from x_0 to x_{11}.

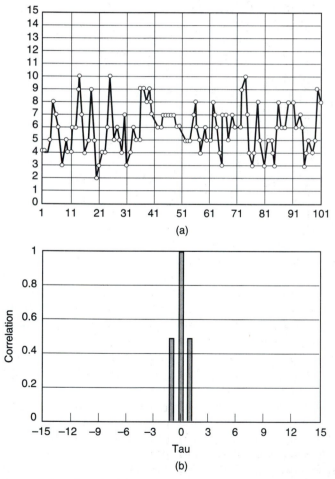

(a)

(b)

Fig. 1.2.6.   The signal $x_6(n)$.

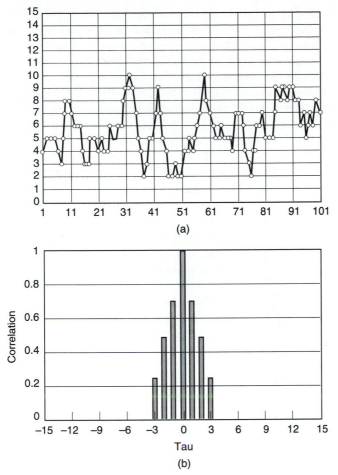

(a)

(b)

Fig. 1.2.7.   The signal $x_9(n)$.

   The predictability of $x_{11}(n)$ is higher than for any of the other signals, meaning that we can guess the next value of the signal with more success. This is also related to the frequency content. Those signals with low frequencies can be predicted with more accuracy than those that fluctuate rapidly.

   Aside from the different rates of fluctuation for these four signals, many characteristics are identical. The expected value of each signal is 6. The total variation of all signals is the same, from about 2 to 10, with values near the mean of 6 occurring more often than those further from the mean. Notice that large deviations from the mean are rare. No signal has a value larger than 10 or less than 2, although it is possible to have values from zero through 12.

With the understanding gained from this example of four signals with different correlation and frequency content, let us ask an important question. How can we use system analysis techniques when we cannot write a formula for the signals that enter and leave systems? Recall your circuit analysis courses. Every technique for system analysis depended on your ability to write a formula for the input and output signal. There are three analysis techniques for linear systems, convolution, transform analysis, and solving difference or differential equations, and all of these depend on a formula for the input signal. So how can we analyze systems with random input signals?

The answer is that we *can* write formulas that describe random signals, but not in the traditional way. The formulas we must use describe the correlation between values of a signal at two different times, or else they describe the power (or energy) density spectrum of the signal. With this type of description for the signal, we must alter our characterization of the system to fit the signal description. Thus, in place of the traditional impulse response $h(n)$, we will use a system description that relates input correlation functions to output correlation functions. And in place of the transfer function $H(z)$, we will use a system description that relates the input spectral density function to the output spectral density function. Once we adapt to this new domain, we will again be able to use many of the techniques of system theory for random signal analysis. The primary purpose of this text is to describe these new procedures.

1.3   Computer Simulation

A computer produced the results of the previous section. No coins were actually tossed; the tosses were only simulated on the computer. A pseudorandom number generator determined "heads" or "tails" with equal probability, and these were used to produce the waveforms and plots of Section 1.2. Our purpose in this section is to describe the various methods used in this process, not so you can duplicate this work, but so you can produce similar results later. The computer is an important tool in investigating random phenomena, and we will use it extensively throughout this text.

To save time and duplication we will introduce pseudocode, meaning code that is neither FORTRAN, C, Pascal, nor any other language, but should be legible to all programmers. Here is a description of pseudocode.

Pseudocode is so named because it resembles languages such as Pascal and C. Unlike actual computer languages, however, which fuss over semicolons, special words, and sometimes upper- and lowercase letters, any verison of pseudocode is acceptable so long as it is unambiguous. The following example will introduce some features of our pseudocode.

All scientific programming depends on looping, a DO loop. The following code segment uses two loops to calculate a sum and then store this sum in an array. Given a 10×12 matrix $A = \{a_{ij}\}$, this algorithm finds the sum of each row and stores this sum in an array labeled b.

```
Procedure rowsum(a)         /* comments appear like this */
define a(10,12), b(10)
   i = 1
   while i ≤ 10    /* loops through the rows */
   begin
      sum = 0
      j = 1
      while j ≤ 12   /* sums each row in array a */
      begin
         sum = sum + a(i,j)
         j = j + 1
      end for j
      b(i) = sum
      i = i + 1
   end for i
   return(b)
end rowsum
```

This routine consists of a title, the input to the procedure (the array a), and the instructions for performing the algorithm. This algorithm consists of a single procedure. The first line of a procedure will consist of the word *procedure,* then the name of the procedure, and then, in parentheses, the parameters supplied to the procedure. These parameters describe the data supplied to the routine. The last line consists of the word *end* followed by the name of the procedure.

The code above translates into FORTRAN as

```
   SUBROUTINE ROWSUM(A,B)
   DIMENSION A(10,12), B(10)
   DO 20 I = 1, 10
   SUM = 0
   DO 21 J = 1, 12
21 SUM = SUM + A(I,J)
   B(I) = SUM
20 CONTINUE
   RETURN
```

In C, the pseudocode becomes

```
rowsum( )
{
    int i,j;    /* we assume float a[i][j] and   */
    float sum;  /* b[i] are defined globally      */

    for(i=0; i<10; i++)
    {
        sum = 0;
        for(j=0; j<12; j++)
            sum += a[i][j];
        b[i] = sum;
        printf("\n%d   %f",i,sum);
    }
}
```

If you program in Pascal or Basic, you are on your own, but the process of translating pseudocode into a workable higher language should be straightforward.

Of the several ways to generate pseudorandom numbers, the simplest and easiest is the linear congruential method:

$$x(n + 1) = [bx(n) + c] \bmod m \qquad (1.3.1)$$

where $x(0)$ = starting value, $x(0) \geq 0$,
  b = the multiplier, $b \geq 0$,
  c = the increment, $c \geq 0$,
  m = the modulus, $m \geq x(0)$, b, and c.

For proper choice of these parameters the sequence of numbers $x(0)$, $x(1)$, $x(2)$, ..., $x(n)$, ... will appear to be random numbers. To see how this works, suppose we start with $x(0) = 757$, $b = 821$, $c = 0$, and $m = 1000$. Then the process is as follows:

  Step 1: Multiply $x(0) * b = 757 * 821 = 621{,}497$
  Step 2: Divide this product by m and set $x(1)$ equal to the remainder, or $x(1) = 497$. (Since $m = 1000$, and since we are using base 10 arithmetic, this is equivalent to simply setting $x(1)$ equal to the last three digits in the product.)
  Step 3: Record the new value of $x = x(1)$ and go to step 1.

The sequence of numbers produced by this algorithm is 757, 497, 37, 377, 517, ... This process can generate any positive number less than 1000 and ending in 7, and all such values seem to be equally likely. We say that the numbers are uniformly distributed over the range (0, 1000), which means that no one number is more likely to be generated than any other number that ends in 7 in this range. The fact that all numbers end in 7 [the same digit as the last digit in the initial $x(0)$] indicates that this is not a good choice of parameters. Associated with this is another problem. There are only 100 different possible numbers less than 1000 and ending in 7. This means that any initial value of $x = x(0)$ must repeat itself in at most 100 iterations of the algorithm. Since the next value of x always depends only on the present value, the sequence of numbers is periodic, with a period of no more than 100.

The properties of the generator depend critically on the parameters, $x(0)$, b, c, and m. The algorithm above is greatly improved if c is changed from 0 to 1. Now the sequence of numbers is 757, 498, 859, 240, 41, 662, ... All numbers in the range from 0 to 999 are possible, not just those ending in 7. If our choice of the other parameters is judicious, the algorithm will have a period of $m = 1000$. The maximum period of any congruential generator is the modulus m.

Notice that we can change the range of the random numbers by dividing by a constant. The range is changed from (0, 1000) to (0, 1) if we divide each random number by m. A program that uses $b = 7821$, $c = 1$, and $m = 65,536$ is shown below, first in pseudocode and followed by versions in FORTRAN and C. They produce numbers in the range $(-0.5, 0.5)$. The two versions are identical except for the differences in language. Here is the pseudocode first. The main program calls the function RAND 20 times to produce 20 random numbers x. The initial $x = x(0)$ is called ISEED in the program.

```
procedure congruential
    iseed = 57
    i = 1
    while i ≤ 20
    begin
        x = rand(iseed) /* call procedure rand */
        print i, x
        i = i + 1
    end for i
end congruential

procedure rand(iseed) /* changes iseed and returns r */
    b = 7821
    m = 65536
    iseed = iseed * b + 1
    r = iseed/m
    return r
end rand
```

Here is the translation of this pseudocode into FORTRAN.

```
      ISEED = 57
      DO 1 I = 1, 20
      X = RAND(ISEED)
  1   WRITE(*,10) I, X
 10   FORMAT(I5,F10.5)
      STOP
      END
C
      FUNCTION RAND(ISEED)
      INTEGER*2 ISEED
C
C     **********************************************
C     * GENERATES RANDOM NUMBERS UNIFORM (-0.5,0.5) *
C     **********************************************
C
      AM = 65536.0
      ISEED = 7821*ISEED + 1
      RAND = FLOAT(ISEED)/AM
      RETURN
      END
```

Notice that ISEED is declared as INTEGER*2. This means that it consists of 16 bits (2 bytes), or 15 binary digits plus the sign bit. Thus when we multiply ISEED by 7821 and store the result in a 2-byte space, we throw away the most significant bits and retain only the 16 least significant bits. The largest possible 16-bit number is 65,535. In binary, this is 16 ones. FORTRAN and C interpret this as 15 ones and the sign bit. Therefore, division of ISEED by 65,536 normalizes the result to a number between -0.5 and $+0.5$. The sequence of numbers produced appears to be random, but of course it is not. With the same initial value of ISEED, the program will produce the same output sequence every time.

The C version of this program is shown below. Integer variables in C are 2 bytes long to begin with, instead of being 4 bytes as in FORTRAN. Hence there is no need to declare iseed as anything except type ''int.'' Otherwise the two programs are as nearly identical as possible under the constraints of different languages. They will produce the same sequence of pseudorandom numbers.

```
#include <math.h>
#include <stdio.h>
```

```
main( )
{
    int i, iseed=57;    /* declare variables */
    float x, rand( );

    for(i=0; i<20; i++)   /* call rand 20 times */
    {
        x = rand(&iseed);
        printf("\n%d  %f",i,x); /* print results */
    }
}


float rand(iseed)
int *iseed;

/************************************************************
*    generates random numbers uniform (-0.5,0.5) *
************************************************************/
{
    int b = 7821;
    float r, m = 65536;
    *iseed = *iseed * b + 1;

    r = *iseed/m;
    return r;
}
```

The period for this program is 65,536. This is the maximum length for this value of $m = 65,536$. We can determine the period by running the program in a loop and looking for the initial value of iseed to reoccur. For a program of maximum length the initial value of iseed is immaterial, because each value can occur only once in the period.

There are other tests one should perform on a random number generator to assure that the sequence is as much like a true random sequence as possible. The output numbers should be uniformly distributed, and successive values in the output sequence should be uncorrelated. If the numbers are uniformly distributed there will be just as many values between 0.2 and 0.3 as there are from 0.3 to 0.4. The chi-square test will reveal when the number of occurrences of equal-length intervals is too far from the expected value, and there are several ways to discover correlation between successive output values. We will discuss one such way in Section 5.2 and calculate the correlation for this number generator there. This particular algorithm does well in these tests, and it is portable from one computer to another. Therefore it should serve us well in the future. If you need to do some serious work involving a random number generator, however, you should use one with a longer period and it should be checked thoroughly before you trust it. Every expert warns that an untested generator, even one supplied with the system, should not be trusted.

Now here is how to generate x_0 and x_{11} from Section 1.2. Recall that x_0 is the signal with no correlation between samples. This means that $x_0(1)$ is the sum of the number of heads in the first 12 tosses, $x_0(2)$ is the sum of the number of heads in the next 12 tosses, etc. There is no overlap as there is for x_{11} or any of the other signals there. Therefore generating x_0 is easy.

Step 1: Let $n = 1$.
Step 2: For $i = 1$ to 12,

```
    {
        generate a random number z.
        if(z > 0) heads
        else tails.
    }
```

Step 3: Let $x_0(n) =$ number of heads in step 2.
Step 4: Let $n = n + 1$ and go to step 2.

Here is the pseudocode to generate 120 values of x_0. It calls procedure rand, so you need to attach that program before running.

```
procedure x0 /* generates 120 values of x₀ */
    iseed = 57
    n = 1
    while n ≤ 120
    begin
        i = 1
        sum = 0
        while i ≤ 12
        begin
            z = rand(iseed) /* calls procedure rand */
            if(z > 0) sum = sum + 1
            i = i + 1
        end for i
        x(n) = sum
        write n, x(n)
        n = n + 1
    end for n
end x0
```

Here is the FORTRAN code to generate 120 values of x_0.

```
      DIMENSION X(120)
      ISEED = 57
      DO 1 N = 1, 120
      SUM = 0.0
      DO 2 I = 1, 12
      Z = RAND(ISEED)
2     IF(Z.GT.0.0) SUM = SUM + 1
      X(N) = SUM
1     WRITE(*,10) N, X(N)
10    FORMAT(I5,F10.5)
      STOP
      END
```

Here is the C version.

```
#include <math.h>
#include <stdio.h>

main( )
{
    int n, i, iseed=57;   /* initialize variables  */
    float x[120], sum, z, rand( );

    for(n=0; n<120; n++) /* generate 120 values of x */
    {
        sum = 0;
        for(i=0; i<12; i++) /* count heads */
        {
            z = rand(&iseed);
            if(z > 0) sum += 1;
        }
        x[n] = sum;       /* x[n] = number of heads */
        printf("\n%d   %f",n,x[n]);
    }
}
```

Generating x_{11} is a bit more complicated. There are several approaches, but we can illustrate one way if you will look back at Fig. 1.2.4. That diagram shows a picture of the approach we will use. We generate a sequence of heads and tails—say, 150 of them. Then we use the first 12 to calculate $x_{11}(1)$, shift one coin toss to the right and use toss 2 through toss 13 to calculate $x_{11}(2)$, and keep going this way until we have calculated 120 values of x_{11}. Here is the pseudocode.

```
procedure x11    /* generates 120 values of x₁₁  */
define toss(150), x(120)
   iseed = 57
   i = 1
   while i ≤ 150 /* generates 150 values of toss(i) */
   begin
      z = rand(iseed)
      toss(i) = 0
      if(z > 0) toss(i) = 1
      i = i + 1
   end for i
   n = 1
   while n ≤ 120
   begin
      sum = 0
      i = n
      while i ≤ n + 12
      begin
         sum = sum + toss(i)
         i = i + 1
      end for i
      x(n) = sum
      print n, x(n)
      n = n + 1
   end for n
end x11
```

Here is the FORTRAN program to do this.

```
      DIMENSION TOSS(150), X(120)
      ISEED = 57
C
C     GENERATE 150 COIN TOSSES
C
      DO 1 I = 1, 150
      Z = RAND(ISEED)
      TOSS(I) = 0.0
      IF(Z.GT.0.0) TOSS(I) = 1.0
    1 CONTINUE
C
C     GENERATE 120 VALUES OF X
C
      DO 2 N = 1, 120
      SUM = 0.0
      DO 3 I = N, N+12
```

```
3   SUM = SUM + TOSS(I)
    X(N) = SUM
2   WRITE(*,10) N, X(N)
10  FORMAT(I5,F10.5)
    STOP
    END
```

Here is the C version of this program.

```c
#include <math.h>
#include <stdio.h>

main( )
{
    int n, i, iseed=57;   /* initialize variables   */
    float toss[150], x[120], sum, z, rand( );

    for(i=0; i<150; i++)   /* generate 150 tosses   */
    {
        z = rand(&iseed);
        if(z > 0) toss[i] = 1;
            else toss[i] = 0;
    }

    for(n=0; n<120; n++)   /* generate 120 values of x */
    {
        sum = 0;
        for(i=n; i<n+12; i++) /* sum 12 tosses */
            sum += toss[i];
        x[n] = sum;
        printf("\n%d %f",n,x[n]);
    }
}
```

Of course you must attach the random number generator to these programs.

With the programs provided for you in this section, plus a basic understanding of how these random number generators perform, you should be prepared to write and run your own programs to simulate the random processes we will encounter in this text. The purpose of this section was twofold: to illustrate how to generate random signals and to provide you with a random number generator for future use.

1.4 Problems

1.1. A histogram represents the number of outcomes that have a specific value, or that fall in bins of specific value. When applied to a random number generator, a histogram of the six outcomes 0.17, −0.16, 0.02, 0.15, 0.32, 0.18 might look like this:

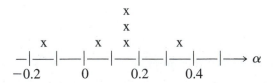

There are three numbers in the bin from 0.1 to 0.2, namely, 0.17, 0.15, and 0.18. There is one number in three other bins, and none in all the others. The idea is that the height of each bin is proportional to the number of outcomes in that bin.

(a) Generate 100 numbers from the subroutine rand (or the generator of your choice) and plot a histogram of the outcomes.

(b) Repeat for 200 numbers from this generator.

Ideally, the height of all bins in the range −0.5 to 0.5 should be equal, but due to the random nature of the generator you should notice that 200 samples gives a more uniform height than 100 samples.

1.2. In Problem 1.1 we used a uniform number generator, meaning that the value of any number in the range −0.5 to 0.5 is equally likely. Here is a way to generate numbers for which this is no longer true. Sum 12 of the numbers from subroutine rand and call this Y_1, sum the next 12 and call this Y_2, continuing in this way until 100 Y values are generated. Thus

$$Y_1 = \sum_{i=1}^{12} X_i$$

$$Y_2 = \sum_{i=13}^{24} X_i$$

$$\vdots$$

$$Y_{100} = \sum_{i=1189}^{1200} X_i$$

Now the largest possible Y value is 6 and the smallest is −6, so these numbers are no longer restricted to the range −0.5 to 0.5. Find and plot a histogram for these 100 Y values. You should make these bins wider than those in Problem 1.1, perhaps of width 0.5. Is the distribution of the Y values uniform, as was that of the X values?

1.3. In Problem 1.2 the Y values have no Xs in common, so they are uncorrelated. Here is a way to generate correlated numbers: Sum the first 12 Xs and call this Z_1, sum 11 of these Xs plus a new X and call this Z_2, continuing in this way until 100 Zs are generated. Thus

$$Z_1 = \sum_{i=1}^{12} X_i$$

$$Z_2 = \sum_{i=2}^{13} X_i$$

$$Z_3 = \sum_{i=3}^{14} X_i$$

$$\vdots$$

$$Z_{100} = \sum_{i=100}^{111} X_i$$

Compare the histogram for Z to the histogram for Y from Problem 1.2. Does being correlated have any effect on the distribution?

1.4. The median is the middlemost or most central item in an ordered set of numbers. For a set of seven numbers, 0.17, −0.16, 0.02, 0.15, 0.32, 0.18, −0.25, we arrange them in ascending order and choose the middle entry as the median:

Item	1	2	3	4	5	6	7
Value	−0.25	−0.16	0.02	0.15	0.17	0.18	0.32

$$\uparrow$$
$$\text{median}$$

When the set contains an even number of items there is no middle item, and we average the two items nearest the middle. For the set of numbers in Problem 1.1, the median is given by

Item	1	2	3	4	5	6
Value	−0.16	0.02	0.15	0.17	0.18	0.32

$$\uparrow$$
$$\text{median} = 0.16$$

Find the median for the Xs, Ys, and Zs from Problems 1.1, 1.2, and 1.3. Compare the median to the mean or average, found from

$$m_x = \frac{1}{N} \sum_{i=1}^{N} X_i$$

1.5. The sequence of numbers Y_n or Z_n from Problems 1.2 and 1.3 could represent one coordinate (say, the altitude) of samples of the flight path of a German Stuka bomber. Norbert Wiener, an American mathematician (1894–1964), was among the first (in 1939) to solve the problem of how best to predict future values of the signal from its past. His solution was known as the "yellow peril" because of its difficult mathematics and yellow cover. It was classified secret until sometime after World War II because of its application in combat. We will study his solution in later chapters, but the theory is no longer perilous. We have discovered better and simpler ways to present the mathematics. In this problem, however, you are asked simply to consider the problem and try several solutions to see which is best. The mean square error (mse) is the average of the squared error terms over several estimates:

$$\text{mse} = \frac{1}{N} \sum_{i=1}^{N} (X_i - \hat{X}_i)^2 \tag{1.4.1}$$

where $\hat{X}_i$ is the estimate of X_i. The criterion used by Wiener was to minimize the mean square error. In this problem we will consider three ways to estimate the next value of the signal, given the present and all past values of the signal. (We should choose the best of these three if we wish to shoot down the German bomber.) The three are

1. $\hat{X}_{n+1} = X_n$ (present value)
2. $\hat{X}_{n+1} = \frac{1}{N} \sum_{i=1}^{N} X_{n-1}$ (sample average)
3. $\hat{X}_{n+1} = \frac{1}{72} (12X_n + 11X_{n-1} + 10X_{n-2} + \cdots + X_{n-11})$
   (weighted average)

Try these three methods of estimating the next value of the signal on the Y sequence and on the Z sequence from Problems 1.2 and 1.3. Average over a large number of estimates (at least 80) in using Eq. 1.4.1 and see which minimizes the mean square error for each signal. You should not necessarily obtain the best results from the same estimator for both signals. From your experimental results you should see that the choice of coefficients for the x_i's determines how good is the estimate. In later chapters we will study the Wiener filtering problem, which explains how best to choose these coefficients.

CHAPTER 2 _____

Probability

2.1 Experiments _____

Preview

The Italian mathematician Gerolamo Cardano (1501–1576) deserves much credit for his accomplishments, for he initiated several common ideas of today. The historian Will Durant said that every custom began as a broken precedent, and Cardano broke several precedents. He was the first person to write down on paper a complex number. He apologized for its use, saying it was obviously illusory and imaginary, nevertheless, he was brave enough to do it. He also solved the problem of finding the roots of cubic and quardic equations by algebraic means.

Cardano was an inveterate gambler, confessing in his autobiography that he gambled every day. Fortunately, he subjected this vice to scientific scrutiny. His *Book on Games of Chance,* published posthumously in 1663, was the first serious text on probability. Blaise Pascal (1623–1662) and Pierre de Fermat (1601–1665) advanced the subject beyond this rudimentary beginning by their correspondence, and Jakob Bernoulli (1654–1705) summarized and consolidated earlier work in his text published posthumously in 1713. His text contains most of the material in present-day introductory probability texts.

Our purpose in this text is to study random signals. We need an approach that is closely related to existing deterministic signal analysis techniques, but one that will provide a valid analysis of random signals. For this we need the background of random processes, which, in turn, depends on the concepts of random variables and probability. We begin with some basic set theory and probability theory.

A *set* is a collection of objects. Other names for a set are a space, a collection, a group, and a class. Often these names are used for sets with special properties. A *group,* for example, is a set with one binary operation (e.g., addition) defined on it. The items that constitute a set are called points, elements, members, and other names with similar meanings. In pure mathematics, a *set* is usually an undefined concept, because it is necessary in any logical system to have some undefined terms, and a set is often one of those.

We will denote sets with capital letters, A, B, C, . . . , X, Y, Z. Braces, { }, are also used to denote sets. For example, $A = \{a, b\}$ means that A is a set consisting of the two letters a and b. We write $a \in A$ to mean that a is an element of the set A. We write $c \notin A$ to mean that c is not an element of A. Two ways to specify a set are

1. The tabular method, whereby the elements of a set are listed within braces.
2. The rule method, whereby the elements are specified by some rule.

$A = \{a, b\}$ is an example of the tabular method. In the rule method, we would say that A is the set consisting of the first two letters of the alphabet. *Set builder notation* combines these two methods for large sets. For example, it is awkward to list the elements in the set

$$B = \{1, 2, 3, 4, 5, 6, 7, 8, 9, 10, 11, 12\}$$

A more convenient notation for this set is

$$B = \{x: 0 < x < 13, \text{ and } x \text{ is an integer}\}$$

The symbol : is sometimes replaced by |, but in either case these symbols are read "such that." Thus B is the set of all integers such that x is between 0 and 13.

We say that B is a subset of A if every element in B is in A, and we write this $B \subset A$, or $A \supset B$. This means that every set is a subset of itself: $A \subset A$. Thus it is not necessary for a subset to be smaller (i.e., contain fewer members) than the original set. However, the subset cannot be larger than the original set. The union $A \cup B$ is the set of all elements in A or B or both. The intersection $A \cap B$ is the set of all elements in both A and B. Consider the following three sets:

$$X = \{a, b, c, e, f\}$$
$$Y = \{a, b, c, d\}$$
$$Z = \{c, d, a, b\}$$

Then $Y \subset Z$ and $Z \subset Y$ (they are the same set). We say that $Y = Z$ if and only if they are the same set. Y is not a subset of X because Y contains d, which is not in X. The union and intersection of X and Y are

$$X \cup Y = \{a, b, c, d, e, f\}$$
$$X \cap Y = \{a, b, c\}$$

The difference $A - B$ is the set of all elements in A but not in B. The complement is a special case of the difference. The complement of A (written $\overline{A}$) is the difference $U - A$, where U is the universal set. The *universal set* is the totality of all things being considered. The complement has no meaning unless the universal set has been defined (or is understood.)

With this rudimentary review of set theory, let us apply these concepts to experiments and probability. The fundamental entity in probability theory is the *experiment,* also called the probability space. Understanding the concept "experiment" will be important for two reasons: It captures the intuitive ideas behind random phenomena in a precise way, and experiments have entropy.

In mathematics an experiment is a triplet (S, F, P), where S is the sample space, F is the field of events, and P is the probability assigned to all events. This concept of experiment embodies the intuition we feel for tossing a coin, spinning a pointer, or trying to pick the winner at Churchill Downs in the eighth race on the first Saturday in May. In these experiments we know that heads is just as likely as tails, the pointer is just as likely to stop in one quadrant as another, and if we could only guess the winner in advance we could get rich. Our task here is not to get rich, but to understand the symbols in the triplet (S, F, P). We discuss each concept in turn, beginning with the sample space.

DEFINITION 2.1.1.   The sample space S is the set of all possible outcomes of an experiment.

When we toss a coin there are two possible outcomes, heads (H) or tails (T). Therefore the sample space is $S = \{H, T\}$.

When we spin a pointer on a circle marked in radians from 0 to 2π (see Fig. 2.1.1), the pointer can stop on any number in the range $(0, 2\pi)$. Hence the sample space is the set of all real numbers between 0 and 2π, $S = \{x: 0 \leq x < 2\pi\}$.

When we try to pick a winner in a horse race, the set of all possible outcomes is the collection of all possible winners. If there are, say, 8 horses labeled $h_1, h_2, \ldots, h_8$, then the sample space is $S = \{h_1, h_2, \ldots, h_8\}$.

Notice that we use braces { } to enclose the elements in the sample space. This is customary and indicates that we are dealing with a set of elements. This set, if it contains all possible outcomes, is the sample space.

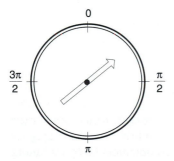

Fig. 2.1.1

Now let us turn to the concept of an event. Intuitively, an event is "something that happens." Formally, an *event* is a subset of the sample space. These two definitions say the same thing, a fact you may better understand after the following discussion and examples.

Every set with n elements has 2^n subsets. The set $S = \{H, T\}$ has $2^2 = 4$ subsets. Therefore the following 4 sets are all the events in the coin-tossing experiment:

$\{H, T\}$
$\{H\}$
$\{T\}$
$\varnothing$

where $\varnothing$ denotes the empty set. (The empty set is a subset of every set, by definition.)

In the horse race there are 8 elements in the sample space, and $2^8 = 256$ subsets of S. A partial list includes the following:

$\{h_1, h_2, h_3, h_4, h_5, h_6, h_7, h_8\}$
$\{h_1, h_2\}$
$\{h_1, h_3\}$
$\varnothing$

The first is just S, the sample space, and this event occurs if any one of the horses wins the race. This is the certain event. The next subset means that the first horse or the second horse won the race. The next means that either the first or the third horse won, and the empty set means there was no winner. Although it is conceivable that a disaster could occur during the race so there would be no winner, that outcome is not allowed. The empty set (no winner) cannot happen. There must be a winner. Therefore the empty set has zero probability.

If the set is continuous, as in our pointer-spinning example, there are an infinite number of elements in the sample space and, therefore, an infinite number of subsets. Here is a list of some of them:

$S = \{x : 0 \leq x < 2\pi\}$
$A = \{x : 0 \leq x < \pi\}$
$B = \{x : (0 \leq x < \pi/2) \cup (\pi \leq x < 3\pi/4)\}$
$\varnothing$

These correspond to the pointer stopping anywhere on the circle, to the pointer stopping in the first or second quadrant, to the pointer stopping in the first or third quadrant, and the empty set corresponds to the pointer not stopping anywhere. This last event is clearly impossible, so the empty set has probability zero. After all, the pointer has to stop somewhere, and it can only stop in the interval between 0 and 2π.

All of these subsets are examples of events, because an event is defined as follows.

DEFINITION 2.1.2.   An event is a subset of the sample space.

Events are important because events and only events have probability. When we speak of the probability of something, that something must be an event.

In the coin-toss experiment there are 4 events, the 4 subsets of the sample space listed above. The probabilities assigned to these events are

$$P(S) = 1$$
$$P(\{H\}) = \tfrac{1}{2}$$
$$P(\{T\}) = \tfrac{1}{2}$$
$$P(\varnothing) = 0$$

In the horse race there are 256 events, with four listed above. Assuming equal ability among the 8 horses, the probabilities are

$$P(S) = 1$$
$$P(\{h_1, h_2\}) = \tfrac{1}{4}$$
$$P(\{h_1, h_3\}) = \tfrac{1}{4}$$
$$P(\varnothing) = 0$$

It is important for you to realize the meaning of the event $\{h_1, h_2\}$. This does *not* mean that both horses win the race. It means that *either* the first *or* the second horse wins the race, but not both at the same time.

The probabilities for the events above in the pointer-spinning experiment are

$$P(S) = 1$$
$$P(A) = \tfrac{1}{2}$$
$$P(B) = \tfrac{1}{2}$$
$$P(\varnothing) = 0$$

In this experiment there are an uncountably infinite number of events, and it is this case that makes the inclusion of the field F in our definition of experiment necessary. You see, we must be careful about which subsets of the continuous interval we allow in the field of events, because there are some subsets to which we cannot assign probability in a meaningful way. Thus the field F is a special field (called a Borel field, or a sigma algebra) that is defined expressly to exclude these bad subsets. The following discussion may help you understand why.

In mathematics, a field is a set together with two binary operations, called addition and multiplication. This concept originated from the real number system with the usual addition and multiplication operations defined there, but we can also define a field on the set of all events with the union and intersection of sets taking the place of addition and multiplication of numbers. We use this idea in the following definition of a Borel field F.

DEFINITION 2.1.3.   Let S be a sample space. A sigma field F of events (also called a Borel field) is any collection of subsets such that the following conditions hold:

1. $\varnothing \in F, S \in F$.
2. If $A_1, A_2, \ldots$ are elements of F, then $A_1 \cup A_2 \cup \cdots$ is an element of F.
3. If $A \in F$, then $S - A \in F$.

The term sigma field is meant to suggest that whenever we sum elements of F (that is, form their union) the resulting subset of S is an element of F. Each element of F is an event of S according to Definition 2.1.3. On the other hand, there may be events of S that are not elements of F.

If $S = \{x_1, x_2, x_3\}$, a particular sigma field F of events is the collection whose elements are all subsets of S, given by $\varnothing$, $\{x_1\}$, $\{x_2\}$, $\{x_3\}$, $\{x_1, x_2\}$, $\{x_1, x_3\}$, $\{x_2, x_3\}$, and S. But a second sigma field F' is the collection whose elements are $\varnothing$, $\{x_1\}$, $\{x_2, x_3\}$, S. Notice that F' satisfies all three properties of the definition, but not all subsets of S are in F'. We explore this point further in the problems.

If we use Borel fields we effectively eliminate those subsets of a continuous sample space that cannot be assigned a probability in a meaningful way. For example, a single point in a continuum has probability 0, but what about an infinite number of isolated points in the continuum? The set of all irrational numbers in the interval $[0, 2\pi)$ is uncountable, yet we don't understand how to assign a probability to them. Therefore we ignore them by using only Borel fields of events in the study of probability.

Now we come to the last item in the triplet (S, F, P). We have already said that probability is assigned to events, and that events and only events have probability. Thus we are dealing with a function.

DEFINITION 2.1.4.   The probability P of an event is a function whose domain is the Borel field F (the set of all events), and whose range is the set of real numbers $0 \le P \le 1$, satisfying the three properties

1. $P(S) = 1$.
2. $0 \le P(A) \le 1$ for all events A.
3. For all events A and B, $P(A \cup B) = P(A) + P(B)$ if $A \cap B = \varnothing$.

Notice that the range is the set of real numbers between 0 and 1, so the probability of an event can never be more than 1 nor less than 0.

One critical mistake often made is to assume that the domain of the function P is the sample space S. This is not so. The domain is the set of all subsets of S, or the set of all events. There is a big difference. This is

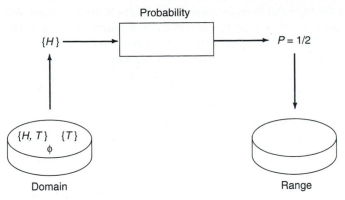

Fig. 2.1.2.   Probability is a function.

pictured in Fig. 2.1.2 for the coin-toss experiment. The input to the function (the box labeled probability) is any one of the sets $\{H, T\}$, $\{H\}$, $\{T\}$, or $\varnothing$. For each of these events we get a number P out of the box, and this number will be either $1, \frac{1}{2}$, or 0. This number P is an element in the codomain, which consists of all the real numbers between 0 and 1. Thus $P\{H, T\} = 1$, $P\{H\} = \frac{1}{2}$, $P\{T\} = \frac{1}{2}$, and $P(\varnothing) = 0$.

Review

You should have learned from the discussion above that we may use the term experiment formally or informally. Formally, the term refers to a triplet (S, F, P), where S is the sample space, F is the Borel field, and P is the probability assigned to the events in F. The informal meaning is the same. It means that we perform some act with a random outcome, but also we know what outcomes are possible, and we know all the probabilities associated with individual outcomes or with combinations of these outcomes. And that is exactly what the triplet (S, F, P) implies: that we know all the facts about which elements are in S, which subsets are in F, and how to assign P to each subset.

2.2   Joint, Conditional, and Marginal Probability ___

Preview

No serious horse player would think of selecting a horse to bet on at random. Horse players study the form charts, much as you study your textbooks, in an effort to determine which horse has a better chance to win. Their estimate of the probability of a particular horse winning the race is conditioned on information gained from events that took place in other races, and from events that took place in the breeding of the particular horse in question. This type of intuitive reasoning on the part of gamblers is reflected in the definition of conditional probability. Knowledge of other events changes the probability of the event in question. The topics of joint, conditional, and marginal probability are the formal way of expressing the gambler's intuitive understanding that events are related. We begin with joint probability, define conditional and marginal probability, and examine Bayes' law.

The intersection $A \cap B$ of two events A and B is called the joint event, and is also written AB. The probability of the joint event is written as $P(A \cap B)$ or $P(AB)$, and also as $P(A, B)$, which can be confusing. Notice that the comma here is read "and," while the comma between two elements of the sample space in the previous section meant "or." When we write $P\{a, b\}$ with a and b as elements in the sample space, we mean that either a or b occurs, but not both at the same time. When we write $P(A, B)$ with A and B as subsets of the sample space, we mean that both occur on the same trial of the experiment. This notation is inconsistent, but it is well established in probability theory so we will just have to get used to it.

The idea of joint probability is pictured in the Venn diagram of Fig. 2.2.1. The probability of A is proportional to the size of the set A, and likewise for B. The probability of AB is proportional to the size of AB, and this is smaller (or at least no bigger) than either A or B. Thus $P(A, B)$ is less than or equal to the smallest of $P(A)$ and $P(B)$.

EXAMPLE 2.2.1.   A single die with faces marked 1 through 6 is thrown. Define the events as follows:

A is the event "the number is even."
B is the event "the number is larger than 3."
Then

$$P(A) = \tfrac{1}{2}$$
$$P(B) = \tfrac{1}{2}$$
$$P(A, B) = \tfrac{1}{3}$$

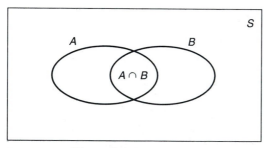

Fig. 2.2.1.   Venn diagram.

where $P(A, B)$ means the probability that both A and B occur. Notice that the probability of the joint event (A, B) is less than either $P(A)$ or $P(B)$. The event (A, B) occurs if either the 4 or 6 face turns up, and this has probability $\frac{2}{6}$, or $\frac{1}{3}$.

Conditional probability is in effect a change in the sample space. Suppose we are told the event B has occurred, and we are asked for the probability of the event A. This is called the probability of A given B and is written $P(A|B)$. Referring to Fig. 2.2.1, the outcome is no longer any element in S; it is now an element in B. Likewise, if A is to occur, it must be an element in $A \cap B$. Thus you can see from Fig. 2.2.1 that

$$P(A|B) = \frac{P(A \cap B)}{P(B)} \tag{2.2.1}$$

Similar reasoning can be used to show that

$$P(B|A) = \frac{P(A \cap B)}{P(A)} \tag{2.2.2}$$

EXAMPLE 2.2.2.   Find the probability of the die face being even, given that the number is greater than 3 in Example 2.2.1.

SOLUTION

$$P(A|B) = \frac{P(A \cap B)}{P(B)} = \frac{\frac{1}{3}}{\frac{1}{2}} = \frac{2}{3}$$

EXAMPLE 2.2.3.   A box contains 5 white balls, 3 red balls, and 2 black ones. What is the probability that two balls drawn from the box will both be red?

SOLUTION: We can make this into a conditional probability problem if we think of drawing one ball at a time. Let R_1 be the event a red ball is drawn on the first draw, and let R_2 be the event a red ball is drawn on the second draw (without replacing the first ball). Then

$$P(R_1, R_2) = P(R_1)P(R_2|R_1)$$
$$= \left(\frac{3}{10}\right)\left(\frac{2}{9}\right) = \frac{1}{15}$$

Subsethood and Probability

Let us return again to Fig. 2.2.1, which pictures the universal set $U = S$ along with the two overlapping subsets A and B. A is a subset of U, and B is a subset of U, but A is not a subset of B, nor is B a subset of A in the usual sense. They do have elements in common, however, because $A \cap B$

is not empty. That is, part of A is in B, and part of B is in A. Thus they are subsets of each other in a partial sense. Bart Kosko has recently added to our understanding of set theory and probability theory by introducing the concept of subsethood. Let $A \subset U$ and $B \subset U$. Then *subsethood* $\mathcal{S}$ is defined by

$$\mathcal{S}(A, B) = \text{Degree}(A \subset B) \tag{2.2.3}$$

This function measures the degree to which A is a subset of B, with $\mathcal{S}(A, B) = 0$ representing no subsethood, and $\mathcal{S}(A, B) = 1$ representing complete subsethood, which is the usual meaning of $A \subset B$. That is, if A is a subset of B in the usual sense, then $\mathcal{S}(A, B) = 1$.

The value of $\mathcal{S}(A, B)$ is determined by the number of elements in $A \cap B$ divided by the number of elements in A:

$$\mathcal{S}(A, B) = \frac{N(A \cap B)}{N(A)} \tag{2.2.4}$$

EXAMPLE 2.2.4.   Let $A = \{f_2, f_4, f_6\}$ and $B = \{f_3, f_4, f_5, f_6\}$, where U is the 6 die faces in Example 2.2.1. Then $A \subset U$ and $B \subset U$ as required in the definition of $\mathcal{S}(A, B)$. Find the subsethoods $\mathcal{S}(A, B)$ and $\mathcal{S}(B, A)$.

SOLUTION: $\mathcal{S}(A, B) = \frac{2}{3}$ because there are two elements in $A \cap B$ and three elements in A. $\mathcal{S}(B, A) = \frac{1}{2}$ because half of B's elements are in A. Note that $\mathcal{S}(A, B) \neq \mathcal{S}(B, A)$.

We can define probability in terms of subsethood as

$$P(A|B) = \mathcal{S}(B, A) \tag{2.2.5}$$

Since $P(A) = P(A|U)$, we see that $P(A) = \mathcal{S}(U, A)$ is the degree to which the universal set (the sample space) is in A.

This scant introduction to subsethood gives you a look at some recent theoretical developments, but we will not use these new ideas in any of the following. We now turn to the most often used formula in probability, Bayes' law.

Bayes' Law

Solving Eqs. 2.2.1 and 2.2.2 for $P(A \cap B)$, we arrive at the fundamental relationship

$$P(A \cap B) = P(A|B)P(B) = P(B|A)P(A) \tag{2.2.6}$$

Bayes' law is derived from Eq. 2.2.6. Bayes' formula, law, or rule states that

$$P(A|B) = \frac{P(B|A)P(A)}{P(B)} \tag{2.2.7}$$

This is probably the most used formula in probability applications. The following example illustrates some features of Bayes' rule.

EXAMPLE 2.2.5.   Three urns, U_1, U_2, and U_3, contain black marbles and white marbles as shown in Fig. 2.2.2. An urn is chosen at random and a marble is drawn from this urn. The marble is black. What is the probability that it came from U_3?

SOLUTION: Define the following events:

B = the marble drawn is black.
U_1 = the marble came from U_1.
U_2 = the marble came from U_2.
U_3 = the marble came from U_3.

Alternatively, these events are subsets of the sample space S, where S is the set of all marbles in the three urns. Then B is the set of all black marbles, U_1 is the set of marbles in urn 1, etc.
    The problem asks for $P(U_3|B)$. By Eq. 2.2.7 this is given by

$$P(U_3|B) = \frac{P(B|U_3)P(U_3)}{P(B)} \qquad (2.2.8)$$

We can easily compute the numerator of this equation as follows:

$$P(B|U_3) = \tfrac{1}{2}$$
$$P(U_3) = \tfrac{1}{3}$$

We now contend that $P(B)$ is given by

$$P(B) = P(B|U_1)P(U_1) + P(B|U_2)P(U_2) + P(B|U_3)P(U_3)$$
$$= (\tfrac{3}{10})(\tfrac{1}{3}) + (\tfrac{4}{10})(\tfrac{1}{3}) + (\tfrac{5}{10})(\tfrac{1}{3}) = 0.4 \qquad (2.2.9)$$

so the solution to our problem is

$$P(U_3|B) = \frac{(\tfrac{1}{2})(\tfrac{1}{3})}{0.4} = \frac{5}{12}$$

Derivation of Eq. 2.2.9.   We want to show the conditions that must hold for Eq. 2.2.9 to be true.

3 B	4 B	5 B
7 W	6 W	5 W
U_1	U_2	U_3

Fig. 2.2.2

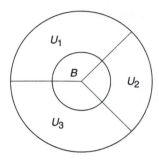

Fig. 2.2.3

Suppose that B can occur only if U_1, U_2, or U_3 occurs. Also suppose that the U_i's are mutually exclusive, that is, $U_i \cap U_j = \varnothing$ for $i \neq j$. Now B can be partitioned into disjoint sets $(B \cap U_1)$, $(B \cap U_2)$, and $(B \cap U_3)$, as shown in Fig. 2.2.3. This means that

$$B = BU_1 \cup BU_2 \cup BU_3 \qquad (2.2.10)$$

By Property 3 of Definition 2.1.4 (probability), we have

$$P(B) = P(BU_1) + P(BU_2) + P(BU_3) \qquad (2.2.11)$$

and from Eq. 2.2.6 this gives the desired result, Eq. 2.2.9.

Bayes' formula is most useful in the form

$$P(U_3|B) = \frac{P(B|U_3)P(U_3)}{\Sigma_i P(B|U_i)P(U_i)} \qquad (2.2.12)$$

which is valid so long as B can occur only if one of the U_i's occurs, and where the U_i's are mutually exclusive.

Review

In this section we introduced conditional probability and used it to derive Bayes' law. If you can apply conditional probability to problems, you are in good shape. Here is a sample problem that you can use to test yourself. The answer is 15/153, or less than 0.1.

A box contains 6 gold coins, 6 silver coins, and 6 copper coins. If two coins are drawn at random, what is the probability that they both will be gold?

2.3   Mutually Exclusive and Independent Events ___

Preview

After completing this section you should be able to define mutually exclusive events, define independent events, and prove that events cannot be both mutually exclusive and independent. It is important to have a clear understanding of these concepts, for they pop up in almost all situations. Knowing that events cannot be both independent and mutually exclusive may save you lots of trouble in the future. We will also introduce the chain rule for calculating conditional probabilities.

Two events are *mutually exclusive* if the occurrence of one prohibits the occurrence of the other. This property, which is captured in the following definition, is illustrated in Fig. 2.3.1. Part (a) shows overlapping events, where it is possible for an outcome to occur in $A \cap B$, and part (b) shows mutually exclusive events.

DEFINITION 2.3.1.   Two events A and B are mutually exclusive if and only if

$$P(A \cup B) = P(A) + P(B) \tag{2.3.1}$$

Note: It is always true that

$$P(A \cup B) = P(A) + P(B) - P(A \cap B)$$

If $P(A \cap B)$ is 0, then the two events are nonoverlapping (Fig. 2.3.1b) and $P(A \cup B)$ is given by Eq. 2.3.1.

Another important property of events is independence. Intuitively, independence means that the occurrence of one event does not affect the occurrence or nonoccurrence of another event. Then two events A and B are *independent* if the probability of A given that B has occurred is equal to the probability of A; that is,

$$P(A|B) = P(A) \tag{2.3.2}$$

But by Eq. 2.2.3 it is always true that

$$P(A \cap B) = P(A|B)P(B) \tag{2.3.3}$$

so we define independence as follows.

DEFINITION 2.3.2.   Two events A and B are statistically independent if and only if

$$P(A \cap B) = P(A)P(B) \tag{2.3.4}$$

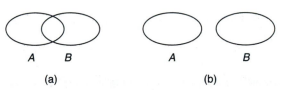

A	B	A	B
(a)		(b)	

Fig. 2.3.1.   (a) A and B are not mutually exclusive. (b) A and B are mutually exclusive.

Note: In view of Eq. 2.3.3, this is equivalent to Eq. 2.3.2. Either condition can be taken as the definition of statistical independence.

Events cannot be both independent and mutually exclusive if their probabilities are not 0. To prove this we will use the method of contradiction; we assume that events A and B are both mutually exclusive and independent, and then show that a contradiction results.

Let events A and B have nonzero probability:

$$P(A) > 0 \tag{2.3.5}$$

$$P(B) > 0 \tag{2.3.6}$$

Assume that A and B are mutually exclusive. By Definition 2.3.1 this means that

$$P(A \cap B) = 0$$

Now assume that A and B are also independent. By Definition 2.3.2 this means that

$$P(A \cap B) = P(A)P(B)$$

But by Eqs. 2.3.5 and 2.3.6 the product $P(A)P(B)$ cannot equal 0. Thus we have a contradiction, and our statement is proved.

The Chain Rule

The concepts of conditional probabilities, dependent events, and mutually exclusive events may be extended to three or more events as follows. Since

$$P(A|B \cap C) = \frac{P(A \cap B \cap C)}{P(B \cap C)} \tag{2.3.7}$$

We can write

$$
\begin{aligned}
P(A \cap B \cap C) &= P(A|B \cap C)P(B \cap C) \\
&= P(A|B \cap C)P(B|C)P(C)
\end{aligned}
\tag{2.3.8}
$$

This is called the *chain rule* and will have wide application in all that follows.

Review

Two mutually exclusive events cannot both occur in one trial of the experiment. Independent events have no effect on each other, meaning they may or may not occur. It is important to know, however, that events cannot be both mutually exclusive and independent (so long as their probability is greater than 0). We used contradiction to prove this. We assumed that two events were mutually exclusive, which meant that $P(A \cap B) = 0$. If they are also independent, then $P(A \cap B) = P(A)P(B)$. If this is 0, then either $P(A) = 0$, or $P(B) = 0$, or both. This is a contradiction because we assumed events with nonzero probabilities to begin with. Since a statement must either be true or false, with no in-between, then events cannot be both mutually exclusive and independent.

2.4 Entropy

Preview

Ludwig Boltzmann's tombstone in the central cemetery of Vienna has on it his name, the dates 1844–1906, and the inscription $S = k \log W$. In this formula S stands for entropy, k is now known as Boltzmann's constant, and W is a measure of the possible states of nature. This is one of the most remarkable formulas of science, for it quantifies the chaos that exists at the atomic level in matter and applies to any system with random parameters. The concept of entropy has application to astronomy, filter design, coding theory, and data fusion, to name a few.

Boltzmann did his work before the existence of atoms was generally accepted, and many of his contemporaries doubted the credibility of his assumptions. Doubt was also cast on his work by those who saw a threat to their beliefs in the purposiveness of nature through his work, just as Galileo and Darwin had been persecuted for their views. Suffering from the scorn of his peers, Boltzmann was overcome by instability and unhappiness and killed himself in 1906. It is ironic that Einstein's paper on thermodynamics in 1905 led the vanguard in accepting Boltzmann's views.

R. V. L. Hartley, a scientist at Bell Laboratories, used Boltzmann's entropy to measure the information storage capacity of a storehouse, or the information transmission capacity of a communication channel, in a paper titled "Transmission of Information" in 1928. The reasoning in Boltzmann's and Hartley's application is essentially the same, and goes as follows. Suppose we have a knob with N click positions, as shown in Fig. 2.4.1. Then we can store any one of N different messages (or symbols) by assigning one message to each click position. Now if

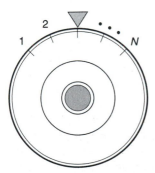

Fig. 2.4.1.   One knob with N click positions.

we double the physical size of the storehouse by using two identical knobs, as shown in Fig. 2.4.2, there are N^2 positions. For each position of the first knob there are N positions of the second knob, giving $N \times N$ positions for the two together. With S the storage capacity, $k = 1$, and W the total number of states, Boltzmann's formula gives

$$S_1 = \log N$$
$$S_2 = \log N^2 = 2 \log N = 2S_1$$

where S_1 is the storage capacity of one knob, and S_2 is the storage capacity of two knobs. Thus Hartley's measure of capacity coincides with intuition. Doubling the size of the storehouse doubles its capacity.

In 1948 Claude Shannon (1916–   ) published his classic work and founded the science of information theory. He expanded Hartley's concept to include not only information storage capacity, but also the information content of a message. He defined the message information as the capacity of the smallest storehouse that will hold the message. His work led to a flurry of activity among researchers to find ways to condense and encode messages, and helped provide the inexpensive worldwide communications networks we enjoy today. It is now our intent to explain the basis for Shannon's work and to demonstrate some of the resulting applications.

We begin with the concept of randomness in experiments. (Randomness is another name for entropy.) Experiments and only experiments have entropy, so anytime we talk about entropy (or randomness, or information content, or uncertainty—all names for entropy) we must be talking about an experiment. Recall that an experiment E is a three-tuple (S, F, P), where S is the sample space, F is the field of events, and P is the probability assignment of F. (See Section 2.1.) This means we have available, or know, the possible outcomes in the experiment and their probabilities. This is the information we need to calculate entropy.

Let us now consider some examples of experiments with random out-

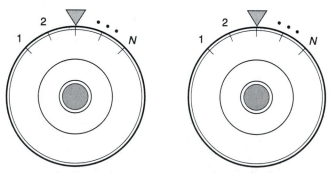

Fig. 2.4.2. Two knobs, each with N click positions.

S_1	H	T
P_1	$\dfrac{1}{2}$	$\dfrac{1}{2}$

Fig. 2.4.3. Experiment E_1.

comes. Figure 2.4.3 shows experiment E_1 where a fair coin is tossed, giving the sample space $S_1 = \{H, T\}$. Figure 2.4.4 shows the sample space and probability assignment for E_2, where a fair die is tossed with possible outcomes of 6 dies faces, each with equal probability. Obviously, E_2 is more random than E_1. There are more possible outcomes in E_2 than in E_1, and it would be easier to guess the outcome in E_1 correctly. But how can we quantify the randomness in these experiments to reflect our intuition, i.e., so that the randomness in E_2 will be larger than the randomness in E_1 by our measure?

As a first attempt, try the reciprocal of the probability. This gives

$$H(E_1) = 2$$
$$H(E_2) = 6$$

where we use the symbol H for entropy. If the outcomes in all experiments were equally likely, this might suffice. But consider the experiment E_3 in Fig. 2.4.5. This is a loaded die, with the one-spot turning up half the time, and the other sides turning up with probability $\frac{1}{10}$. Intuition says that the entropy of this experiment should be between $H(E_1)$ and $H(E_2)$. There are more possible outcomes in E_3 than in E_1, and we can correctly guess the outcome in E_3 with more luck than in E_2. Thus the randomness in E_3 should be more than in E_1 but less than in E_2.

The reciprocal of probability won't work here, and neither will the average. The average of $1/P_i$ in Fig. 2.4.5 gives 6. But if we introduce the logarithm into this by finding the average of the $\log(1/P_i)$, we get

$$H(E) = \sum_i P_i \log\left(\frac{1}{P_i}\right) \qquad (2.4.1)$$

S_2	⚀	⚁	⚂	⚃	⚄	⚅
P_2	$\dfrac{1}{6}$	$\dfrac{1}{6}$	$\dfrac{1}{6}$	$\dfrac{1}{6}$	$\dfrac{1}{6}$	$\dfrac{1}{6}$

Fig. 2.4.4. Experiment E_2.

S_3	⚀	⚁	⚂	⚃	⚄	⚅
P_3	$\frac{1}{2}$	$\frac{1}{10}$	$\frac{1}{10}$	$\frac{1}{10}$	$\frac{1}{10}$	$\frac{1}{10}$

Fig. 2.4.5.   Experiment E_3.

This is Shannon's measure of entropy. For our three experiments this gives (using base 2 logarithms)

$$H(E_1) = 0.5 \log 2 + 0.5 \log 2 = 1 \text{ bit}$$
$$H(E_2) = 6[(\tfrac{1}{6}) \log 6] = 2.58 \text{ bits}$$
$$H(E_3) = 0.5 \log 2 + 5(0.1 \log 10) = 2.16 \text{ bits}$$

which agrees with intuition.

Of course we may use any base for the logarithm. The most commonly used are base 10, base e, and base 2. The unit of entropy has a different name for each base as follows:

Base 10—hartley
Base e—nat (for natural units)
Base 2—bit (for binary unit)

A useful formula to convert from one unit to another is

$$\log_2(x) = \frac{\log_e(x)}{\log_e(2)} = \frac{\log_{10}(x)}{\log_{10}(2)} \tag{2.4.2}$$

One advantage of using the logarithm in the definition of entropy is that combined experiments will have entropy equal to the sum of the individual entropies. This is illustrated in the following example.

EXAMPLE 2.4.1.   Let E_4 be the combined experiment $E_1 \times E_2$ as shown in Fig. 2.4.6. Then E_4 has entropy given by

$$H(E_4) = \sum_{i=1}^{12} P_i \log \left(\frac{1}{P_i} \right) = 3.58 \text{ bits}$$

Notice that this is $H(E_1) + H(E_2)$.

S_4	$H, ⚀$	$H, ⚁$	...	$H, ⚅$	...	$T, ⚅$
P_4	$\frac{1}{12}$	$\frac{1}{12}$	...	$\frac{1}{12}$	...	$\frac{1}{12}$

Fig. 2.4.6.   The combined experiment $E_1 \times E_2$.

We said that experiments and only experiments have entropy, and this is true, but you will see in the literature that entropy applies to partitions and random variables. This seems to conflict with our claim that experiments and only experiments have entropy, but it really doesn't. We will describe entropy in random variables in the next chapter, so let us now discuss entropy in partitions.

A *partition* of the sample space is a collection of subsets that are mutually exclusive and exhaustive. Let A_1 and A_2 be subsets of the sample space S. Then A_1 and A_2 are mutually exclusive if they have no elements in common, i.e., if

$$A_1 \cap A_2 = \varnothing \qquad (2.4.3)$$

The sets $A_1, A_2, \ldots, A_N$ are exhaustive if their union includes all elements of S, i.e., if

$$S = A_1 \cup A_2 \cup \cdots \cup A_N \qquad (2.4.4)$$

EXAMPLE 2.4.2.   Toss a die, so the sample space is the set of 6 die faces $S = \{f_1, f_2, f_3, f_4, f_5, f_6\}$. We form the following three collections of sets:

$$A_1 = \{f_1, f_5\} \qquad A_2 = \{f_2, f_4\} \qquad A_3 = \{f_3\} \qquad A_4 = \{f_6\}$$
$$B_1 = \{f_1, f_5\} \qquad B_2 = \{f_2\} \qquad B_3 = \{f_3\} \qquad B_4 = \{f_6\}$$
$$C_1 = \{f_1, f_5\} \qquad C_2 = \{f_2, f_4\} \qquad C_3 = \{f_3, f_4\} \qquad C_4 = \{f_6\}$$

Then $\{A_i\}$ is a partition of S, because each element of S occurs exactly once in the collection of subsets. $\{B_i\}$ is not a partition because f_4 is left out, and $\{C_i\}$ is not a partition because f_4 occurs in two different subsets.

Since the A_i's form a partition, we can find the entropy by thinking of this as a new experiment with outcomes A_1, A_2, A_3, A_4. This gives

$$H = - \sum_i P(A_i) \log P(A_i) = 2 \left(\frac{1}{3} \log 3 \right) + 2 \left(\frac{1}{6} \log 6 \right) = 1.918 \text{ bits}$$

EXAMPLE 2.4.3.   Toss a die. If 6 die faces form the sample space, then odd spots and even spots form a partition:

$$A_1 = \{f_1, f_3, f_5\} \qquad A_2 = \{f_2, f_4, f_6\}$$

Note that $A_1 \cap A_2 = \varnothing$ and $A_1 \cup A_2 = S$. Therefore the sets A_1, A_2 form a partition of S. Since the probability of each part is $\frac{1}{2}$, the entropy is the same as the coin toss in Fig. 2.4.3. But we could also define a new experiment, call it E_5, where the sample space is $S = \{$even, odd$\}$ and arrive at the same entropy. This new experiment has two elements in the sample space, each with equal probability, but it is derived by tossing a die rather than a coin.

This illustrates that partitioning an existing sample space is equivalent to defining a new experiment with sample space equal to the partitions of the original space. Therefore entropy applies to partitions of the sample space.

Properties of Entropy

Here are several important properties of entropy. Most have been illustrated in one way or another by our examples, so they should seem reasonable.

1. $H(E)$ is a continuous function of p_i.
2. $H(E)$ is maximum for equally likely outcomes. If there are N elements in the sample space with $p_1 = p_2 = \cdots = p_N = 1/N$, no other set of probabilities gives larger entropy.
3. If $p_1 = p_2 = \cdots = p_N = 1/N$, then $H(E)$ is an increasing function of N.
4. $H(E_1, E_2) \le H(E_1) + H(E_2)$ with equality if and only if E_1 and E_2 are independent.

Property 1 means that a small change in the probability of some outcome will not cause a large change in the entropy of the experiment. Consider an experiment E with two possible outcomes occurring with probabilities p and $1 - p$. The entropy is

$$H(E) = -p \log p - (1 - p) \log(1 - p)$$

The plot of $H(E)$ versus p shown in Fig. 2.4.7 illustrates the first two properties. This plot is continuous in p, and the maximum value occurs when all outcomes are equally likely, i.e., when $p = 1 - p$. Property 3 is believable from our interpretation of entropy as a measure of randomness. An increase in the number of equally likely outcomes increases the entropy. Finally, Example 2.4.1 illustrated that the combination of two independent experiments has entropy equal to the sum of their individual entropies. When the experiments are dependent, the entropy of their combination is less than the sum of each entropy.

These properties will prove useful to us in a number of applications. Here

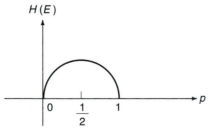

Fig. 2.4.7. Entropy as a function of p.

is an example illustrating one important application (the maximum entropy method).

Suppose that we have a suspect die, one that is loaded so the one-spot turns up more often than any other. Upon tossing the die 900 times, the one-spot turns up 400 times. The experimenter failed to record the number of times the other spots turned up, and we don't want to repeat the experiment. We would like to make some reasonable estimate for the probabilities of the other 5 spots on the die turning up.

The maximum entropy method (MEM) maximizes the entropy in the experiment subject to any constraints. The only known constraints (in the absence of any other information about the die) are that all probabilities sum to 1, and that $P(f_1) = \frac{4}{9}$. Property 2 then implies that for maximum entropy all other probabilities should be equal, giving

$$P(f_2) = P(f_3) = \cdots = P(f_6) = \frac{1}{9}$$

This is called the forward problem of MEM. We are given the sample space and some constraints. The problem is to choose the unknown probabilities so as to maximize the entropy and to satisfy the constraints. The inverse problem in MEM is to find the constraints that lead to the given probabilities. Here we are given the sample space and the probabilities, and are asked to specify the constraints that give the probabilities if the maximum entropy method is used to find them. In our example, we are told that the sample space consists of the 6 die faces, and that their probabilities are $P(f_1) = \frac{4}{9}$, with all others equal to $\frac{1}{9}$. The inverse problem is to find a reasonable set of constraints that lead to this solution, so we will probably choose those constraints given in the original problem. See the last chapter in reference 4 for a good discussion of the maximum entropy method.

Review

Experiments and only experiments have entropy. This means that we must use all the sample space in calculating entropy. When we use less than the entire sample space we have either (perhaps tacitly) defined a new experiment with a reduced sample space, or we have made a mistake. Otherwise, there is nothing to finding entropy for one experiment: Simply find the average of $\log(1/P_i)$.

2.5 Observation Systems _____

Preview

Signal processing systems fall into two categories, those that make decisions and those that make estimates. In response to an input signal, decision makers produce a declaration, decision, or number as their output; estimators produce a number or a waveform. Entropy is one of the few measures of system performance that applies to both types.

Most criteria measure only system performance, ignoring the source. In this section we demonstrate how entropy measures performance in the system's intended application, taking into account both the source and the system. For that reason, entropy is the criterion of choice in many applications.

Before tossing a fair die, the outcome is uncertain and the randomness in the experiment is log 6. After observing the outcome, however, the randomness is 0. There is no longer any uncertainty about the outcome. We have changed the entropy from log 6 to 0 by observing the outcome. This can be expressed by the formula

(change in entropy) = (initial randomness) − (final randomness)   (2.5.1)

where we can use any of the terms entropy, randomness, ignorance, and uncertainty on the right side of this formula. Information gain and mutual information are other names for the change in entropy on the left side.

In this example there is no question about the observer's reliability. But many observers are unreliable, giving the correct outcome only on occasion. In such a case the final entropy is not 0 after the experiment is performed, although it may be reduced. For example, the three urns in Fig. 2.5.1 contain two marbles each. Urn 1 has two black marbles, urn 2 has one white and one black marble, and urn 3 has two white marbles. Let E_1 represent the experiment of selecting an urn, with each urn equally likely. Then E_1 has entropy given by

$$H(E_1) = \sum_i P_i \log \left(\frac{1}{P_i}\right) = \frac{1}{3}\log 3 + \frac{1}{3}\log 3 + \frac{1}{3}\log 3 = 1.585 \text{ bits}$$

Now we add a wrinkle to the experiment. After selecting an urn, a ball is drawn from the urn and shown to you. You are allowed to see the ball, but you are not told which urn it came from. Thus you have incomplete information about E_1. This has the effect of reducing the randomness in experiment E_1, but not completely.

Let us label the experiment of selecting a marble E_2, where the experimen-

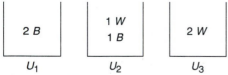

Fig. 2.5.1.   Two experiments: E_1, select an urn; E_2, select a marble.

tal outcomes are $\{M_1, M_2\}$, representing the two possible marbles, $M_1 =$ black and $M_2 =$ white. We wish to find the entropy remaining in E_1 after seeing the outcome of E_2. This should have the effect of reducing the entropy in E_1, because knowing the outcome of E_2 gives us some information about E_1. This remaining entropy must be related to the conditional probabilities $P(U_i|M_j)$ in the same way that the initial entropy in E_1 is related to the probabilities $P(U_i)$. There we had $H(E_1)$ equal to the average of $\log[1/P(U_i)]$. Here we find the average of $\log[1/P(U_i|M_j)]$. We label this remaining entropy $H(E_1|E_2)$:

$$H(E_1|E_2) = -\sum_{i,j} P(U_i, M_j) \log P(U_i|M_j) \qquad (2.5.2)$$

This is the entropy remaining in E_1 after observing E_2, and it is called the conditional entropy in E_1 given E_2.

In order to calculate the entropy remaining in E_1 after performing experiment E_2 in Fig. 2.5.1, we need the joint probabilities $P(U_i, M_j)$ and the conditional probabilities $P(U_i|M_j)$. These are displayed in Fig. 2.5.2. Substituting into Eq. 2.5.2 gives

$$H(E_1|E_2) = -2[(\tfrac{1}{3}) \log(\tfrac{2}{3}) + (\tfrac{1}{6}) \log(\tfrac{1}{3})] = 0.918 \text{ bits}$$

This quantity $H(E_1|E_2)$ represents the last term in Eq. 2.5.1, and $H(E_1)$ represents the initial randomness. The change in entropy is the information gain, and it is labeled $I(E_1, E_2)$. Therefore we can rewrite Eq. 2.5.1 in symbols as

$$I(E_1, E_2) = H(E_1) - H(E_1|E_2) \qquad (2.5.3)$$

This quantity is symmetrical, meaning that $I(E_1, E_2) = I(E_2, E_1)$, or

$$I(E_1, E_2) = H(E_2) - H(E_2|E_1)$$

This is also called the mutual information because it is symmetrical. The mutual information between E_1 and E_2 in our example is

$$I(E_1, E_2) = 1.585 - 0.918 = 0.667 \text{ bits}$$

The concepts of mutual information and remaining entropy are useful as a criterion in system design. The following discussion of observation systems introduces some of these ideas.

E_2 \ E_1	U_1	U_2	U_3
$M_1 = B$	$\tfrac{1}{3}$	$\tfrac{1}{6}$	0
$M_2 = W$	0	$\tfrac{1}{6}$	$\tfrac{1}{3}$

E_2 \ E_1	U_1	U_2	U_3
$M_1 = B$	$\tfrac{2}{3}$	$\tfrac{1}{3}$	0
$M_2 = W$	0	$\tfrac{1}{3}$	$\tfrac{2}{3}$

Fig. 2.5.2. The probabilities $P(U_i, M_j)$ and $P(U_i|M_j)$.

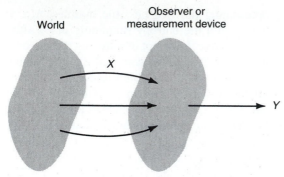

Fig. 2.5.3.   A model for an observation system.

Generally speaking, some error is possible when we make an observation or measurement. Two voltmeters may read different values of the same voltage. Eye witnesses are often mistaken about details in their observations. False alarms and missed targets occur in radar. In short, our methods of measuring and observing the world are often unreliable.

A general model for observations with randomness is shown in Fig. 2.5.3. One or more connections between the world and the observer are denoted by X. The observer output is called Y. Our urn–marble selection experiment is a specific example of this general model, where E_1 corresponds to X and E_2 corresponds to Y.

If the observation Y in Fig. 2.5.3 is to be of use to us, we must have some knowledge about how Y is related to X. The general form of an observation system is shown in Fig. 2.5.4. This unit selects Y from X, but this selection is not unique. Therefore the relationship between X and Y cannot be a function, for a function has a unique relationship between X and Y. In a function, each X produces exactly one Y. Here, X may produce a particular Y on one trial, and another Y on a second trial. The output is selected with probability $P(Y|X)$. For all possible pairs (X, Y), the function $P(Y|X)$ is the probability that a particular Y will be selected when the input

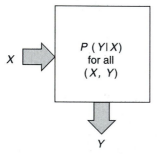

Fig. 2.5.4.   A mathematical model for the observation system.

is X. Knowing $P(X)$ we can obtain $P(Y)$, the probability of the output, by the formula

$$P(Y) = \sum_{i=1}^{N} P(X_i) P(Y|X_i) \tag{2.5.4}$$

Also, the joint probability is given by

$$P(X, Y) = P(X) P(Y|X) \tag{2.5.5}$$

If not all randomness is removed, i.e., if there is some remaining uncertainty about X after observing Y, then that is how much randomness X can exhibit if Y is already known. In Figs. 2.5.3 and 2.5.4 neither X nor the internal operation of the system is observable. Only Y is observable. A measure of how much good it did to construct the system, or how reliable was the witness, is determined by how much randomness is removed by the system. That is, how close does $P(Y|X)$ come to a unique mapping? The system is a mapping with randomness. We observe its output and guess its input. The more unique the mapping, the better the guess will be. With $H(X)$ denoting the randomness in X, the entropy that remains after observing Y is $H(X|Y)$, given by

$$H(X|Y) = - \sum_{x,y} P(X, Y) \log P(X|Y) \tag{2.5.6}$$

(This is just Eq. 2.5.2 restated in terms of X and Y.) Knowing $P(Y)$ and $P(X, Y)$, we can calculate $P(X|Y)$ to substitute into Eq. 2.5.6 from

$$P(X|Y) = \frac{P(X, Y)}{P(Y)} \tag{2.5.7}$$

EXAMPLE 2.5.1.   Suppose that the bandpass filter in Fig. 2.5.5a has input signal $x(t)$ with frequency components in the neighborhood of 3 kHz and 8 kHz, as shown in Fig. 2.5.5b. If this signal is supplied to a bandpass filter (BPF) with transfer function $G(f)$ passing 8 kHz but rejecting 3 kHz, then the output $Y(f)$ has the frequency components shown in Fig. 2.5.5c.

Initially, X has entropy $H(X)$. We are allowed to observe Y, so the entropy remaining in X after observing Y is $H(X|Y)$, and this is contained in the frequencies around 3 kHz, as shown in Fig. 2.5.5d. The mutual information $I(X, Y)$ is contained in the frequency components around 8 kHz that pass through the filter.

Information-Destroying Processes

Any operation performed on a signal cannot create or add information about the source. Information can only be destroyed or, at best, preserved. Recall that information is another name for entropy, so we are calling the randomness

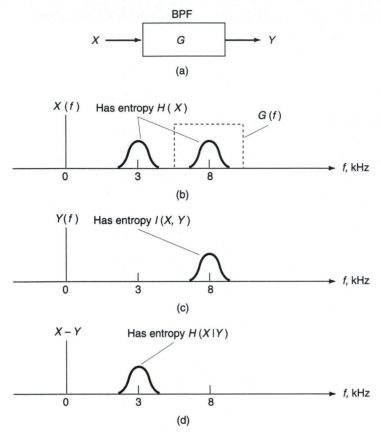

Fig. 2.5.5. Mutual information in a bandpass filter.

in the experiment information. For system A in Fig. 2.5.6, if the original signal X cannot be recovered exactly from Y, then obviously something has been lost. Any nonreversible operation destroys information.

The purpose of any information processing system is to either preserve or destroy information. Systems often destroy unwanted randomness (noise) while preserving desired information (the signal). Examples of systems that destroy information are filters, detectors, and declarations:

Filters destroy information about the noise but, hopefully, destroy little information about the signal. The aim is to recover the desired signal with high fidelity.

Detectors destroy a lot of information about the signal. The output of a detector is typically a number. There is no hope of recovering the input signal (a waveform) from the detector output.

The threshold in *a declaration* circuit destroys even the detector output and provides a "yes" or "no" decision about the presence of a target

Fig. 2.5.6.   System A performs a reversible operation if X can be recovered from Y.

in a sonar or radar system. This threshold unit destroys a great deal of information.

On the other hand, the purpose of transmission systems is to preserve information. If this is accomplished, then it should be possible to recover the input signal from the transmission system output. To illustrate, let the transmission system have known impulse response. In Fig. 2.5.6 we are given the impulse response of system A and the output signal $y(n)$ at point Y in the diagram. From this information we are to recover the signal $x(n)$ at point X. A practical application is to design a filter to counteract the distortion introduced by the transmission system.

Given a transmission system with known impulse response $h(n)$, one way to do this is to construct an inverse system with impulse response $g(n)$ such that the overall impulse response of the cascaded pair in Fig. 2.5.6 is $\delta(n)$. This implies that

$$h(n) * g(n) = \delta(n) \qquad (2.5.8)$$

where the symbol $*$ stands for convolution. If both $h(n)$ and $g(n)$ are causal, then this implies that the corresponding transfer functions are reciprocals of one another.

EXAMPLE 2.5.2.   Suppose that $H(z)$, the transfer function for system A in Fig. 2.5.6, is given by

$$H(z) = \frac{z - \frac{1}{2}}{z - 1}$$

Find the system impulse response $h(n)$ and the causal inverse system impulse response $g(n)$. Show by direct convolution that Eq. 2.5.8 is satisfied.

SOLUTION: We first calculate $h(n)$ by using partial fraction expansion on $H(z)/z$:

$$\frac{H(z)}{z} = \frac{z - \frac{1}{2}}{z(z - 1)} = \frac{\frac{1}{2}}{z} + \frac{\frac{1}{2}}{z - 1}$$

or

$$H(z) = \frac{1}{2} + \frac{0.5z}{z - 1}$$

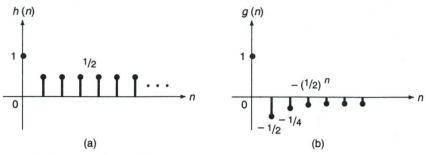

Fig. 2.5.7.   The reciprocal impulse responses (a) $h(n)$ and (b) $g(n)$.

Thus

$$h(n) = 0.5\,\delta(n) + 0.5u(n)$$

where $u(n)$ is the unit step function. This is the impulse response of the original system. Now set $G(z) = 1/H(z)$ and repeat these steps to get

$$g(n) = 2\,\delta(n) - (0.5)^n u(n)$$

The functions $h(n)$ and $g(n)$ are plotted in Fig. 2.5.7. You should be able to show that convolving $h(n)$ with $g(n)$ satisfies Eq. 2.5.8.

Review

In this section we introduced two concepts: mutual information and information-destroying processes. These can be used to measure the effectiveness of a given system in its intended application, something most other measures of performance cannot do. The values of these performance measures depend on both the system itself and the nature of the input signal. Examples 2.6.7 and 2.6.8 in the next section further illustrate these points.

Every system must destroy information if it is to be useful in any information processing task. A system that makes decisions must destroy substantial information, while a transmission system should destroy only unwanted noise. In order for a system to destroy no information it must be invertible, i.e., one must be able to recover the input from the output.

2.6   Examples

Preview

This section provides additional examples to further illustrate the concepts in this chapter. They are provided for those who may need further explanation for some of the concepts. The examples are keyed to their relevant section.

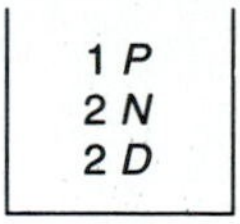

```
1 P
2 N
2 D
```

Fig. 2.6.1.   A box of coins.

Section 2.1.   Experiments

EXAMPLE 2.6.1.   A box contains 1 penny, 2 nickels, and 2 dimes as shown in Fig. 2.6.1. You reach in and select at random one coin from the box. What is the (formal) experiment? That is, what is the sample space, what are the events, and what is the probability of each event?

SOLUTION: The action is to select one coin. Therefore it is possible to select 1 penny, or 1 nickel, or 1 dime. It is not possible to select 2 dimes, for example, because the action is to select one coin. Therefore the sample space is

$$S = \{p, n, d\}$$

There are $2^3 = 8$ events as listed below along with their associated probabilities:

$$P\{p, n, d\} = 1$$
$$P\{p, n\} = \tfrac{3}{5}$$
$$P\{p, d\} = \tfrac{3}{5}$$
$$P\{n, d\} = \tfrac{4}{5}$$
$$P\{p\} = \tfrac{1}{5}$$
$$P\{n\} = \tfrac{2}{5}$$
$$P\{d\} = \tfrac{2}{5}$$
$$P\{\varnothing\} = 0$$

Section 2.2.   Joint, Conditional, and Marginal Probability

EXAMPLE 2.6.2.   A box contains 5 white balls, 5 red balls, and 5 black ones. What is the probability that two balls drawn from the box will both be red?

SOLUTION: Suppose we draw one ball at a time. Let R_1 be the event a red ball is drawn on the first draw, and let R_2 be the event a red ball is drawn on the second draw (without replacing the first ball). Then

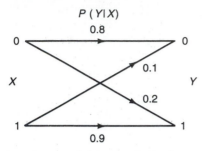

$P(Y|X)$

Fig. 2.6.2.   A binary channel.

$$P(R_1, R_2) = P(R_1)P(R_2|R_1)$$
$$= \left(\frac{5}{15}\right)\left(\frac{4}{14}\right) = \frac{2}{21}$$

EXAMPLE 2.6.3.   A model for a binary communication channel is shown in Fig. 2.6.2. It is called binary because one of two symbols, $X = 0$ or $X = 1$, is supplied to the channel input. The output Y is then either 0 or 1. The probabilities on the diagram represent $P(Y|X)$. Thus if $X = 0$, Y will equal 0 with probability 0.8, and Y will equal 1 with probability 0.2. Suppose that a symbol for X is selected with equal probability and transmitted over the channel. If $Y = 0$, find the probability that the transmitted symbol was $X = 0$.

SOLUTION: From Eq. 2.2.7 we have

$$P(X = 0|Y = 0) = \frac{P(Y = 0|X = 0)P(X = 0)}{P(Y = 0)}$$

where $P(Y = 0) = P(Y = 0|X = 0)P(X = 0) + P(Y = 0|X = 1)P(X = 1)$. Plugging these numbers into Eq. 2.2.12 gives

$$P(X = 0|Y = 0) = \frac{0.8(0.5)}{0.8(0.5) + 0.1(0.5)} = \frac{8}{9}$$

Notice that this is an urn problem. If we identify U_1 with $X = 0$ and U_2 with $X = 1$, then the diagram in Fig. 2.6.2 can be replaced by the one in Fig. 2.6.3. The problem statement now would ask for $P(U_1|W)$.

Section 2.3.   *Mutually Exclusive and Independent Events*

EXAMPLE 2.6.4.   Consider an ordinary deck of playing cards. Define the events A, B, and C as

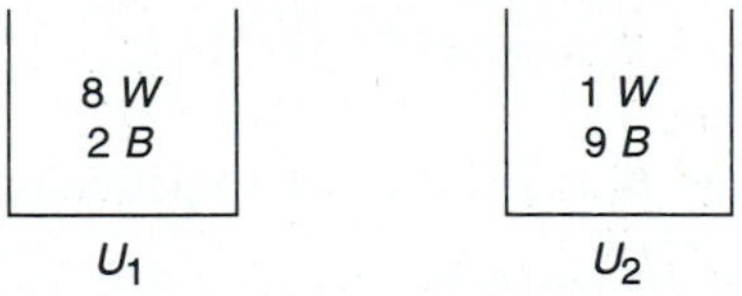

8 W		1 W
2 B		9 B

$$U_1 \qquad\qquad U_2$$

Fig. 2.6.3. An equivalent urn problem.

$$A = \text{an ace is selected}$$
$$B = \text{a heart is selected}$$
$$C = \text{a club is selected}$$

Clearly A and B are not mutually exclusive. Therefore

$$P(A \cup B) = P(A) + P(B) - P(A \cap B)$$
$$= \left(\frac{1}{13}\right) + \left(\frac{1}{4}\right) - \left(\frac{1}{52}\right) = \frac{16}{52}$$

This says that 1 of every 13 cards is an ace, and 1 of every 4 cards is a heart, but in counting the possibilities of both you should not count the ace of hearts twice.

Events B and C are mutually exclusive. Therefore

$$P(B \cup C) = P(B) + P(C) = \left(\frac{1}{4}\right) + \left(\frac{1}{4}\right) = \frac{1}{2}$$

This says that since no card is both a heart and a club, their probabilities add with no danger of counting the same card twice.

EXAMPLE 2.6.5.   Let us define A, B, and C as in Example 2.6.4, where event A means an ace is drawn and B means a heart is drawn. Consider drawing the ace of hearts, which is $A \cap B$. Then

$$P(A \cap B) = P(A|B)P(B) = \left(\frac{1}{13}\right)\left(\frac{1}{4}\right) = \frac{1}{52}$$

But $P(A) = \frac{4}{52}$, so

$$P(A)P(B) = \left(\frac{4}{52}\right)\left(\frac{1}{4}\right) = \frac{1}{52}$$

Since $P(A \cap B) = P(A)P(B)$, we conclude that the events A and B are statistically independent.

Now consider events B and C. We find

$$P(B|C) = 0$$

(It is impossible to select a heart and a club on the same draw.)

$$P(B) = \frac{1}{4}$$

Since $P(B|C)$ is not equal to $P(B)$, events B and C are not statistically independent.

EXAMPLE 2.6.6.   Referring once again to the events in Example 2.6.4, where A means an ace is drawn, B means a heart is drawn, and C means a club is drawn, we have the following relationships.

A and B are independent but not mutually exclusive. They are independent because $P(A) = P(A|B)$, but they are not mutually exclusive because both can happen in the drawing of one card.

B and C are dependent and mutually exclusive. They are dependent because $P(B) = \frac{1}{4}$ and $P(B|C) = 0$, so $P(B) \neq P(B|C)$. They are mutually exclusive because it is impossible to draw a heart and a club in one draw.

Section 2.5.   Observation Systems

EXAMPLE 2.6.7.   A radar system has a false alarm rate of 0.1 and probability of miss equal to 0.2. Assuming that a target is present with probability 0.1, find the initial entropy, the final entropy, and the mutual information.

SOLUTION: With reference to Fig. 2.5.3, the information in X is binary: Either a target is present or it is not present. Thus we have a binary world, either $X = 1$ (target present) or $X = 0$ (target absent). The probabilities are

$$P\{X = 1\} = 0.1$$
$$P\{X = 0\} = 0.9$$

A false alarm rate of 0.1 means that the radar will report a target present, when in fact there is no target, with probability 0.1. A probability of miss equal to 0.2 means that the radar will report no target when one is present 20% of the time. This gives the system probabilites $P(Y|X)$ as follows:

$P(Y\|X)$	$Y = 0$	$Y = 1$
$X = 0$	0.9	0.1
$X = 1$	0.2	0.8

This summarizes the information given in the problem statement in the form of $P(X)$ and $P(Y|X)$. $H(X)$ is calculated from $P(X)$, but for $H(X|Y)$

we need $P(X, Y)$ and $P(X|Y)$. Therefore we proceed as follows: First, we know $P(X)$ and $P(Y|X)$, which is needed to calculate $P(X, Y)$ in Eq. 2.5.5. Using this information we get

$P(X, Y)$	$Y = 0$	$Y = 1$
$X = 0$	0.81	0.09
$X = 1$	0.02	0.08

Now, to calculate $P(X|Y)$ to use in Eq. 2.5.6 we first need $P(Y)$. From Eq. 2.5.4 we have

$$P(Y = 0) = P(Y = 0|X = 0) \cdot P(X = 0) + P(Y = 0|X = 1) \cdot P(X = 1)$$
$$= (0.9)(0.9) + (0.2)(0.1) = 0.83$$

Also,

$$P(Y = 1) = P(Y = 1|X = 0) \cdot P(X = 0) + P(Y = 1|X = 1) \cdot P(X = 1)$$
$$= (0.1)(0.9) + (0.8)(0.1) = 0.17$$

Using these values along with the values from above for $P(X, Y)$ in Eq. 2.5.7, we obtain

$P(X|Y)$	$Y = 0$	$Y = 1$
$X = 0$	0.9759	0.5294
$X = 1$	0.0241	0.4706

We now have the information needed to calculate $H(X)$ and $H(X|Y)$:

$$H(X) = -0.1 \log(0.1) - 0.9 \log(0.9) = 0.469 \text{ bits}$$
$$H(X|Y) = -0.08 \log(0.4706) - 0.02 \log(0.0241)$$
$$-0.81 \log(0.9759) - 0.09 \log(0.5294)$$
$$= 0.3056 \text{ bits}$$

The entropy remaining in X is a large portion of the initial entropy. Therefore the mutual information is small, given by

$$I(X, Y) = H(X) - H(X|Y) = 0.1634 \text{ bits}$$

The relative merits of this system can be judged only in relation to other systems, but an information gain of only 0.1634 bits out of an initial entropy of 0.469 bits seems small. These numbers will change if the false alarm rate and probability of miss are changed. Another factor that affects system performance is the environment that the system is supposed to observe, especially the prior probabilities for the occurrence of X. The next example illustrates this point.

EXAMPLE 2.6.8.   Figure 2.6.4 illustrates a system where the entries in the table are $P(Y|X)$ for each pair (X, Y). Recall that our system model is the probability $P(Y|X)$ for all possible (X, Y) pairs, and this is provided in Fig. 2.6.4a. In working through this example we will see that this is a good unit if X is primarily 1 or 2, but a poor unit if X tends to be 3 or 4. Figures 2.6.4b and 2.6.4c show signal probabilities for two different signals, labeled "good signal" and "poor signal." The system transfers information better for the good signal. This is because the system is a unique mapping from input to output if X is 1 or 2, but it is completely random if X is 3 or 4. To see how much better the system is for each type of input, calculate the entropy in X given Y for each input in Fig. 2.6.4.

SOLUTION: In order to apply Eqs. 2.5.2 and 2.5.3, we must calculate $P(X, Y)$ and $P(X|Y)$ from the given data. From Eq. 2.5.5 we find $P(X, Y)$,

$P(Y|X)$

	4	0.25	0.25	0.25	0.25
	3	0.25	0.25	0.25	0.25
X	2	0	1	0	0
	1	1	0	0	0
		1	2	3	4

Y

(a) The system

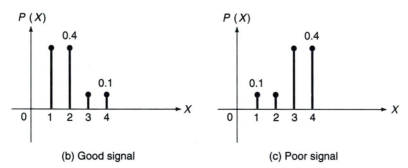

(b) Good signal   (c) Poor signal

Fig. 2.6.4.   This system is good if X is primarily 1 or 2, but poor otherwise.

Good signal

X				
4	0.025	0.025	0.025	0.025
3	0.025	0.025	0.025	0.025
2	0	0.4	0	0
1	0.4	0	0	0
	1	2	3	4

Y

Poor signal

X				
4	0.1	0.1	0.1	0.1
3	0.1	0.1	0.1	0.1
2	0	0.1	0	0
1	0.1	0	0	0
	1	2	3	4

Y

Fig. 2.6.5. $P(X, Y)$ for both signals.

first for the good signal, then for the poor signal, given by the tables of Fig. 2.6.5.

Now we need $P(X|Y)$, which we find from

$$P(X|Y) = \frac{P(X, Y)}{P(Y)}$$

Good signal

$P(Y)$	0.45	0.45	0.05	0.05
	1	2	3	4

Poor signal

$P(Y)$	0.3	0.3	0.2	0.2
	1	2	3	4

$P(X|Y)$

X				
4	0.0556	0.0556	0.5	0.5
3	0.0556	0.0556	0.5	0.5
2	0	0.889	0	0
1	0.889	0	0	0
	1	2	3	4

Y

$P(X|Y)$

X				
4	1/3	1/3	1/2	1/2
3	1/3	1/3	1/2	1/2
2	0	1/3	0	0
1	1/3	0	0	0
	1	2	3	4

Y

Fig. 2.6.6

where

$$P(Y) = \sum_{i=1}^{4} P(Y|X_i)\, P(X_i)$$

These quantities are displayed in Fig. 2.6.6.

Having calculated all necessary quantities, we plug into Eq. 2.5.6 to obtain

Good signal: $H(X|Y) = 0.653$ bits

Poor signal: $H(X|Y) = 1.351$ bits

The entropy remaining in X after observing Y is much lower for the good signal, being approximately half that obtained for the poor signal.

Finally, we calculate the mutual information between X and Y using Eq. 2.5.3. Note that $H(X)$ is the same for both the good signal and the bad signal, given by $H(X) = 1.722$ bits. Then Eq. 2.5.3 gives

Good signal: $I(X, Y) = 1.722 - 0.635 = 1.069$ bits

Poor signal: $I(X, Y) = 1.722 - 1.351 = 0.371$ bits

The mutual information between X and Y is larger for the good signal than it is for the poor signal, indicating that more information about experiment X is conveyed by the system when the good signal is used.

As this example illustrates, the utility of an observation system is a function of its intended use, and not solely a function of its internal construction. A good system in one application may be bad in another setting. When used properly, entropy measures the utility of a system in its intended setting.

2.7   Problems

2.1. Define (a) sample space, (b) event, (c) probability, (d) experiment. Be sure to include any relevant properties in your definitions.

2.2. A box contains 5 red, 3 black, and 2 white marbles. The experiment is to select one marble from the box.
(a) List the sample space.
(b) How many events are there?
(c) List all events and assign a probability to each.

2.3. Suppose that an experiment consists of spinning a pointer on a dial marked from 0 to 2π and noting where it stops.
(a) What is the sample space?
(b) Since the sample space is infinite, there are an uncountably infinite number of subsets, so it makes no sense to list the set of all events.

But a few events are listed below. Determine the probability of each event listed.

$\{x : 0 < x < \pi/2\}$
$\{x : 0 < x < 1\}$
$\{x : 0 < x < \pi \text{ or } \pi/2 < x < 3\pi/2\}$
$\{x : (0 < x < \pi) \text{ and } (\pi/2 < x < 3\pi/2)\}$

2.4. The game of craps uses a pair of dice, and the only outcome of interest is the sum of spots on the two dice. Suppose we toss a clear die and a red die. Answer the following questions.
(a) What is the experiment?
(b) List the sample space.
(c) List all possible outcomes corresponding to the event $r + c = 5$, where $r = $ number of spots on the red die and $c = $ number of spots on the clear die.
(d) Given that $r + c = 5$, what is the probability that $r = 3$?

2.5. Let $S = \{x_1, x_2, x_3\}$ be the sample space for an experiment. Then $F' = \{\emptyset, \{x_1\}, \{x_2, x_3\}, S\}$ is a Borel field. Show that although some subsets of S are not in F', we can assign a probability to the events in F' that satisfies all properties of Definition 2.1.4.

2.6 Let $S = \{a, b, c, d\}$ be the sample space for an experiment. Which of the following are legitimate Borel fields?

$F = \{\emptyset, \{a, b, c\}, \{c, d\}, S\}$
$F = \{\emptyset, \{a\}, \{b\}, \{c, d\}, S\}$
$F = \{\emptyset, \{a, b\}, \{c, d\}, S\}$

Can you assign a probability that satisfies Definition 2.1.4 in each case, or must we use legitimate fields?

2.7 The set of all subsets of a set A is sometimes called the power set and is denoted by the symbol 2^A. Suppose the sample space has 5 elements, so $S = \{x_1, x_2, x_3, x_4, x_5\}$. List a Borel field other than the power set 2^S.

2.8. The general definition of a field is given as follows: Let A be a set consisting of the elements $\{a, b, c, \ldots\}$. The field $(A, +, \cdot)$ is a set together with two operations, $+$ and $\cdot$. These operations, called "addition" and "multiplication," together satisfy the following nine conditions.
(i) $a \cdot (b + c) = a \cdot b + a \cdot c$ for all a, b, c in A.
(ii) $a + (b + c) = (a + b) + c$ for all a, b, c in A.
(iii) $a \cdot (b \cdot c) = (a \cdot b) \cdot c$ for all a, b, c in A.
(iv) $a + b = b + a$ for all a, b in A.
(v) $a \cdot b = b \cdot a$ for all a, b in A.
(vi) There exists an element 0 in A such that $a + 0 = a$ for all a in A.
(vii) There exists an element 1 in A such that $a \cdot 1 = a$ for all a in A.

(viii)   To each element a in A there corresponds an element $-a$ in A such that $a + (-a) = 0$.

(ix)   To each $a \neq 0$ in A there corresponds an a^{-1} in A such that $a \cdot a^{-1} = 1$.

(a) Show that the addition and multiplication defined by

$+_2$	0	1
0	0	1
1	1	0

$\cdot_2$	0	1
0	0	0
1	0	1

makes the system $(\{0, 1\}, +, \cdot)$ a field. This is called the binary field.

(b) Give the tables $+$ and $\cdot$ to make $(\{0, 1, 2\}, + \cdot)$ a field.

(c) Give the tables $+$ and $\cdot$ to make $(\{0, 1, 2, 3\}, + \cdot)$ a field. You should be warned that finding the operations to make a set with a composite (rather than a prime) number of elements is not easy. The magic word here is "Galois," because Galois field theory tells us how to construct these fields. This is an example of theory that pure mathematicians said would never be applied. It now constitutes the theoretical basis for many of the sophisticated codes used in communication theory.

2.9.   Prove that events with nonzero probability cannot be both independent and mutually exclusive.

2.10.   Let $S = \{f_1, f_2, f_3, f_4, f_5, f_6\}$, the set of all possible die-toss outcomes. Define events A, B, and C as

$$A = \{f_2, f_4, f_6\}$$
$$B = \{f_4, f_5, f_6\}$$
$$C = \{f_1, f_2, f_3, f_4, f_6\}$$

Evaluate $P(ABC)$ two ways:

(a) Find the intersection set ABC and evaluate its probability.

(b) Use the chain rule, Eq. 2.3.8.

Show that you get the same answer both ways.

2.11. Resistors coming off the production line are tested to see if they meet specifications. The probability of success is 0.95. If 5 parts are tested, what is the probability that all 5 will meet specifications? That 4 of 5 will meet specifications?

2.12. In a rocket there are 10 components. The probability of failure is 0.1 for any component, and the failure of any one is independent of another. If any component fails, the rocket will fail. Find the probability of a successful flight.

2.13. Here is an urn-type problem. Your friend chose at random one of three states A, B, or C to visit. The probability of rain in A was $\frac{1}{3}$, in B it was $\frac{1}{4}$, and in C it was $\frac{1}{6}$. He returned with mud on his car. What is the probability that he visited state C?

2.14. A deck of 52 ordinary playing cards consists of 13 cards in each of 4 suits. The suits are named spades, hearts, diamonds, and clubs. The 13 cards in each suit are numbered from 1 to 13, although labels of jack, queen, and king are attached to cards numbered 11, 12, and 13, respectively. Suppose that an experiment consists of drawing one card at random from the deck. The events A, B, and C are defined as follows:

$$A = \text{draw a club}$$

$$B = \text{draw an ace}$$

$$C = \text{draw the 7 of clubs}$$

(a) Are events A and B mutually exclusive? Independent?
(b) Are events B and C mutually exclusive? Independent?
(c) Are events A and C mutually exclusive? Independent?

2.15. Wilt Chamberlain is one of the most gifted athletes ever to have played professional basketball. During his career he had the coordination of a small guard on a 7-foot, 1-inch frame, and his ability made him the highest-scoring player in history at one time. He was a horrible free-throw shooter, however, making only about 50% of his free throws.

   Suppose that he is shooting free throws in his back yard while his neighbor's small child counts his successes. After shooting the ball 20 times, the child reports 8 baskets. Given that the child can count to 8 successfully 60% of the time, what is the probability that the number of baskets is in fact 8?

2.16. Three urns contain red and white marbles as shown in Fig. 2.7.1. An urn is selected, a ball drawn, and the color noted. It is white. If the prior probability of selecting each urn is $\frac{1}{3}$, what is the probability that the white ball came from urn 2?

2.17. Repeat Problem 2.16 if $P(U_1) = \frac{1}{2}$, $P(U_2) = \frac{1}{3}$, $P(U_3) = \frac{1}{6}$.

2.18. Find the entropy in the experiment in Problem 2.2. Express your answer in (a) hartleys, (b) nats, and (c) bits.

3 W	3 W	8 W
3 R	7 R	2 R
U_1	U_2	U_3

Fig. 2.7.1

2.19. In communication theory a binary channel has two input symbols (x_1, x_2) and two output symbols (y_1, y_2). The model for a binary channel consists of the transition probabilities $\Pr[y|x]$, as shown in Fig. 2.7.2. Thus for this channel,

$$\Pr[Y = y_1 | X = x_1] = 0.9$$
$$\Pr[Y = y_2 | X = x_1] = 0.1$$
$$\Pr[Y = y_1 | X = x_2] = 0.2$$
$$\Pr[Y = y_2 | X = x_2] = 0.8$$

Assume that the input probabilities are equal, $\Pr[x_1] = \Pr[x_2] = 0.5$.
(a) Find the entropy in X.
(b) Find the entropy in X given Y.
(c) Find the mutual information $I(X, Y)$.
(d) The channel capacity in bits per symbol is defined as the maximum mutual information $I(X, Y)$ obtained by varying the input probabilities. Suppose that we are able to vary $\Pr[x_1]$ and $\Pr[x_2]$, which is equal to $1 - \Pr[x_1]$. Calculate $I(X, Y)$ for input probabilities $\Pr[x_1]$ of 0.49, 0.5, 0.51, 0.52, 0.53, and 0.54. What is the best estimate of the channel capacity from these results?

2.20 Repeat parts (b), (c), and (d) of Problem 2.19 for the binary symmetric channel in Fig. 2.7.3.

2.21. Figure 2.7.4 shows a matrix of probabilities $P(Y|X)$ for a particular observation system. It should be better if the input signal x tends to have value 1 or 2, such as the distribution $P_1(x)$. Show that this is true by calculating the conditional entropies and mutual information for (a) $P_1(x)$,

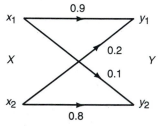

Fig. 2.7.2

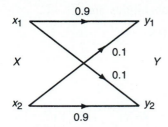

Fig. 2.7.3

and (b) $P_2(x)$.

2.22. Three urns contain red and white marbles as follows:

Urn 1: 4 red, 1 white
Urn 2: 3 red, 3 white
Urn 3: 1 red, 3 white

(a) Find the entropy in the experiment of selecting an urn at random, i.e., with equal probability.
(b) Find the entropy remaining in the urn selection experiment after you know the color of the marble selected from the urn.

2.23. Repeat Problem 2.22 if the urns have prior probabilities $P(U_1) = \frac{1}{2}$, $P(U_2) = \frac{1}{4}$, $P(U_3) = \frac{1}{4}$.

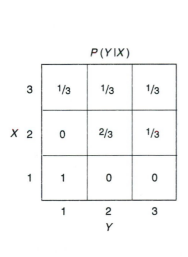

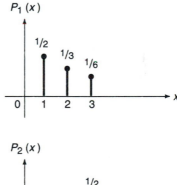

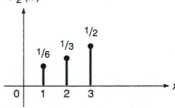

Fig. 2.7.4

2.24. Repeat Problem 2.19 if the transition probabilities are

$$\Pr[Y = y_1 | X = x_1] = 0.8$$
$$\Pr[Y = y_2 | X = x_1] = 0.2$$
$$\Pr[Y = y_1 | X = x_2] = 0.2$$
$$\Pr[Y = y_2 | X = x_2] = 0.8$$

2.25.   (a) Find the mutual information $I(X, Y) = H(X) - H(X|Y)$ for the ternary symmetric channel of Fig. 2.7.5. All inputs are equally likely.
(b) What is the channel capacity? Label your answer in either bits, nats, or hartleys per symbol.

2.26.   The following statements are either correct, partially correct, or wrong. Change each statement as needed to make it correct.
(a) If events A and B are mutually exclusive, then $P(A \cap B) = 0$.
(b) Probability is a function whose domain is the sample space, and whose range is the set of numbers $0 \le P \le 1$.
(c) An event is a subset of the sample space.
(d) The sample space is the set of all events.
(e) If a coin is tossed with possible outcomes H and T, then there are two events.
(f) Probability must satisfy three properties. $P(X \cup Y) = P(X) + P(Y)$ is one of these properties.
(g) If two events A and B are independent, then $P(A \cap B) = 0$.
(h) If a die is tossed there are 6 possible experimental outcomes. This means there are 64 events.
(i) A box contains 6 red and 4 black balls. If the experiment is to select one ball at random, the sample space is $S = \{B, R\}$.
(j) A box contains 6 red and 4 black balls. If the experiment is to select one ball at random, there are 2 events.

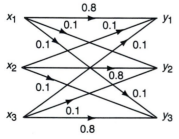

Fig. 2.7.5.   A ternary channel.

Further Reading

1. RICHARD W. HAMMING, *The Art of Probability for Scientists and Engineers,* Addison-Wesley, Reading, MA, 1991.
2. ROBERT G. GALLAGER, *Information Theory and Reliable Communication,* John Wiley & Sons, New York, 1968.
3. ATHANASIOS PAPOULIS, *Probability, Random Variables, and Stochastic Processes,* 3rd ed., McGraw-Hill, New York, 1991.
4. ATHANASIOS PAPOULIS, *Probability and Statistics,* Prentice Hall, Englewood Cliffs, NJ, 1990.
5. BART KOSKO, *Neural Networks and Fuzzy Systems,* Prentice Hall, Englewood Cliffs, NJ, 1992.

CHAPTER 3 _____

Random Variables

3.1 Definition of Random Variable _____

Preview

Gambling motivated the development of probability theory by Blaise Pascal and Pierre de Fermat in the seventeenth century. The French nobility had the time and money to devote to games of chance, so there was great interest in obtaining an edge in the competition. This motivated development of the basic theory that we enjoy today.

Most gambling games are repetitious, with the score or payoff changing frequently. In craps the player tosses the dice until he loses his turn, and then another player tries his luck at casting the dice. Money changes hands at times determined by the experimental outcomes. In bridge the player's score may or may not change on each deal of the cards. In horse racing the tote board displays the payoff for each possible experimental outcome. In each of these experiments there is a variable (the payoff) associated with each experimental outcome. Furthermore, the variability of the outcome is displayed by the repetitious nature of most gambling games—hence the name *random variable*.

Notice that in these examples the payoff is associated with the experimental outcome. Recall that the sample space is the set of all experimental outcomes. In the examples above there is a number (the payoff) associated with each experimental outcome, or a mapping from the sample space to the set of real numbers. A random variable converts an experimental outcome into a number. This makes random phenomena easier to work with. In this section we will formalize this definition and illustrate it with several examples. We also discuss entropy for a random variable.

DEFINITION 3.1.1.   A (real-valued) random variable X is a function with domain S (the same space) and codomain R (the set of real numbers), written $X: S \rightarrow R$. In addition, this function must satisfy the following two properties:

1. $P\{X = \infty\} = 0$, $P\{X = -\infty\} = 0$.
2. The set $\{X \leq \alpha\}$ is an event for every real number α.

Another function whose domain is related to the sample space is probability (see Definition 2.1.4). There are several important differences: (1) The domain for random variables is just S, the sample space. The domain for probability is all subsets of S (all events). Hence, their domains are different. (2) The range for a random variable can be any set of real numbers, while the range for probability is the interval $[0, 1]$. (3) There is an important conceptual difference. Probability is assigned before the experiment is performed. (After all, if we knew the experimental outcome, we would assign probability 1 to any event that included the actual outcome, and probability 0 to all other events.) A random variable is a function that assigns numbers to the experimental outcomes, and it is immaterial whether the experiment has been performed or not.

We use a capital letter (usually X, Y, or Z) to denote a random variable, and the corresponding lowercase letter (x, y, or z, and sometimes α) to denote an element in the range, i.e., a particular value of the random variable. The Greek letter ζ denotes an element in the domain (a possible experimental outcome). Some examples will illustrate these concepts.

EXAMPLE 3.1.1.   Suppose you match coins with a friend, winning \$1.00 if the two coins match and losing \$1.00 if the coins do not match. A random variable may be defined as the mapping from the sample space {HH, HT, TH, TT} to the numbers $+1$ and -1. We identify the elements in the domain with the Greek letter ζ_i as follows:

$$\zeta_1 = HH, \quad \zeta_2 = HT, \quad \zeta_3 = TH, \quad \zeta_4 = TT$$

If our random variable is labeled X, then

$$X(\zeta_1) = 1, \quad X(\zeta_2) = -1, \quad X(\zeta_3) = -1, \quad X(\zeta_4) = 1$$

The following chart displays this function:

X	1	-1	-1	1
ζ	HH	HT	TH	TT

Notice that this random variable satisfies properties 1 and 2 in Definition 3.1.1. There are only two possible values for X, $+1$ and -1, so X can never be infinite and property 1 is satisfied. To show that property 2 is

Y	-2	-1	0	1	2	3
ζ						

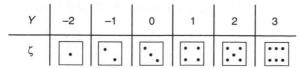

Fig. 3.1.1.   The random variable Y.

satisfied, choose a value for α—say, $\alpha = 0.6$. Then $X \leq \alpha$ if $\zeta = $ HT
or if $\zeta = $ TH, that is, if the event {HT, TH} occurs. Therefore $\alpha = 0.6$
determines an event. Let $\alpha = -10$. The event determined is $\varnothing$, because
there is no experimental outcome that produces a value of $X \leq -10$. This
is perfectly okay, however, because $\varnothing$ counts as an event. For $\alpha > 1$ the
event is the sample space. Thus for every α we have an event, and property
2 is satisfied.

This is an example of a discrete random variable because there are
only two elements in the range, -1 and $+1$.

EXAMPLE 3.1.2.   Figure 3.1.1 illustrates the random variable Y. A
single die is thrown. The random variable assigns the numbers -2, -1,
0, 1, 2, and 3 to the six possible outcomes ζ_i. Figure 3.1.2 illustrates
another random variable Z. The random variable Z is different from Y,
nevertheless, it is a perfectly good random variable.

Property 2 of the definition for random variable says that the set $\{Y \leq \alpha\}$
must be an event for every real number α. Let $\alpha = 0.5$. Then if $Y \leq 0.5$,
the event {⊡ ⊡ ⊡} has occurred. If $Y \leq -5.6$, the event $\varnothing$ has occurred.
You can see that every number α identifies a subset of the sample space,
and $\{Y \leq \alpha\}$ is an event for all α. This is a discrete random variable, for
there are a countable number of elements in the range, -2, -1, 0, 1, 2,
and 3.

A *discrete* random variable has a countable number of elements in the
range. A *continuous* or *mixed* random variable has an uncountably infinite
number of elements in the range, and therefore in the domain. Here is an
example.

EXAMPLE 3.1.3.   The pointer-spinning experiment in Fig. 2.1.1 can
produce either a continuous or a discrete random variable. Recall that a

Z	4	1	0	1	4	9
ζ						

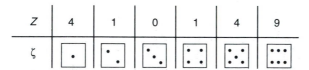

Fig. 3.1.2.   The random variable Z.

pointer is spun on a circle marked off in radians, from 0 to 2π. The experimental outcome is the position where the pointer stops, so the sample space is the set of all angles, written $S = \{\beta: 0 \le \beta < 2\pi\}$. Now here are two random variables, one continuous and one mixed.

CONTINUOUS RANDOM VARIABLE. Let $X = \beta/2\pi$. By this we mean that the value of the random variable is determined by where the pointer stops on a trial of the experiment, and this value is between 0 and 1. Then there are an uncountably infinite number of values of X, all numbers α between 0 and 1.

MIXED RANDOM VARIABLE. Let $Y = \beta/2\pi$ for $0 \le \beta < \pi$. Let $Y = 1$ for other β. Then Y takes on values between 0 and $\frac{1}{2}$ when the pointer stops on the right side of the circle, and it equals 1 when the pointer stops anywhere on the left side of the circle. This is called mixed because there are an uncountable number of elements in the range plus some countable values (in this case only one countable value, $Y = 1$).

Both X and Y satisfy the two properties for a random variable in Definition 3.1.1. For example, Y has no values greater than 1 nor less than 0, so $P\{Y = \infty\} = 0$, and $P\{Y = -\infty\} = 0$. Also, every value of α in $\{Y \le \alpha\}$ defines an event. Let $\alpha = 0.82$. The corresponding event is $\{0 \le \beta < \pi\}$.

In random signal processing we are mostly concerned with waveforms, be they voltage waveforms, current waveforms, digital signals, or acoustic (speech) waveforms. Suppose we wish to design a system to transmit telephone conversations over a satellite link. This means we are to design analog-to-digital (A/D) and digital-to-analog (D/A) converters, filters, modulators, demodulators, and antenna systems. These designs will be influenced by the characteristics of the signals processed by the system. Therefore we need to know the maximum and minimum values, the power, and the dc value of the signals. We will see that if we identify the values taken by the waveforms with a random variable, where the domain (sample space) may or may not be known, then it will be possible to find these and other characteristics of the signal. Therefore this concept of random variable is more than just an interesting anomaly. We will illustrate the use of this concept in the next few sections.

Entropy of Random Variables

In Section 2.4 we said two things: First, that experiments and only experiments have entropy; and, second, that partitions and random variables have entropy. We demonstrated that a partition of an existing sample space is equivalent to defining a new experiment, and now we wish to show that a random variable can partition a sample space. Since a random variable can partition a sample space, and since this is equivalent to defining a new experiment, it follows that one should look at the entropy of a random variable as the entropy in this new experiment.

We will use the random variable Z in Fig. 3.1.2 for this discussion. That random variable can assume four different values, 0, 1, 4, and 9. The equivalent subsets of the sample space S that correspond to these values are

$$0—\{f_3\}$$
$$1—\{f_2, f_4\}$$
$$4—\{f_1, f_5\}$$
$$9—\{f_6\}$$

where f_i represents the die face with i spots. This means that

If $Z = 0$, then f_3 occurred with probability $P_1 = \frac{1}{6}$.
If $Z = 1$, then either f_2 or f_4 occurred ($P_2 = \frac{1}{3}$)
If $Z = 4$, then either f_1 or f_5 occurred ($P_3 = \frac{1}{3}$)
If $Z = 9$, then f_6 occurred ($P_4 = \frac{1}{6}$)

Now we have a set of probabilities, so we can calculate entropy:

$$H(Z) = -\sum_i P_i \log P_i = \tfrac{2}{3}\log 3 + \tfrac{1}{3}\log 6 = 1.9183 \text{ bits}$$

Therefore entropy can be assigned to random variables, but you can avoid mistakes if you remember that this is possible only because the random variable forms a partition of the sample space, and this is equivalent to defining a new experiment. This is important because it prevents you from calculating the entropy in less than the entire experiment. It is meaningless to discuss the entropy in the values 0 and 1 of the random variable Z, because this ignores the other possible outcomes that produce values $Z = 4$ and $Z = 9$. One must use the entire sample space (and all probabilities) in calculating entropy, because entropy is defined as an average.

Review

The idea of a random variable started with gambling. For a given trial, the experimental outcome determines how much money you win or lose. Thus an experimental outcome determines a number. This is a random variable. Hopefully we can put this number to better use than gambling.

The outcome can be from a continuum, or it can be from a finite (discrete) set. The outcomes in the coin-toss and die-toss experiments are discrete. The outcome in the pointer-spinning experiment is from a continuum. Thus there are three possible types of domain for random variables, discrete, continuous, and mixed, depending on whether the sample space is discrete or continuous.

3.2   Distribution Functions ————————

Preview

We now wish to introduce two functions, the cumulative distribution function (cdf) and the probability density function (pdf). A random variable has already been introduced as a function whose domain is the sample space and whose codomain is the set of real numbers, but this is only an interesting concept until we relate the values of the random variable to a particular application. For us, this means finding a relation to the power, dc value, autocorrelation function, and other parameters of a random waveform. This is accomplished through the cdf and the pdf.

DEFINITION 3.2.1.   The probability of the event $\{X \le \alpha\}$, denoted by

$$F_X(\alpha) = P\{X \le \alpha\} \qquad (3.2.1)$$

is called the cumulative distribution function (cdf) of the random variable X.

By the symbol $\{X \le \alpha\}$ we mean the event "the value of the random variable X is less than or equal to α," where α is just some number. The cdf is therefore a function of α. The larger α is, the more likely is the event $\{X \le \alpha\}$. The subscript X on F indicates that this is the cdf of the random variable X. The cdf for the random variable Y would be written F_Y.

The domain of F_X is the range of the random variable X. See Fig. 3.2.1. The random variable X is a mapping from the sample space S to the range R_X. The function F_X is a mapping from R_X to the range of F_X, namely, $[0, 1]$. The function F_X is a probability, so its domain must be a set of events, and it is. Recall that property 2 of the definition for random variable said $\{X \le \alpha\}$ is an event for every α. It is these numbers α that form the domain of the function F.

EXAMPLE 3.2.1.   In the experiment of tossing a coin there are two possible outcomes, heads or tails. Let $X = 1$ if heads is the outcome, and

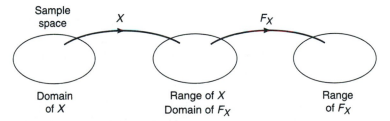

Fig. 3.2.1.   Illustrating the relationship between the random variable X and its cdf.

$X = -1$ if the outcome is tails. Then the random variable X shown in Fig. 3.2.2 has the cdf shown there. To see this, choose a value for α and find $P\{X \leq \alpha\}$. Let $\alpha = 10$. Then the probability is 1 that $X \leq 10$ because every experimental outcome gives a value of $X \leq 10$. You can see that this holds for any $\alpha \geq 1$, because every experimental outcome gives a value of $X \leq 1$.

Let $\alpha = 0.6$. Now the probability of $\{X \leq \alpha\}$ is $\frac{1}{2}$, because this event occurs if tails comes up, but does not occur if heads comes up. Thus $F_X(\alpha) = 0.5$ for $-1 \leq \alpha < 1$. For $\alpha < -1$, the value of $F_X(\alpha)$ is 0, because this event is the empty set (no experimental outcome gives a value of $X < -1$).

Properties of $F_X(\alpha)$

1. $0 \leq F_X(\alpha) \leq 1$ [since $F_X(\alpha)$ is a probability].
2. $F_X(\infty) = 1$, $F_X(-\infty) = 0$.
3. $F_X(\alpha_1) \leq F_X(\alpha_2)$ if $\alpha_1 < \alpha_2$.

These properties follow from the properties of probability. It will be instructive to prove property 3.

Let $\alpha_1 < \alpha_2$. We can write

$$P\{X \leq \alpha_2\} = P\{X \leq \alpha_1\} + P\{\alpha_1 < X \leq \alpha_2\}$$

Rearranging terms, we obtain

$$P\{\alpha_1 < X \leq \alpha_2\} = P\{X \leq \alpha_2\} - P\{X \leq \alpha_1\}$$
$$= F_X(\alpha_2) - F_X(\alpha_1) \qquad (3.2.2)$$

but $P\{\alpha_1 < X \leq \alpha_2\} \geq 0$ since it is a probability. From this and Eq. 3.2.2 we obtain property 3.

Our next important topic is the probability density function, the pdf. The term *density* has a special meaning in engineering and physics: It means that you must add or integrate the density function. In mechanics, the mass density function for a beam must be integrated to find the mass of the beam. In signal processing, the power density function must be integrated to find the power in the signal. Here we have the probability density function, which must be integrated to find the probability. Here is the definition.

DEFINITION 3.2.2.   The derivative with respect to α of the cdf is called the probability density function (pdf). It is given by

$$f_X(\alpha) = \frac{d}{d\alpha} F_X(\alpha) \qquad (3.2.3)$$

EXAMPLE 3.2.2.   The random variable X in Fig. 3.2.2 has the pdf shown in Fig. 3.2.3. The derivative of $F_X(\alpha)$ is 0 everywhere except at

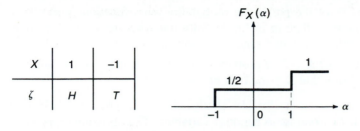

Fig. 3.2.2.   The random variable X and its cdf.

$\alpha = -1$ and $\alpha = 1$. There the derivative is a delta function of area 0.5. Notice that if we integrate the pdf we obtain the cdf, and the value of the cdf is a probability for each value of α. Thus integrating the density function gives a probability.

Since the pdf f_X is the derivative of the cdf F_X, the cdf must be the integral of the pdf. Therefore we have the relationship

$$F_X(\alpha) = \int_{-\infty}^{\alpha} f_X(\lambda)\, d\lambda \qquad (3.2.4)$$

Properties of f_X

1. $f(\alpha) \geq 0$ for all α.

2. $\int_{-\infty}^{\infty} f_X(\alpha)\, d\alpha = 1$.

3. $F_X(\alpha) = \int_{-\infty}^{\alpha} f_X(\lambda)\, d\lambda$.

These properties follow from the properties of the cdf. Property 1 follows from the monotonically nondecreasing property of the cdf (property 3). Since the cdf is nondecreasing, the derivative cannot be negative. Property 2 of the pdf follows directly from property 2 of the cdf. And property 3 of the pdf follows from the relationship between the cdf and the pdf.

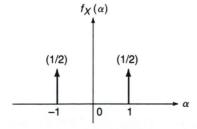

Fig. 3.2.3.   The probability density function for the random variable X.

A close relative of the pdf is the probability mass function (pmf). It is easier to add numbers than to integrate delta functions, so we use the pmf in place of the pdf for discrete random variables.

DEFINITION 3.2.3.   The probability mass function is, for any number α, the probability that $X = \alpha$. It is given by

$$p_X(\alpha) = P(X = \alpha) \tag{3.2.5}$$

The pmf is 0 for continuous random variables. Thus it is appropriate to use the pmf only for discrete random variables. The pdf can be used for both discrete and continuous random variables, but the advantage of the pmf, where appropriate, is that we can use sums instead of integrals in the formulas that follow.

EXAMPLE 3.2.3.   The cdf and pmf for the random variable Y in Fig. 3.2.4a are shown in Figs. 3.2.4b and 3.2.4c. Find the probability that $\{1.5 < Y \le 4.2\}$.

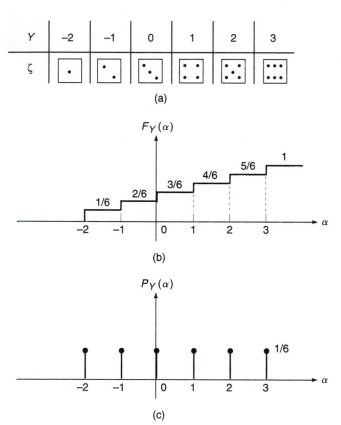

(a)

(b)

(c)

Fig. 3.2.4.   The cdf and pmf for Y.

First solution: Here we use the cdf:

$$P\{1.5 < Y \le 4.2\} = P\{Y \le 4.2\} - P\{Y \le 1.5\}$$
$$= F_Y(4.2) - F_Y(1.5) = 1 - \tfrac{2}{3} = \tfrac{1}{3}$$

Second solution: Here we use the pmf:

$$P\{1.5 < Y \le 4.2\} = p_Y(2) + p_Y(3) = \tfrac{1}{3}$$

Review

The cdf is a probability, the probability that the value of the random variable will be less than or equal to a given number α. We choose α in advance and perform the experiment. The experimental outcome determines the value of the random variable, and the cdf is the probability that this value is less than or equal to α.

The pdf is the derivative of the cdf with respect to α. Since the cdf is monotonically nondecreasing, the pdf is greater than or equal to 0. The pdf is a density function, which indicates that it must be integrated to produce probability.

3.3   Signal Analysis

Preview

Our purpose in this section is to explain how to find the cdf and the pdf for periodic waveforms, and to describe how to measure the cdf and the pdf for random waveforms. The procedure for finding the cdf is simple, straightforward, and hard to grasp. At least, many students fail to understand it. But if you pay particular attention to Eq. 3.3.2 and Fig. 3.3.7, you should be able to grasp the idea.

Figure 3.3.1 shows a periodic continuous-time square wave $v(t)$ that has values 3 and 1. If we choose at random a point on the time axis in the interval $(0, T)$, the value of the function for that particular time will be a number, either 3 or 1. Here we have the ingredients of an experiment with

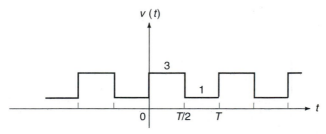

Fig. 3.3.1.   The periodic function $v(t)$.

a random variable defined on a sample space. The sample space consists of all values of time that might be chosen, and the values of the random variable are 3 and 1. Let us call the random variable so defined V. Then the cdf is the probability that V is less than or equal to some number α.

$$F_V(\alpha) = P\{V \le \alpha\} \tag{3.3.1}$$

The number α is a value on the vertical axis in Fig. 3.3.1.

To find the cdf we choose a value of α and find the probability that $V \le \alpha$. For $\alpha \ge 3$, the event $\{V \le \alpha\}$ is certain (the voltage must be 1 or 3 and either value is less than or equal to α). Therefore $P\{V \le \alpha\} = 1$ for $\alpha \ge 3$. For values of α in the range $1 \le \alpha < 3$, the probability $P\{V \le \alpha\} = \frac{1}{2}$. The random variable V can assume only the values 1 or 3. If $V = 3$, then it is not true that $V \le \alpha$. If $V = 1$, it is true. These occur with equal probability, so the cdf has value $\frac{1}{2}$ for $1 \le \alpha < 3$. Finally, for $\alpha < 1$, it can never happen that $V \le \alpha$. Therefore $P\{V \le \alpha\} = 0$ for $\alpha < 1$. The function $F_V(\alpha)$, for all values of α, is given by

$$F_V(\alpha) = P\{V \le \alpha\} = \begin{cases} 1, & 3 \le \alpha \\ \frac{1}{2}, & 1 \le \alpha < 3 \\ 0, & \alpha < 1 \end{cases}$$

The cdf and pdf are plotted in Fig. 3.3.2. We differentiate $F_V(\alpha)$ with respect to α to find the pdf $f_V(\alpha)$. The pmf looks like the pdf, except that the delta functions are replaced by numbers. We could have found the pmf directly, because $P\{V = 1\} = \frac{1}{2}$, and $P\{V = 3\} = \frac{1}{2}$.

The discrete-time signal shown in Fig. 3.3.3 has exactly the same cdf and pdf as the continuous-time signal in Fig. 3.3.1. The experiment is to select at random a time n, and the random variable defined on this experiment has a value of either 1 or 3. Hence the pdf and cdf are those shown in Fig. 3.3.2.

In these examples we determined the cdf (and pdf) from the waveform. Here is a different example. Toss a coin once each second, beginning at time $n = 1$. Let the random variable V assume the value 1 if heads occurs at time n, and let V assume the value 3 if tails occurs. A typical sequence so generated is shown in Fig. 3.3.4. (Keep in mind that this is only one of

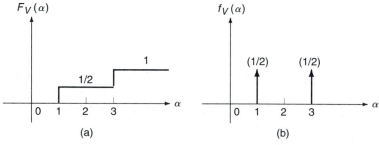

Fig. 3.3.2.   (a) The cdf and (b) the pdf for $v(t)$.

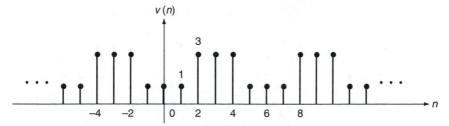

Fig. 3.3.3.   The discrete-time function $v(n)$.

an infinite number of possible sequences, assuming we continue to toss the coin forever.) For $n > 0$ this function has the cdf and the pdf in Fig. 3.3.2.

Let us now consider some waveforms that have a different distribution. (The word ''distribution'' is used to describe the frequency with which the various values of a random variable occur. Therefore either or both the cdf and the pdf are referred to as the distribution.)

Figure 3.3.5 illustrates another voltage waveform with the corresponding cdf and pdf. Again, the random variable is discrete, with possible values of 1 and 3. The probabilities of these two values are no longer equal, however. From the diagram of the waveform $v(t)$ it is evident that $P\{V = 3\} = \frac{2}{3}$ and $P\{V = 1\} = \frac{1}{3}$. The associated cdf and pdf should be compared to those of Fig. 3.3.2 to illustrate the effect of these different probabilities. The cdf is given by (Fig. 3.3.5)

$$F_V(\alpha) = \begin{cases} 1, & 3 \leq \alpha \\ \frac{1}{3}, & 1 \leq \alpha < 3 \\ 0, & \alpha < 1 \end{cases}$$

For another example, toss a die once each second, beginning at $n = 1$. Let the random variable V assume the value 1 if the one-spot or two-spot occurs. Let $V = 3$ if any of the other spots occurs. Then this sequence might look like $v(n)$ in Fig. 3.3.6. This signal has the cdf and the pdf shown in Fig. 3.3.5.

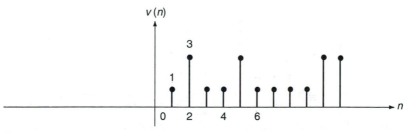

Fig. 3.3.4.   A coin-toss experiment.

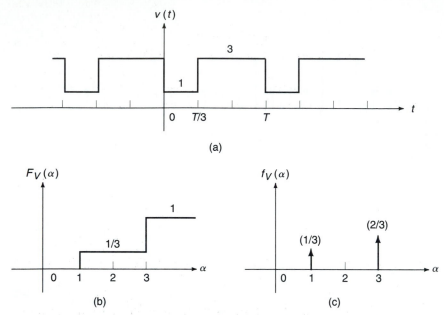

(a)

(b)   (c)

Fig. 3.3.5.   (a) Another waveform and (b) its cdf and (c) its pdf.

There is a fundamental difference between these last two examples. The probabilities for the cdf are obtained from the waveform in Fig. 3.3.5, while the probabilities in the die-toss example are obtained from the underlying experiment. Future values of the waveform are predictable in all our examples involving periodic waveforms, while future values are not predictable in the coin-toss and die-toss experiments. We will find all applications of interest to be of the second type; future values of the waveform will be unpredictable with complete accuracy. These are the only types of signals that transmit information. After all, if the waveform is predictable, there is no need to transmit it. Simply tell the receiver how to predict future values.

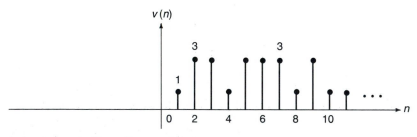

Fig. 3.3.6.   A die-toss experiment.

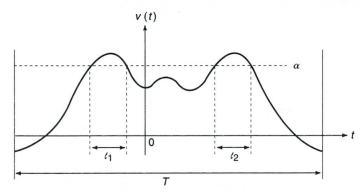

Fig. 3.3.7. Measuring the empirical cdf.

In many cases we will not know enough about the underlying physical experiment to be able to derive the cdf directly. We must measure the cdf from the waveform. The method is to fix a threshold α and measure the percentage of time the waveform has a value less than or equal to this threshold. This percentage is the value of the estimated cdf for that value of α. This is illustrated in Fig. 3.3.7, where the estimated cdf for the given value of α is

$$F_V(\alpha) = \frac{T - l_1 - l_2}{T} \tag{3.3.2}$$

This empirical cdf is estimated by measuring time intervals. It is the ratio of time the function has a value less than or equal to α over the total time T. This procedure applies to any waveform. For discrete-time functions the number of samples with a value less than or equal to α over the total number of samples replaces the time intervals.

Let us apply this concept to the periodic sawtooth signal $v(t)$ in Fig. 3.3.8. The maximum value of $v(t)$ is 2, so for any value of $\alpha \geq 2$ the cdf is 1.

$$F_V(\alpha) = 1, \qquad \alpha \geq 2$$

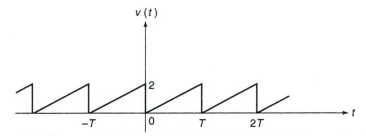

Fig. 3.3.8. A sawtooth waveform.

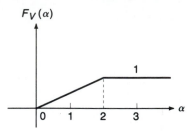

$F_V(\alpha)$

Fig. 3.3.9

For $\alpha < 0$, we again have an easy problem because the cdf is 0. But for values of α between 0 and 2, we need to apply the concept in Eq. 3.3.2. To be specific, let $\alpha = 1$. Now we ask: "For what percentage of the period is $v(t) \leq 1$?" The answer is $\frac{1}{2}$. When $0 \leq t < T/2$, the voltage is less than 1. For the other half of the period, the voltage is greater than 1. Therefore

$$F_V(\alpha) = 0.5, \qquad \alpha = 1$$

Using similar reasoning for $\alpha = 0.5$ and 1.5, we have

$$F_V(\alpha) = 0.25, \qquad \alpha = 0.5$$
$$= 0.75, \qquad \alpha = 1.5$$

Because of the continuous nature of $v(t)$ in one period you can see that the resulting cdf will have the form shown in Fig. 3.3.9.

Review

To find the empirical cdf for continuous-time signals we measure time intervals. For discrete-time signals we count pulses. The two procedures are as follows.

1. Continuous-time signals:

$$F_V(\alpha) = \frac{\text{length of time } v(t) \leq \alpha}{\text{total time considered}}$$

2. Discrete-time signals:

$$F_V(\alpha) = \frac{\text{number of pulses with value } v(n) \leq \alpha}{\text{total number of pulses considered}}$$

3.4 Statistical Averages

Preview

A short sample of one random signal does not give accurate results when we compute the mean square value, or the correlation function, or any other quantity derived from the signal. Improved accuracy results from averaging over several signals. This is not true, of course, for deterministic signals; but for random signals, the more signals the better the results.

   This suggests that to achieve complete accuracy we should average over all possible cases, and that is precisely what it means to find the statistical average. The purpose of this section is to introduce statistical averages and to indicate their importance to signal processing. The most important of these are the mean, the variance, and the mean square value. We will indicate what we can do with statistical averages that we cannot do with time averages. In the process we will find need for further theory—specifically, the concept of stochastic processes. We begin with the average or the expected value of a random variable.

The expected value of the random variable X is given by

$$E(X) = \int_{-\infty}^{\infty} \alpha f_X(\alpha) \, d\alpha \tag{3.4.1a}$$

$$= \sum_{\alpha} \alpha p_X(\alpha) \tag{3.4.1b}$$

where Eq. 3.4.1a is used for continuous or mixed distributions, and Eq. 3.4.1b is used for discrete distributions. Other names for $E(X)$ are average value, mean value, and statistical average. The only information needed to calculate the expected value of a random variable is its pdf (or pmf). No other information is needed.

   We often find the average value of a function of X—say, $g(X)$—where $g(X)$ may be X^2, or $[X - E(X)]^2$, or any other function of the random variable X. The definition in Eq. 3.4.1 may be extended to any function of X by

$$E[g(X)] = \int_{-\infty}^{\infty} g(\alpha) f_X(\alpha) \, d\alpha \tag{3.4.2a}$$

$$= \sum_{\alpha} g(\alpha) p_X(\alpha) \tag{3.4.2b}$$

This formula is known as the fundamental theorem of statistics. It will serve us well in the future, for it is probably the single most important formula in this or any other book on statistics. As before, part b of this formula is to be used only for discrete random variables. In the future we will drop this second form, it being understood.

Moments

The average value of X is called the first moment m_1. If we let $g(X) = X^2$ in Eq. 3.4.2, this is the second moment m_2. In general, the nth moment is given by

$$m_n = E(X^n) = \int_{-\infty}^{\infty} \alpha^n f_X(\alpha) \, d\alpha \tag{3.4.3}$$

Another important term is *central moment*. If we first subtract m_1 (the mean) from X and then find the average value of $g(X) = (X - m_1)^n$, we get the nth central moment. The second central moment is so important that it rates a special name, *variance,* and a special label, σ^2. The variance of X is given by

$$E[(X - m_1)^2] = \sigma^2 = \int_{-\infty}^{\infty} (\alpha - m_1)^2 f_X(\alpha)\, d\alpha \qquad (3.4.4)$$

The square root of the variance is called the *standard deviation, σ.* The variance (or standard deviation) is a measure of the spread of the values of the random variable. You might say that it is a measure of the randomness of the random variable, but that would be confusing since variance and entropy are different. The variance is a measure of how much the values of a random variable can vary, and entropy is a measure of how nearly equal are the probabilities in an experiment (see Section 2.4).

Expected values have the following useful properties:

1. Let c be a constant. Then $E(c) = c$.
2. Let X be a random variable and c be a constant. Then $E(cX) = cE(X)$.
3. Let X and Y be two random variables. Then $E(X + Y) = E(X) + E(Y)$.
4. Let X and Y be independent. Then $E(XY) = E(X)E(Y)$.

The Use of Moments in Waveform Analysis

For random variables related to waveforms (such as those in Section 3.3), the first moment is the dc value, the second moment is the average power, and the variance is the ac power. These relationships among parameters for waveforms are tabulated as follows:

$$
\begin{aligned}
m_1 &= \text{dc value} \\
m_1^2 &= \text{dc power} \\
\sigma^2 &= \text{ac power} \\
m_2 &= \text{total power} = \sigma^2 + m_1^2
\end{aligned}
\qquad (3.4.5)
$$

where the power terms are on a 1-Ω basis. That is, if the voltage waveform is impressed across a 1-Ω resistor, the average power will be given by the terms in Eq. 3.4.5. The last relationship between ac, dc, and total power can be demonstrated as follows. By definition,

$$\sigma^2 = E[(V - m_1)^2] = E(V^2 - 2m_1 V + m_1^2)$$

Then

$$\sigma^2 = E(V^2) - 2m_1 E(V) + m_1^2 = E(V^2) - m_1^2 = m_2 - m_1^2 \qquad (3.4.6)$$

The relationships in Eq. 3.4.5 simply state that if the waveform is decomposed by the subtraction of the dc level, then m_1^2 is the power in a dc signal of that amplitude, the variance σ^2 is the power in the signal after the dc level is removed, and the second moment m_2 is the power in the original waveform. We assume, of course, that we are dealing with a power signal, i.e., a signal that has finite but nonzero average power.

For the waveforms discussed in Section 3.3, we can compute the dc power, ac power, and total power by the techniques of circuit analysis, or we can use this new (statistical) technique. This gives us a second method with application to random signals. In many cases we cannot describe the waveform by a formula, and therefore we cannot use circuit analysis techniques to calculate these quantities. But if the cdf or the pdf is known for the waveform, we can compute these parameters by statistical means. Below are some sample calculations for the waveforms given in Section 3.3 that illustrate both methods of calculation.

EXAMPLE 3.4.1.   The waveform in Fig. 3.3.1 has the pdf shown in Fig. 3.3.2. The average value is given by

$$m_1 = \sum_\alpha \alpha p_V(\alpha) = 1(0.5) + 3(0.5) = 2$$

so the dc power is 4. The variance (ac power) is

$$\sigma^2 = \sum_\alpha (\alpha - m_1)^2 p_V(\alpha) = (1 - 2)^2(0.5) + (3 - 2)^2(0.5) = 1$$

And the mean square value (total power) is

$$m_2 = \sum_\alpha \alpha^2 p_V(\alpha) = 1^2(0.5) + 3^2(0.5) = 5$$

Notice that $m_2 = \sigma^2 + m_1^2$.

We can calculate these same numbers directly from the waveform using circuit analysis techniques. The average value is given by

$$m_1 = \frac{1}{T} \int_0^T v(t)\, dt = \frac{1}{T} \left(\int_0^{T/2} 3\, dt + \int_{T/2}^T 1\, dt \right) = 2$$

The average power is given by

$$P = \frac{1}{T} \int_0^T v^2(t)\, dt = \frac{1}{T} \left(\int_0^{T/2} 3^2\, dt + \int_{T/2}^T 1^2\, dt \right) = 5$$

And, of course, the ac power is the total (average) power minus the power in the dc component, or $5 - 4 = 1$. These same quantities apply to the discrete waveform in Fig. 3.3.3, since the pdf is the same.

Review

When we apply statistics to waveforms, the mean is the dc value, the square of the mean is the power in the dc component, and the variance is the power in the ac component. The sum of the dc power and the ac power is the total power, also called the mean square value. This means we now have two ways to calculate power, the old way using formulas from circuit theory, and the new way using the pdf to calculate the mean, the variance, and the mean square value.

3.5 Characteristic Functions _____

Preview

The characteristic function of a random variable is the Fourier transform of the pdf. We customarily calculate the Fourier transform for a signal, so it may seem strange to calculate the Fourier transform of the pdf. But that is what the characteristic function is.

Knowing that the Fourier transform has many useful properties, perhaps there is some advantage to applying the Fourier transform to the pdf. And of course there is, for otherwise we would not even mention the subject. In this section we define the characteristic function, list its properties, and show some of its applications.

The usual Fourier transform used in signal analysis is

$$V(\Omega) = \int_{-\infty}^{\infty} v(t)e^{-j\Omega t}\, dt \tag{3.5.1}$$

$$V(t) = \frac{1}{2\pi}\int_{-\infty}^{\infty} v(\Omega)e^{j\Omega t}\, d\Omega \tag{3.5.2}$$

where Eq. 3.5.1 is the forward transform used to transform from the time domain to the frequency domain, and Eq. 3.5.2 is the inverse transform used to go from the frequency domain to the time domain. Let us begin with an example to remind you how to apply the formula in Eq. 3.5.1.

EXAMPLE 3.5.1.   Find and plot the Fourier transform of the function in Fig. 3.5.1.

SOLUTION: Using $v(t)$ in Eq. 3.5.1 gives

$$V(\Omega) = \int_{0}^{1}(1)e^{-j\Omega t}\, dt = \frac{1}{j\Omega}(1 - e^{-j\Omega}) = \left[\frac{\sin(\Omega/2)}{\Omega/2}\right]e^{-j\Omega/2}$$

This is graphed in Fig. 3.5.2. Since $V(\Omega)$ is complex-valued, it takes either one three-dimensional plot or two two-dimensional plots to display

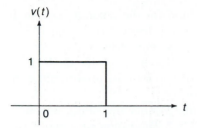

Fig. 3.5.1. The function $v(t)$.

the function. The two plots in Fig. 3.5.2 display the amplitude and phase angle. The amplitude $A(\Omega)$ is the $\sin(x)/x$ term in brackets in the solution above, and the phase is the exponent $\theta(\Omega) = -\Omega/2$.

Although the characteristic function is a Fourier transform, there is a difference, which we will explain shortly. The characteristic function has the following definition.

DEFINITION 3.5.1. The characteristic function of the random variable X is the expected value of e^{jvX}:

$$\phi_X(v) = \int_{-\infty}^{\infty} f_X(\alpha)e^{jv\alpha}\, d\alpha \qquad (3.5.3)$$

We use the symbol ϕ_X to denote the characteristic function in the same way we used the symbol f_X to denote the probability density function for the

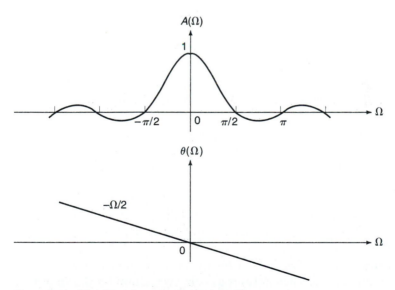

Fig. 3.5.2. The Fourier transform of $v(t)$.

random variable X. The characteristic function for the random variable Y would be ϕ_Y. Equation 3.5.3 is the fundamental theorem of statistics (Eq. 3.4.2) applied to the function $g(X) = e^{jvX}$. We have interchanged the positions of f_X and $g(X)$ in Eq. 3.4.2, but that is so it will look more like the Fourier transform. Tradition dictates that the symbols α and v be used in place of t and Ω, which may be more familiar to you.

The characteristic function differs from the Fourier transform. Equation 3.5.3 is not identical to either Eq. 3.5.1 or 3.5.2. The exponent should have a minus sign to make Eq. 3.5.3 equal to 3.5.1, or there should be a $1/2\pi$ factor in front for it to equal Eq. 3.5.2. Nevertheless, once Eq. 3.5.3 has been applied to a pdf to find the characteristic function, the inverse operation to recover the pdf is given by

$$f_X(\alpha) = \frac{1}{2\pi} \int_{-\infty}^{\infty} \phi_X(v) e^{-jv\alpha} \, d\alpha \qquad (3.5.4)$$

This also has either the wrong sign on the exponent or else it has an extra $1/2\pi$ factor for it to be identical to either Eq. 3.5.1 or 3.5.2. In any event, Eqs. 3.5.3 and 3.5.4 form a Fourier transform pair that is just as good as the original pair. (Yogi Berra would say that they are just as good as a good one.) The next example contrasts the Fourier transform with the characteristic function for the waveform in Fig. 3.5.1.

EXAMPLE 3.5.2.   Find the characteristic function for the pdf in Fig. 3.5.1.

SOLUTION: Applying Eq. 3.5.3 to $f_X(\alpha)$ gives

$$\phi_X(v) = \int_0^1 (1)e^{jv\alpha} \, d\alpha = \frac{1}{jv}(e^{jv} - 1) = \left[\frac{\sin(v/2)}{v/2}\right] e^{jv/2}$$

The only difference between this and the Fourier transform in Fig. 3.5.2 is the sign on the phase angle. Here the phase angle has a slope of $v/2$.

The characteristic function has many uses, and we will use it in discussing the central limit theorem, in evaluating moments, and in special cases of transformation of random variables. All the properties of the Fourier transform apply to the characteristic function. The negative phase angle has no essential effect on the properties. It is the properties of any transform that make it useful, and there are many properties of the Fourier transform. Hence there are many uses for ϕ_X. Here are some more examples.

EXAMPLE 3.5.3.   Let X be a binary random variable with the pdf shown in Fig. 3.5.3a. Then $\phi_X(v)$ is a cosine waveform, as shown in Fig. 3.5.3b, since

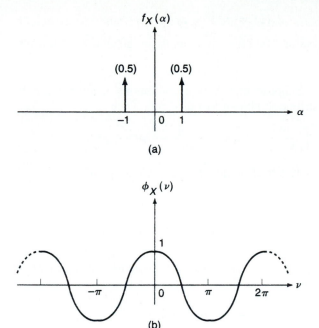

Fig. 3.5.3.   The pdf and the characteristic function for a binary random variable.

$$\phi_X(v) = \int_{-\infty}^{\infty} [0.5\delta(\alpha + 1) + 0.5\,\delta(\alpha - 1)]e^{jv\alpha}\,d\alpha$$

$$= 0.5e^{-jv} + 0.5e^{jv} = \cos v$$

EXAMPLE 3.5.4.   As a generalization of the previous example, suppose that X is a binary random variable with the pdf shown in Fig. 3.5.4. Then the characteristic function is given by

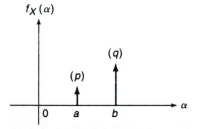

Fig. 3.5.4.   Another binary random variable.

$$\phi_X(v) = pe^{jav} + qe^{jbv}$$

which is a complex-valued function of v, as is the general case for any Fourier transform.

EXAMPLE 3.5.5.   Suppose that X is a continuous random variable with the uniform distribution shown in Fig. 3.5.5. Then

$$\phi_X(v) = \int_a^b \frac{1}{b-a} e^{jv\alpha} \, d\alpha = \frac{1}{jv(b-a)} (e^{jbv} - e^{jav})$$

Once again the characteristic function is a complex-valued function of the variable v.

Use of Characteristic Function in Evaluating Moments

One of the more common and useful applications of the characteristic function is evaluating moments. Recall from Eq. 3.4.3 that the nth moment of a random variable is given by

$$m_n = E(X^n) = \int_{-\infty}^{\infty} \alpha^n f_X(\alpha) \, d\alpha$$

The only information needed to calculate this number is the pdf $f_X(\alpha)$. The characteristic function is equivalent to the pdf in an information theory sense, because the Fourier transform is a reversible operation. Therefore we should be able to use the characteristic function to calculate the nth moment.

To do this we use the differentiation property of Fourier transforms. Differentiating Eq. 3.5.3 with respect to v gives

$$\frac{d}{dv} \phi_X(v) = \int_{-\infty}^{\infty} j\alpha f_X(\alpha) e^{jv\alpha} \, d\alpha$$

Evaluating this derivative at $v = 0$, we get

$$\frac{d}{dv} \phi_X(v)\Big|_{v=0} = j \int_{-\infty}^{\infty} \alpha f_X(\alpha) \, d\alpha = jm_1$$

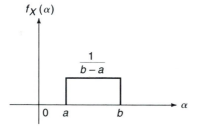

Fig. 3.5.5.   A continuous random variable.

Or, dividing by j, we have

$$m_1 = \frac{1}{j} \frac{d}{dv} \phi_X(0) \qquad (3.5.5)$$

Thus to find the first moment (the mean) we differentiate the characteristic function, set $v = 0$, and divide by j.

The nth moment of X is found by using the nth derivative of $\phi_X(v)$ evaluated at $v = 0$ and dividing by j to the nth power.

$$m_n = \frac{1}{j^n} \frac{d^n}{dv^n} \phi_X(0) \qquad (3.5.6)$$

EXAMPLE 3.5.6.   Find the first two moments for the random variable in Example 3.5.3 in two ways: (a) Use the probability density function; (b) use the characteristic function.

SOLUTION: (a) Using the pdf we get

$$m_1 = \sum_\alpha \alpha p_X(\alpha) = (-1) \cdot \tfrac{1}{2} + (1) \cdot \tfrac{1}{2} = 0$$

$$m_2 = \sum_\alpha \alpha^2 p_X(\alpha) = (-1)^2 \cdot \tfrac{1}{2} + (1)^2 \cdot \tfrac{1}{2} = 1$$

(b) Using the characteristic function gives

$$m_1 = \frac{1}{j} \frac{d}{dv} \cos(v)\big|_{v=0} = \frac{1}{j} \sin(0) = 0$$

$$m_2 = \frac{1}{j^2} \frac{d^2}{dv^2} \cos(v)\big|_{v=0} = \cos(0) = 1$$

EXAMPLE 3.5.7.   Find the first two moments of the random variable in Fig. 3.5.4 using the characteristic function.

SOLUTION: The characteristic function is given by

$$\phi_X(v) = pe^{jav} + qe^{jbv}$$

Differentiating, we get

$$\frac{d}{dv} \phi_X(v) = jape^{jav} + jbqe^{jbv}$$

Upon setting $v = 0$ and dividing by j we get

$$m_1 = ap + bq$$

The second derivative is given by

$$\frac{d^2}{dv^2} \phi_X(v) = -a^2 pe^{jav} - b^2 qe^{jbv}$$

After setting $\nu = 0$ and dividing by $(j)^2 = -1$, we get

$$m_2 = a^2 p + b^2 q$$

These examples illustrate the characteristic function for evaluating moments. It may seem superfluous first to calculate the characteristic function and then to take derivatives in order to find moments, when it is much simpler to use the pdf directly. There are two reasons for introducing this roundabout way of calculating moments. For one, you may be given the characteristic function but not the pdf in a particular application. And second, the more you know, the better able you are to apply concepts. The applications of theory will grow steadily more complex as we progress, and you will find applications in unexpected places.

Review

The characteristic function is the Fourier transform of the probability density function. We are accustomed to finding the Fourier transform of signals, so this application may seem strange to you. But the Fourier transform has many good properties that we can use in this new application, and the pdf for any random variable, discrete or continuous, satisfies the requirements for having a Fourier transform. Therefore the characteristic function finds wide application in statistics.

We have demonstrated only one application for the characteristic function in finding moments, but we promise more later in the text. We will use it in proving the central limit theorem, and in other applications.

3.6   Some Important Distributions

Preview

There are a million and one distributions, but some are more important than others. We have seen examples of some of these, and now we explore some more.

In Section 3.3 we found the distribution of waveforms, and we saw that different waveforms determined different distributions. In this section we discuss two continuous and two discrete distributions that are likely to be encountered in practice. These are the binary, binomial, uniform, and Gaussian distributions.

Binary Distribution

The *binary distribution* is also called the *Bernoulli distribution*. The random variable in Example 3.2.1 (tossing a coin) has binary distribution. There are two numbers in the range of a binary random variable. In Example 3.2.1 these numbers are -1 and $+1$.

For a binary distribution the random variable partitions the sample space into two distinct subsets—say, A and B. All elements of one subset, A, are mapped into one number—say, a—and all elements of the other (complementary) set, B, are mapped into another number—say, b. That is,

$$X(\zeta) = a \qquad \text{if } \zeta \in A$$
$$\quad\;\; = b \qquad \text{if } \zeta \in B \tag{3.6.1}$$

Figure 3.6.1 is a graphic representation of a binary random variable. Each element in the sample space S is mapped into one of two numbers, a or b.

The ''success'' or ''failure'' random variable is binary. An experiment is performed. If it results in success, the random variable X assigns the number 1; if it results in failure, $X = 0$. We can toss a coin and assign the number 1 to heads and the number 0 to tails. Or, on a production line, we can test a part to see if it meets specifications. If so, $X = 1$; if not, $X = 0$. Similarly, we can transmit a binary (on–off) message sequence. If ''on,'' $X = 1$; if ''off,'' $X = 0$.

The symbol p denotes the probability of success, and $q = 1 - p$ denotes the probability of failure. Thus

$$P\{X = 1\} = p$$
$$P\{X = 0\} = q \tag{3.6.2}$$

The mean and variance of this random variable are p and pq, respectively.

Binomial Distribution

The *binomial distribution* is called ''repeated Bernoulli trials.'' Suppose we perform the binary experiment n times with outcomes $X_1, X_2, \ldots, X_n$, where X_i is either 0 or 1. Then there are n random variables $X_1, X_2, \ldots, X_n$, where X_i is associated with the ith trial. Now we define the random variable Y as the sum of the X_i's:

$$Y = \sum_i X_i \tag{3.6.3}$$

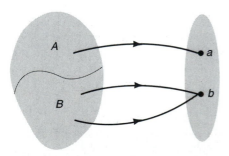

Fig. 3.6.1.   A binary random variable.

The random variable Y has binomial distribution. Since the range of each X_i is 0 or 1, the range of Y is the set of integers $\{0, 1, \ldots, n\}$. That is, Y can assume any one of the numbers $\{0, 1, \ldots, n\}$, and the probability that Y will equal a particular number is determined by the two probabilities $P\{X = 1\}$ and $P\{X = 0\}$, and by the number n. Given that $P\{X = 1\}$ is p and $P\{X = 0\}$ is q, we wish to find the cdf and the pdf of the random variable Y. To do this we need Pascal's triangle.

Pascal's triangle contains the coefficients in the binomial expansion $(p + q)^n$. Figure 3.6.2 shows Pascal's triangle for values of n from 1 to 6. If we wish to find the binomial expansion of $(p + q)$ to the fourth power, we use the fourth row in Fig. 3.6.2 and obtain

$$(p + q)^4 = p^4 + 4p^3q + 6p^2q^2 + 4pq^3 + q^4$$

This triangle is constructed in the following way. Each row begins and ends with the number 1. The numbers in between are determined by adding the two numbers in the row just above the desired number. For example, the numbers in the fourth row are $4 = 1 + 3$, $6 = 3 + 3$, and $4 = 3 + 1$. In this manner the seventh row could easily be added to Fig. 3.6.2. It would be 1, 7, 21, 35, 35, 21, 7, 1.

Another way to construct Pascal's triangle is to use the binomial coefficient $\binom{n}{k}$, which is the symbol for the following operation:

$$\binom{n}{k} = \frac{n!}{k!(n - k)!} \tag{3.6.4}$$

where 0! is defined to be 1. The first four rows of Pascal's triangle in terms of this new notation are shown in Fig. 3.6.3. Note that the center coefficient in the fourth row is given by

$$\binom{4}{2} = \frac{4!}{2!2!} = \frac{4 \cdot 3 \cdot 2 \cdot 1}{(2 \cdot 1)(2 \cdot 1)} = 6 \tag{3.6.5}$$

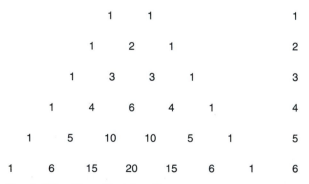

Fig. 3.6.2. Pascal's triangle.

$$\binom{1}{0} \quad \binom{1}{1} \qquad\qquad 1$$

$$\binom{2}{0} \quad \binom{2}{1} \quad \binom{2}{2} \qquad\qquad 2$$

$$\binom{3}{0} \quad \binom{3}{1} \quad \binom{3}{2} \quad \binom{3}{3} \qquad 3$$

$$\binom{4}{0} \quad \binom{4}{1} \quad \binom{4}{2} \quad \binom{4}{3} \quad \binom{4}{4} \quad 4$$

Fig. 3.6.3.   Binomial coefficients in Pascal's triangle.

Pascal's triangle and the binomial coefficients are useful because they enable us to find the number of ways to arrange n things taken k at a time. If we toss a coin four times and want to know the number of ways we can obtain two heads, the answer is given by Eq. 3.6.5; there are six ways to obtain two heads if a coin is tossed four times. Thus $\binom{n}{k}$ is the number of ways we can obtain k successes in n trials of the experiment.

In order for the random variable Y of Eq. 3.6.3 to equal k, where k is one of the numbers $\{0, 1, \ldots, n\}$, the experiment must result in k successes and $(n - k)$ failures. If the trials are independent, the probability of k successes and $(n - k)$ failures in any given order is the product of k p's and $(n - k)$ q's. That is, this probability is equal to $p^k q^{(n-k)}$. Since there are $\binom{n}{k}$ different ways to obtain k successes and $(n - k)$ failures, the distribution of Y is given by

$$P\{Y = k\} = \binom{n}{k} p^k q^{(n-k)} \tag{3.6.6}$$

The mean and the variance of the binomial random variable are np and npq, respectively.

EXAMPLE 3.6.1.   For the experiment of tossing a coin, let $X = 1$ if heads occurs and $X = 0$ if tails occurs. Let Y be the number of heads in three tosses of the coin. Find the cdf and the pdf for Y.

SOLUTION: From Eq. 3.6.6,

$$P\{Y = 0\} = \binom{3}{0}\left(\frac{1}{2}\right)^0\left(\frac{1}{2}\right)^3 = \frac{1}{8}$$

$$P\{Y = 1\} = \binom{3}{1}\left(\frac{1}{2}\right)^1\left(\frac{1}{2}\right)^2 = \frac{3}{8}$$

$$P\{Y = 2\} = \binom{3}{2}\left(\frac{1}{2}\right)^2\left(\frac{1}{2}\right)^1 = \frac{3}{8}$$

$$P\{Y = 3\} = \binom{3}{3}\left(\frac{1}{2}\right)^3\left(\frac{1}{2}\right)^0 = \frac{1}{8}$$

The cdf and the pdf for the random variable Y of this example are plotted in Figs. 3.6.4a and 3.6.4b.

The probabilities in the Bernoulli distribution are functions of only two parameters, n and p. The next example illustrates the effect of changing p.

EXAMPLE 3.6.2.   Suppose in the example above that values of X are chosen by tossing a die rather than by tossing a coin. We do this to obtain a different value for p. Let $X = 1$ if any of the first four die faces turn up, and let $X = 0$ if the five-spot or six-spot turns up. Now $p = \frac{2}{3}$ and $q = \frac{1}{3}$. Find and plot the distribution of Y if Y equals the sum of the X values in three trials of the experiment.

SOLUTION;   Now, with $p = \frac{2}{3}$, Eq. 3.6.6 gives

$$P\{Y = 0\} = \binom{3}{0}\left(\frac{2}{3}\right)^0\left(\frac{1}{3}\right)^3 = \frac{1}{27}$$

$$P\{Y = 1\} = \binom{3}{1}\left(\frac{2}{3}\right)^1\left(\frac{1}{3}\right)^2 = \frac{6}{27}$$

$$P\{Y = 2\} = \binom{3}{2}\left(\frac{2}{3}\right)^2\left(\frac{1}{3}\right)^1 = \frac{12}{27}$$

$$P\{Y = 3\} = \binom{3}{3}\left(\frac{2}{3}\right)^3\left(\frac{1}{3}\right)^0 = \frac{8}{27}$$

The cdf and the pdf for this random variable are shown in Figs. 3.6.5a and 3.6.5b.

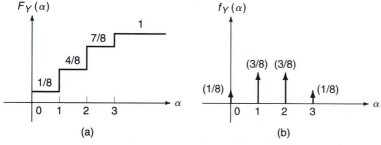

Fig. 3.6.4.   The cdf and the pdf for Example 3.6.1.

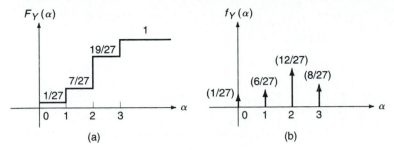

Fig. 3.6.5.   The cdf and the pdf for Example 3.6.2.

EXAMPLE 3.6.3.   Now that we have introduced the binomial distribution, we are in a position to derive the correct (statistical) cdf and pmf for the waveform $x_0(n)$ of Section 1.2. The values of $x_0(n)$ have a binomial distribution with parameters $n = 12$ and $p = q = \frac{1}{2}$. Substituting into Eq. 3.6.6 gives

$$P\{X = 12\} = \binom{12}{0}\left(\frac{1}{2}\right)^{12} = 2.44(10)^{-4}$$

$$P\{X = 11\} = \binom{12}{1}\left(\frac{1}{2}\right)^{12} = 2.93(10)^{-3}$$

$$P\{X = 10\} = \binom{12}{2}\left(\frac{1}{2}\right)^{12} = 0.0161$$

$$P\{X = 9\} = \binom{12}{3}\left(\frac{1}{2}\right)^{12} = 0.0537$$

$$P\{X = 8\} = \binom{12}{4}\left(\frac{1}{2}\right)^{12} = 0.1208$$

$$P\{X = 7\} = \binom{12}{5}\left(\frac{1}{2}\right)^{12} = 0.1934$$

$$P\{X = 6\} = \binom{12}{6}\left(\frac{1}{2}\right)^{12} = 0.2256$$

There are 13 values possible for the waveform at any one time, and we are finding the probabilities of those values here. There are seven shown above, and the other six have probabilities equal to the first six above, but in reverse order. These values are shown in Fig. 3.6.6, and they should be compared to the values in Fig. 3.8.8.

Uniform Distribution

The binary and binomial distributions are both discrete; that is, the random variable can assume only discrete values. We now introduce two continuous

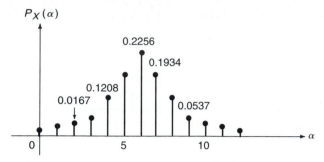

Fig. 3.6.6.   The binomial distribution of $x_0(n)$.

distributions: the uniform and Gaussian distributions.

A random variable X is *uniformly distributed* if the pdf is given by

$$f(\alpha) = \begin{cases} \dfrac{1}{(b - a)} & \text{if } a < \alpha < b \\ 0 & \text{otherwise} \end{cases} \tag{3.6.7}$$

We then say that X is uniformly distributed in the interval (a, b). Figures 3.6.7a and 3.6.7b illustrate the cdf and the pdf for a random variable that is uniformly distributed between values a and b. Since the total area under $f_X(\alpha)$ must be 1, the height of the uniform pdf is determined by its width.

The mean and the variance of the uniformly distributed random variable are $(\frac{1}{2})(b + a)$ and $(\frac{1}{12})(b - a)^2$, respectively.

Gaussian Distribution

In our mathematical models for random phenomena the *Gaussian* (also called *normal*) *distribution* occurs most often. This is because of two factors: the central limit theorem, and the ease with which the Gaussian function can be manipulated. The central limit theorem says that if values of a random

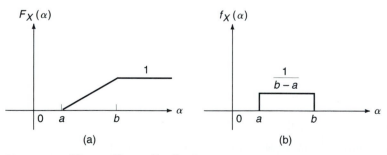

Fig. 3.6.7.   The uniform distribution.

variable are determined by the cumulative effect of a large number of small variables, then under certain conditions the random variable has a Gaussian distribution. These certain conditions are often satisfied in practice, so the Gaussian distribution is prevalent in applications. Because of the exponential form of the pdf, the ratio and the product of Gaussian pdf's are also exponential, and we can often use logarithms to our advantage. Hence the Gaussian form is easy to manipulate in mathematical expressions.

A random variable X has Gaussian distribution if the pdf is given by

$$f_X(\alpha) = \frac{1}{\sqrt{2\pi\sigma_X^2}} \exp\left(-\frac{(\alpha - m_X)^2}{2\sigma_X^2}\right) \qquad (3.6.8)$$

The factors m_X and σ_X^2 are the mean and the variance of the random variable X. The associated cdf is then given by

$$F_X(\alpha) = \int_{-\infty}^{\alpha} f_X(\lambda)\, d\lambda \qquad (3.6.9)$$

The Gaussian pdf is one of those functions for which the integral cannot be evaluated in closed form. This is the reason an explicit formula is not given in Eq. 3.6.9. The pdf and the cdf are illustrated in Figs. 3.6.8a and 3.6.8b. Notice that the mean m_X locates the position of the center of the pdf, and that the standard deviation σ_X determines the width of the curve. The standard deviation is the distance from the mean to the inflection point of the curve.

One of the distinguishing features of the Gaussian pdf is that its Fourier transform is also Gaussian. For a normal distribution with zero mean and unit variance [written $N(0, 1)$], the characteristic function is given by

$$\phi_X(\nu) = \exp\left(\frac{-\nu^2}{2}\right) \qquad (3.6.10)$$

From this let us derive the general form by applying properties of the Fourier transform. First consider the effect of changing the mean. This is illustrated in Fig. 3.6.9. In Fig. 3.6.9b the pdf is shifted so that the center is at m_X

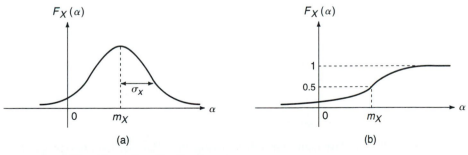

(a)   (b)

Fig. 3.6.8.   The Gaussian distribution.

instead of 0. By the modulation property of Fourier transforms, the characteristic function is multiplied by e^{jvm_X}. That is,

$$\phi_X(v) = e^{jvm_X}e^{-v^2/2} \qquad (3.6.11)$$

Next consider the effect of changing the variance, as shown in Fig. 3.6.9c.
By the scaling property of Fourier transforms, if the variance of X is changed
from 1 to σ_X^2, the characteristic function becomes

$$\phi_X(v) = e^{-v^2\sigma_X^2/2} \qquad (3.6.12)$$

If we combine these two effects, the characteristic function for the general
normal random variable $N(m_X, \sigma_X^2)$ becomes

$$\phi_X(v) = e^{jvm_X}e^{-v^2\sigma_X^2/2} \qquad (3.6.13)$$

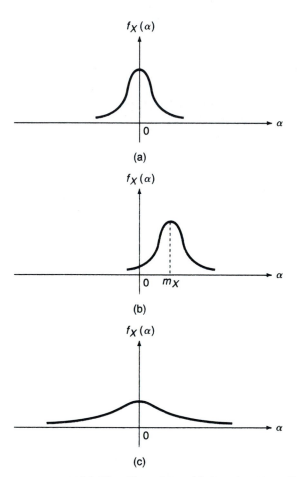

(a)

(b)

(c)

Fig. 3.6.9. (a) The Gaussian pdf showing the effect of (b) changing the
mean and (c) changing the variance.

Notes.   (1) The mean m_X locates the Gaussian curve along the α axis, and the variance σ_X^2 is related to the width of the pdf. (2) The first two moments, m_X and σ_X^2, completely characterize the Gaussian random variable. That is, if the mean and the variance are known, then the formula for the pdf can be written.

Since the pdf cannot be integrated in closed form, most statistics and probability texts provide a table of values for the cdf $F_X(\alpha)$. Otherwise, approximation techniques such as trapezoidal integration must be used. We provide a table of normal curve areas in Appendix A, and the following examples illustrate its use.

EXAMPLE 3.6.4.   Suppose that a random waveform $v(t)$ can be sampled at times $\{t_i\}$ and that these values have Gaussian distribution with zero mean and unit variance. This is written $f_V(\alpha) \sim N(0, 1)$, which means that the random variable V is normal with zero mean and unit variance. (a) Find the probability that a particular value $v(t_i)$ will exceed its root mean square (rms) value.
(b) Find the probability that $\{|v(t_i)| < 0.5\}$.

SOLUTION: From Eq. 3.4.5, the rms value is just the standard deviation σ (since $m_1 = 0$). Thus part (a) asks for $P\{v(t_i) > 1\}$. This is the shaded

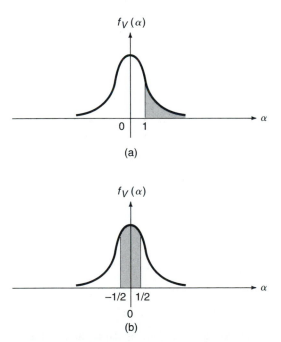

(a)

(b)

Fig. 3.6.10.   The areas in Example 3.6.4.

area in Fig. 3.6.10a, and is given by $1 - F_X(1)$. From the table of normal curve areas, $P\{v(t_i) > 1\} = 1 - 0.8413 = 0.1587$.

Part (b) asks for the shaded area of Fig. 3.6.10b. The area under the curve from 0 to 0.5 is 0.1915. Therefore the area under f_V from -0.5 to 0.5 is $2(0.1915) = 0.3830$, and this is $P\{|v(t_i)| < 0.5\}$.

EXAMPLE 3.6.5.   A Gaussian waveform has dc power of 4 and total power of 5. Find the probability that a sample of this waveform will exceed 3 V.

SOLUTION: The dc value is either $+2$ or -2, since the dc power is 4. If we assume that $m_1 = +2$, then the pdf is shown in Fig. 3.6.11, where the shaded area represents the probability that $v(t) > 3$. The width of the curve here must be the same as the width of the curve in Fig. 3.6.10a, because σ is the same in both cases, and therefore the shaded areas must be the same. Hence $P\{v(t) > 3\} = 0.1587$.

EXAMPLE 3.6.6.   A Gaussian waveform has dc value of 2 and total power of 13. Find the probability that a sample of this waveform will exceed 5 V.

SOLUTION: Here we have reached a situation where our intuition needs some assistance. The Gaussian distribution for this waveform has mean 2 and variance given by

$$\sigma^2 = m_2 - m_1^2 = 13 - 4 = 9$$

This is illustrated in Fig. 3.6.12. The problem asks for the area in the shaded portion to the right of $\alpha = 5$. This area is given by

$$P[v(t) > 5] = \int_5^\infty \frac{1}{\sqrt{2\pi 9}} \exp\left[\frac{-(\alpha - 2)^2}{18}\right] d\alpha$$

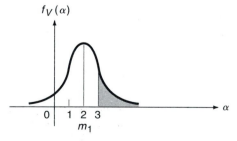

Fig. 3.6.11.   The curve in Example 3.6.5.

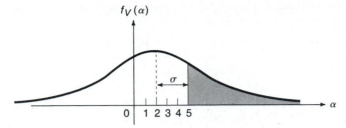

Fig. 3.6.12.   The curve for Example 3.6.6.

By a change of variable,

$$\lambda = \frac{\alpha - m_1}{\sigma} \tag{3.6.14}$$

This becomes

$$P[v(t) > 5] = \int_1^\infty N(0, 1)\, d\lambda = 0.1587$$

Notice that Eq. 3.6.14 is the magic formula to use in transforming Gaussian curves that have distribution $N(m_1, \sigma^2)$ to $N(0, 1)$.

Review

We have introduced only four important distributions in this section, but there are others. The Rayley, Weibull, hypergeometric, and exponential are among the many other distributions you will encounter in signal processing applications. However, our primary need in this text will be for the four in this section.

One purpose of this section is to show you how to find areas under the Gaussian distribution. If you understand how to find the area under a one-dimensional Gaussian distribution with arbitrary mean and variance between two given limits, you are in business.

3.7   Transformation of Variables _____

Preview

Let X be a random variable with known distribution, and suppose we operate on the value of X in some way to produce—say, $y = g(x)$. Then the value y is a sample of a random variable Y with some distribution. The problem we address in this section is to find the distribution of Y given the distribution of X. The method we develop is quite general and applies to any functional relationship $g(x)$, linear or not.

We introduce the concept of a composite function in the following way. Let $f: A \rightarrow B$ be a function with domain A and range B. We say that f is a *map* from A to B. Let $g: B \rightarrow C$ be a map from B to C. Then the *composition* (product) mapping $gf: A \rightarrow C$ takes an element in A and first applies f, producing an element in B, then applies g, producing an element in C. For example, if $y = f(x) = x + 1$ and $z = g(y) = 2y$, then the composition gf produces $gf(x) = 2(x + 1)$. The composition fg produces $fg(x) = 2x + 1$.

Notice two things: $gf \neq fg$, and the order of application for the two maps is turned around. The composite function gf first applies f, the second symbol in the list gf, and then applies g. So you should read the order from right to left.

EXAMPLE 3.7.1.   Let $A = \{x, y\}$, $B = \{1, 2, 3\}$, and $C = \{a, b\}$, and consider the maps $f: A \rightarrow B$ and $g: B \rightarrow C$ shown in Fig. 3.7.1. Notice that the range of f is a subset of the domain of g. In order to define and use composition maps properly, it is necessary that the range of f be a subset (proper or not) of the domain of g. You can see that the composition map in Fig. 3.7.1 is given by

$$gf(x) = g[f(x)] = g(1) = a$$
$$gf(y) = g[f(y)] = g(2) = b$$

We will apply the concept of composition maps to transformation of variables in the following way. Remember that a random variable X is a function with domain S, sample space, and range R_X, which is a subset of the real numbers. If we then operate on values of the random variable with a system or other operation g, the result is a composite function gX, where X is the random variable and g is the second operation. This is pictured in Fig. 3.7.2. According to our discussion above, any function g with domain $D_g \supset R_X$ can be a function of the random variable X, $g(X)$. The domain of the composite function $y = g(X)$ is the sample space S, and therefore y can be correctly labeled a random variable. For this reason we write $Y = g(X)$, using an uppercase Y in place of the lowercase y.

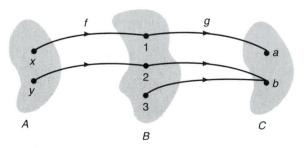

Fig. 3.7.1.   The composite function gf.

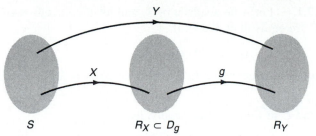

Fig. 3.7.2.   The composite function $Y = gX$.

Suppose that we know the cdf (and hence the pdf) of the random variable X, and suppose that there is a relationship of the form $Y = g(X)$. The problem labeled "transformation of variables" is to find the cdf (and the pdf) for the random variable Y. The reader can appreciate that the solution has wide application in system theory. The application is direct for systems without memory, for in this case the input x and the output y are related by a formula of the form $y = g(x)$. If the input is a random variable with known cdf, then the cdf of the output can be computed. The problem is much more complicated for systems with memory, but the present discussion will form a first step toward the solution of this problem.

Let A be a subset of the sample space, $A \subset S$. Then A is an event. Suppose that the experiment is performed and that the outcome ζ is an element of A, $\zeta \in A$. The point ζ is mapped to the number $X(\zeta)$ by the random variable X, and if there is a relation of the form $Y = g(X)$, then the number $X(\zeta)$ is mapped into $g[X(\zeta)]$. See Fig. 3.7.3. Each element ζ of A is mapped into some element $X(\zeta)$ of B by the random variable X, which in turn is mapped into some element $g[X(\zeta)]$ of C by the function g. Since sets A, B, and C all occur together, they must have the same probability of occurrence. With this in mind, let us define the important concept of equivalent events.

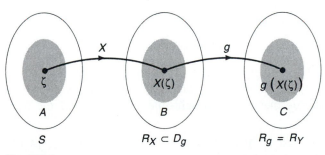

Fig. 3.7.3

DEFINITION 3.7.1.   Let A be a subset of S, and let B be a subset of R_X. Then A and B are equivalent events if

$$A = \{\zeta \in S: X(\zeta) \in B\} \qquad (3.7.1)$$

In words, A is the set of elements ζ in S such that $X(\zeta)$ is an element of B.

Likewise, sets B and C in Fig. 3.7.3 are equivalent if

$$B = \{b \in R_X: g(b) \in C\}$$

The sets A, B, and C occur together. That is, the occurrence of an element in one of the three sets implies the occurrence of an element in each of the other two sets.

Do not confuse the concept of equivalent events with the more common (in mathematics) concept of equivalent sets. Two sets are equivalent if their elements can be put into one-to-one correspondence. (For finite sets, this means that they contain the same number of elements.) In contrast, in our definition of equivalent events the correspondence between two sets is specified by the function.

Recall that an event is a set. That is, an event is a subset of the sample space S. What we have done here is find the corresponding set in the range R_X. We can now think of the corresponding set B (equivalent to A) as an event, and of the range R_X as a sample space. We will find this convenient, for R_X is a set of numbers and is inherently easier to work with than the sample space S.

If A and B are equivalent events, then $P(A) = P(B)$. This important property is the basis for finding the cdf of $Y = g(X)$. You should keep in mind that A and B are in different sets: A is a subset of the sample space S and B is a subset of the range R_X.

Finding the cdf of $Y = g(X)$

We assume that the cdf for X is known, the relationship $Y = g(X)$ is known, and we wish to find the cdf for the random variable Y. For a particular value y_1 the function F_Y defines an event in R_Y, namely, $\{Y \leq y_1\}$. There is an equivalent event in R_X; call it B. Therefore

$$F_Y(y_1) = P\{Y \leq y_1\} = P\{X \in B\} \qquad (3.7.2)$$

and we can express the probability $P\{X \in B\}$ in terms of F_X.

The way we do this depends on the relation $y = g(x)$. To begin with a simple situation, suppose that g is a straight-line relationship of the form $ax + b$:

$$Y = aX + b \qquad (3.7.3)$$

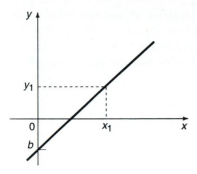

Fig. 3.7.4

There are two cases to consider, because they lead to different results: $a > 0$ and $a < 0$. First consider the case $a > 0$ (Fig. 3.7.4). For arbitrary x_1 and y_1, as pictured in Fig. 3.7.4, the events $\{Y \le y_1\}$ and $\{X \le x_1\}$ are equivalent (because when one occurs, so does the other). Therefore we have

$$F_Y(y_1) = P\{Y \le y_1\} = P\{X \le x_1\} = F_X(x_1)$$

But $x_1 = (y_1 - b)/a$, so

$$F_Y(y_1) = F_X\left(\frac{y_1 - b}{a}\right) \tag{3.7.4}$$

Since y_1 is arbitrary, this relation holds for any $y = y_1$.

EXAMPLE 3.7.2.   Suppose that X is uniform $(3, 5)$ and that Y is related to X by $Y = 2X + 4$. Find the cdf for Y.

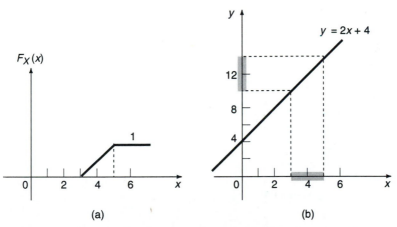

(a)   (b)

Fig. 3.7.5.   (a) The cdf $F_X(x)$ and (b) the relation $y = 2x + 4$.

SOLUTION: Figure 3.7.5 displays the cdf for X and the relationship $y = 2x + 4$. From these diagrams we can see that if $X \le 5$, then $Y \le 14$. If $X \le 3$, then $Y \le 10$. Therefore Y should be uniform (10, 14). We should be able to get these same results from the formulas. The cdf for X is given by

$$F_X(x) = \begin{cases} 0, & x < 3 \\ \frac{1}{2}(x - 3), & 3 < x < 5 \\ 1, & x > 5 \end{cases}$$

From Eq. 3.7.4 with $a = 2$ and $b = 4$ we get

$$F_Y(y) = \begin{cases} 0, & y < 10 \\ \frac{1}{4}(y - 10), & 10 < y < 14 \\ 1, & y > 14 \end{cases}$$

Next consider the case $a < 0$ (Fig. 3.7.6). Now the events $\{Y \le y_1\}$ and $\{X \ge x_1\}$ are equivalent. Therefore

$$F_Y(y_1) = P\{Y \le y_1\} = P\{X \ge x_1\} = 1 - F_X(x_1)$$

Again, $x_1 = (y_1 - b)/a$, so for arbitrary $x = x_1$ and $y = y_1$ we have

$$F_Y(y) = 1 - F_X\left(\frac{y - b}{a}\right) \tag{3.7.5}$$

EXAMPLE 3.7.3.   Suppose that X is uniform (3, 5) and that Y is related to X by $Y = -2X + 4$. Find the cdf for Y.

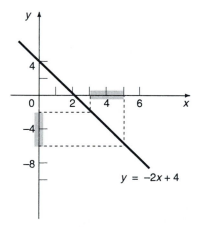

Fig. 3.7.6.   The relation $y = -2x + 4$.

SOLUTION: For this relation $y = -2x + 4$, when $x = 3$, $y = -2$. When $x = 5$, $y = -6$. So Y should be distributed between -6 and -2. To see if this works out, plug into Eq. 3.7.5 to get

$$F_Y(y) = \begin{cases} 0, & y < -6 \\ \frac{1}{4}(y + 6), & -6 < y < -2 \\ 1, & y > -2 \end{cases}$$

Now consider a more general (and more complicated) case. Instead of the straight-line relationship between x and y, suppose that $y = g(x)$ is the curve shown in Fig. 3.7.7. For $y = y_1$, the sets $\{Y \le y_1\}$ and $\{X \le x_1\}$ are equivalent. Therefore

$$F_Y(y_1) = P\{Y \le y_1\} = P\{X \le x_1\} = F_X(x_1) \qquad (3.7.6)$$

For $y = y_2$ there are three solutions to $y = g(x)$, namely, x_2', x_2'', and x_2'''. From the graph we see that the equivalent events are

$$\{Y \le y_2\} \sim \{X \le x_2'\} \cup \{x_2'' < X \le x_2'''\} \qquad (3.7.7)$$

(The symbol $\sim$ denotes equivalent events.) Hence

$$F_Y(y_2) = P\{Y \le y_2\} = P\{X \le x_2'\} + P\{x_2'' < X \le x_2'''\}$$
$$= F_X(x_2') - F_X(x_2'') + F_X(x_2''') \qquad (3.7.8)$$

Finally, for $y < 0$ (say, $y = y_3$), there is no solution to the equation $y_3 = g(x)$. This means that there is no event in R_X equivalent to $\{Y \le y_3\}$ in R_Y. Therefore

$$F_Y(y_3) = P\{Y \le y_3\} = 0 \qquad (3.7.9)$$

The fact that equivalent events have equal probabilities is the basis for the procedure above. This procedure is quite general and also straightforward. Notice that it also allows us to find the pdf, because we can always differentiate the cdf of the random variable Y to find the pdf. Here are some examples to illustrate this procedure.

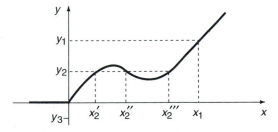

Fig. 3.7.7

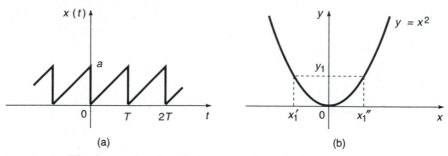

Fig. 3.7.8   The input signal and the square law device characteristic.

EXAMPLE 3.7.4.   Square law device. Suppose that a system has the input–output characteristic $y = x^2$. Figure 3.7.8a shows the input signal, and Fig. 3.7.8b shows the square law device characteristic. If the voltage $v(t)$ is sampled at an arbitrary time t, then we can define a random variable X whose range is the possible values of $x(t)$, that is, whose range is the set of numbers $[0, a]$. The random variable X is uniformly distributed $(0, a)$. Find F_Y and f_Y.

SOLUTION: Let $y = y_1$, where y_1 is any number greater than 0, as shown in Fig. 3.7.8b. Then the equivalent sets are

$$\{Y \le y_1\} \sim \{x_1' \le X \le x_1''\}$$

where x_1' and x_1'' are solutions to the equation $y = x^2$. That is,

$$x_1' = -\sqrt{y_1} \quad \text{and} \quad x_1'' = \sqrt{y_1}$$

Then

$$F_Y(y_1) = P(Y \le y_1) = P(-\sqrt{y_1} < X < \sqrt{y_1})$$
$$= F_X(\sqrt{y_1}) - F_X(-\sqrt{y_1}) \qquad y_1 > 0 \qquad (3.7.10)$$

with the cdf for X given by

$$F_X(x) = \begin{cases} 0, & x < 0 \\ \dfrac{x}{a}, & 0 < x < a \\ 1, & x > a \end{cases}$$

Substituting this into Eq. 3.7.10 gives the cdf (see Fig. 3.7.9)

$$F_Y(y) = \begin{cases} 0, & y < 0 \\ \dfrac{1}{a}\sqrt{y}, & 0 < \sqrt{y} < a \quad \text{or} \quad 0 < y < a^2 \\ 1, & y > a^2 \end{cases}$$

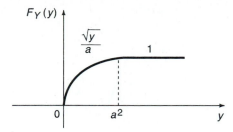

Fig. 3.7.9

We differentiate the cdf to find the pdf for Y:

$$f_Y(y) = \begin{cases} 0, & y < 0 \\ \dfrac{1}{2a\sqrt{y}}, & 0 < y < a^2 \\ 0, & y > a^2 \end{cases}$$

This is a good time to mention that there is a corresponding method for finding the pdf f_Y from the pdf f_X directly, without using the cdf's. The procedure is as follows. Let $y = g(x)$ be the relationship between y and x. Then the pdf for Y is given in terms of the pdf for X by

$$f_Y(y) = \frac{f_X(x_1)}{|g'(x_1)|} + \frac{f_X(x_2)}{|g'(x_2)|} + \cdots + \frac{f_X(x_n)}{|g'(x_n)|} \tag{3.7.11}$$

where $x_1, x_2, \ldots, x_n$ are all the real roots of $y = g(x)$, and

$$g'(x) = \frac{dg(x)}{dx}$$

Obviously, this procedure is applicable only when there are a finite number of roots of $y = g(x)$.

EXAMPLE 3.7.5. Let $X = a \sin \Theta$ where Θ is a random variable uniformly distributed over the inverval $0 < \theta < 2\pi$. The cdf of Θ is shown in Fig. 3.7.10a, and the relationship between X and Θ is shown in Fig. 3.7.10b. Find F_X in terms of F_Θ.

SOLUTION: For $x > a$, the event $\{X \le x\}$ is certain. Therefore

$$F_X(x) = 1, \qquad x > a$$

For $x = x_1$, where $0 < x_1 < a$ (see Fig. 3.7.10b), there are two solutions to the equation $x_1 = a \sin \theta$, namely, θ_1' and θ_1'', as shown in Fig. 3.7.10b. The equivalent events are

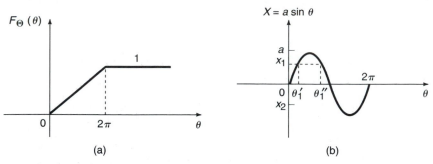

Fig. 3.7.10   The cdf of Θ and the relationship between X and Θ.

$$\{X \le x_1\} \sim \{\Theta \le \theta_1'\} \cup \{\theta_1'' \le \Theta\}$$

and

$$F_X(x_1) = P\{X \le x_1\} = P\{\Theta \le \theta_1'\} + P\{\theta_1'' \le \Theta\} = F_\Theta(\theta_1') + 1 - F_\Theta(\theta_1'')$$

The two solutions θ_1' and θ_1'' are given by

$$\theta_1' = \sin^{-1}\left(\frac{x_1}{a}\right) \qquad \text{and} \qquad \theta_1'' = \pi - \sin^{-1}\left(\frac{x_1}{a}\right)$$

Substituting these values into the expression for F_X, where F_Θ is as given in Fig. 3.7.10, we obtain

$$F_X(x_1) = \frac{1}{2} + \frac{1}{\pi}\sin^{-1}\frac{x_1}{a}, \qquad 0 < x_1 < a \qquad (3.7.12)$$

This completes the derivation for the cdf of X for $0 < x_1 < a$. For negative values of $x = x_2$, where $-a < x_2 < 0$, we can arrive at the same answer. That is, Eq. 3.7.12 is valid for all values of x between a and $-a$.

Finally, for $x < -a$, $F_X(x) = 0$. Figures 3.7.11a and 3.7.11b illustrate the cdf and the pdf of X, where the pdf is found by differentiating the cdf with respect to x.

EXAMPLE 3.7.6.   Let us combine the two previous examples by supplying a sinusoidal waveform to a square law device and calculate the distribution of the output. Assume that a sinusoidal signal of amplitude a is supplied to the square law device in Fig. 3.7.8b. Find the output cdf.

SOLUTION: For $y = y_1$, where $y_1 > 0$ (see Fig. 3.7.8b), the equivalent events are

$$\{Y \le y_1\} \sim \{x_1' \le X \le x_1''\}$$

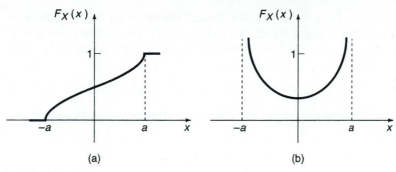

Fig. 3.7.11

Therefore

$$F_Y(y_1) = P\{Y \le y_1\} = P\{x_1 \le X \le x_1''\} = F_X(x_1'') - F_X(x_1')$$

$$= \frac{2}{\pi} \sin^{-1} \frac{\sqrt{y_1}}{a}, \qquad 0 < y_1 < a^2 \qquad (3.7.13)$$

Of course, for $y > a^2$, $F_Y = 1$, and for $y < 0$, $F_Y(y) = 0$.

EXAMPLE 3.7.7.   Limiter. Suppose that the relationship between y and x is that shown in Fig. 3.7.12, rather than the square law relationship of the previous example. The input to the limiter is a sine wave of amplitude a. Find the cdf and pdf of the output $y(t)$.

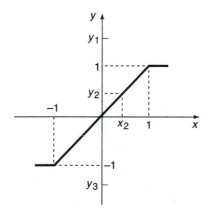

Fig. 3.7.12

SOLUTION: First, for $a < 1$, the limiter has no effect on the input $x(t)$, and the cdf and the pdf are shown in Fig. 3.7.11. But suppose that $a > 1$. For $y = y_1$, where $y > 1$ as shown in Fig. 3.7.12, the event $\{Y \leq y_1\}$ is certain. Hence

$$F_Y(y_1) = P\{Y \leq y_1\} = 1, \qquad y_1 > 1$$

For $y = y_2$, where $-1 < y_2 < 1$, the equivalent events in R_X and R_Y are given by

$$\{Y \leq y_2\} \sim \{X \leq x_2\}$$

Therefore,

$$F_Y(y_2) = P\{Y \leq y_2\} = P\{X \leq x_2\} = F_X(x_2), \qquad -1 < y_2 < 1$$

Finally, for $y = y_3$, where $y_3 < -1$, the event $\{Y \leq y_3\}$ can never occur. Therefore the cdf of Y is given by

$$F_Y(y) = \begin{cases} 1, & y > 1 \\ \dfrac{1}{2} + \dfrac{1}{\pi}\sin^{-1}\dfrac{y}{a}, & -1 < y < 1 \\ 0, & y < -1 \end{cases} \qquad (3.7.14)$$

The cdf and the pdf are shown in Figs. 3.7.13a and 3.7.13b.

Review

The procedure for solving these problems is always the same. Find an event in R_X that is equivalent to the event $\{Y \leq y\}$ in R_Y. Once this equivalent event has been determined, the problem is straightforward. We express the cdf for Y in terms of the cdf for X, that is, in terms of $P\{X \leq x\}$ for one or more x values.

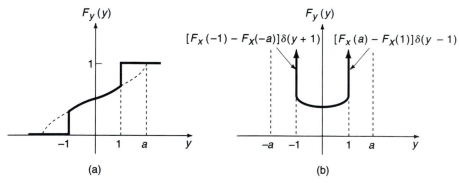

Fig. 3.7.13

3.8   Examples

Section 3.2.   Distribution Functions

EXAMPLE 3.8.1.   The random numbers generated by the linear congruential method in Section 1.3 range in value from -0.5 to 0.5, with no value being more likely than any other. This means that they have the pdf shown in Fig. 3.8.1a. Find the cdf.

SOLUTION: First change the variable from α to λ, for use in Eq. 3.2.4. This gives

$$f_X(\lambda) = \begin{cases} 0, & \lambda < -0.5 \\ 1, & -0.5 < \lambda < 0.5 \\ 0, & 0.5 < \lambda \end{cases}$$

Therefore F_X is given by

$$\begin{aligned} F_X(\alpha) &= \int_{-\infty}^{\alpha} 0 \, d\lambda = 0, & \alpha < -0.5 \\ &= \int_{-\infty}^{-0.5} 0 \, d\lambda + \int_{-0.5}^{\alpha} 1 \, d\lambda = \alpha + 0.5, & -0.5 < \alpha < 0.5 \\ &= \int_{-\infty}^{-0.5} 0 \, d\lambda + \int_{-0.5}^{0.5} 1 \, d\lambda + \int_{0.5}^{\alpha} 0 \, d\lambda = 1, & 0.5 < \alpha \end{aligned}$$

which is shown in Fig. 3.8.1b. The detail in this example illustrates that the integration is from $-\infty$ to α for every value of α. We even displayed the integral of 0 just to make sure you understood that the integration is always from $-\infty$ to α.

EXAMPLE 3.8.2.   Let the experiment consist of spinning a pointer on a circle marked from 0 to 2π, giving a sample space

$$S = \{\beta: 0 \le \beta < 2\pi\}$$

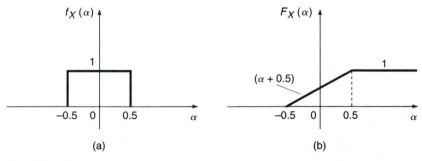

(a)   (b)

Fig. 3.8.1

Define the random variables X and Y as

$$X = \beta/2\pi, \qquad 0 \leq \beta < 2\pi$$

$$Y = \begin{cases} \beta/2\pi, & 0 \leq \beta < \pi \\ 1, & \pi \leq \beta < 2\pi \end{cases}$$

(these are the same as in Example 3.1.3). Find and plot the cdf and the pdf for each random variable.

SOLUTION: For every experimental outcome, the value of X is between 0 and 1. Therefore the probability that X has a value less than or equal to α for $\alpha > 1$ is 1. For values of α between 0 and 1, $P\{X \leq \alpha\}$ is proportional to α. And for values of α less than 0, $P\{X \leq \alpha\} = 0$. The cdf and the pdf for X are shown in Figs. 3.8.2a and 3.8.2b.

The random variables X and Y are equal whenever the pointer stops on the right side of the circle, i.e., for $\beta < \pi$. Therefore the only difference between the cdf's for X and Y occurs on the left side of the circle. The cdf for Y acts like the cdf for the coin-tossing experiment in Fig. 3.2.2 in the neighborhood of $\alpha = 1$. This is shown in Figs. 3.8.3a and 3.8.3b.

Section 3.3   Signal Analysis

EXAMPLE 3.8.3.   The periodic discrete signal in Fig. 3.8.4 has values of 0, 2, 4, and 6. Find the cdf and the pdf for this waveform:

SOLUTION: For $\alpha > 6$ the cdf has value 1. For values of α between 4 and 6 the cdf is equal to $\frac{5}{6}$. We have tried to illustrate this in Fig. 3.8.5a, where you can count the number of pulses that are less than α in one period. Five of the six pulses are less than or equal to α, so $F_V(\alpha) = \frac{5}{6}$. For α between 2 and 4 the cdf is $\frac{3}{6}$, because three of the six pulses are less than or equal to α in Fig. 3.8.5b. Likewise, in Fig. 3.8.5c we show α between 0 and 2 with the corresponding cdf. Finally, for $\alpha < 0$ the

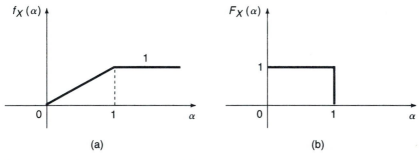

Fig. 3.8.2.   The cdf and the pdf for X.

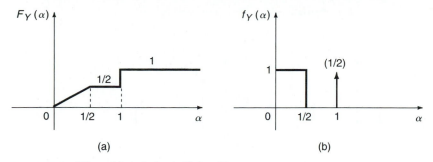

Fig. 3.8.3.   The cdf and the pdf for Y.

cdf has value 0. The complete cdf and the corresponding pdf are shown in Figs. 3.8.6a and 3.8.6b.

EXAMPLE 3.8.4.   Let us use this method to estimate the cdf for the waveform $x_0(n)$ of Section 1.2. This is the waveform with no correlation between successive values shown in Fig. 1.2.2. We can begin to show the connection between time averages and statistics here, because we know the underlying experiment from which the waveform is generated. We found the true cdf from the experiment in Example 3.6.3. You should compare the statistical cdf from that example to the empirical cdf we find here.

SOLUTION: Figure 3.8.7 shows the method used to calculate the empirical cdf for $x_0(n)$. Choose a value of α, draw a horizontal line across the graph at α, and count the number of samples with value less than or equal to α. The cdf is then that number divided by the total number of samples, which is 100 in the graph. The values of the empirical cdf, taken directly from Fig. 3.8.7 are given in the following chart and are plotted in Fig. 3.8.8a, along with the empirical pmf in Fig. 3.8.8b.

α	cdf
10.5	1
9.5	96/100
8.5	91/100
7.5	78/100
6.5	64/100
5.5	34/100
4.5	25/100
3.5	5/100
2.5	1/100
1.5	0

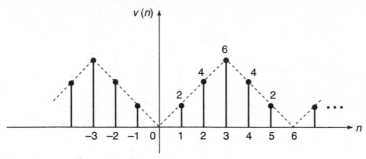

Fig. 3.8.4   The periodic discrete signal.

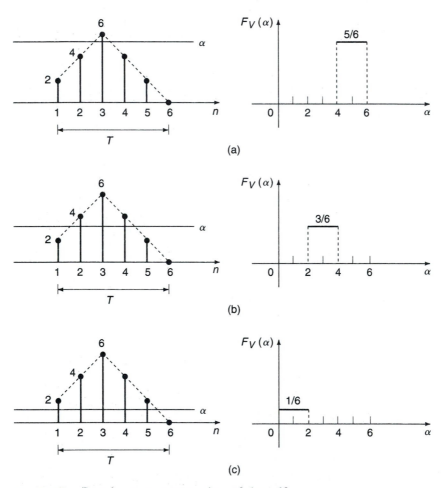

Fig. 3.8.5.   Step-by-step construction of the cdf.

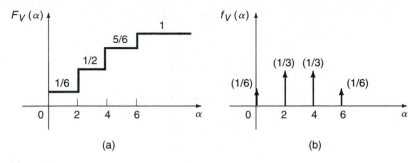

Fig. 3.8.6

Section 3.4.   Statistical Averages

EXAMPLE 3.8.5.   Find the expected value of the random variable X whose pdf is shown in Fig. 3.8.9.

SOLUTION: From Eq. 3.4.1b we have

$$E(X) = (-1)(\tfrac{1}{2}) + (1)(\tfrac{1}{2}) = 0$$

EXAMPLE 3.8.6.   Find the expected value of the random variable X whose pdf is shown in Fig. 3.8.10.

SOLUTION: From Eq. 3.4.1a we have

$$E(X) = \int_{-0.5}^{0.5} \alpha \, d\alpha = 0$$

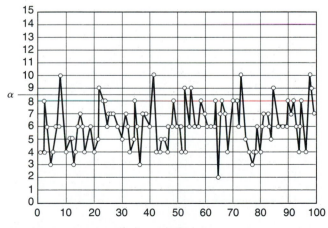

Fig. 3.8.7. $x_0(n)$ with $\alpha = 0.85$.

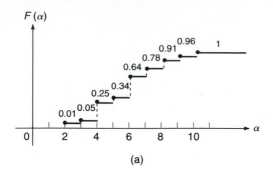

(a)

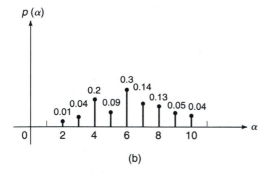

(b)

Fig. 3.8.8.   The empirical cdf and the empirical pmf for x_0.

EXAMPLE 3.8.7.   Repeat for the random variable Y in Fig. 3.8.11.

SOLUTION

$$E(Y) = (\tfrac{1}{6})(-2 - 1 + 0 + 1 + 2 + 3) = 0.5$$

EXAMPLE 3.8.8.   Repeat for the random variable Y in Fig. 3.8.12.

SOLUTION

$$E(Y) = \int_0^{0.5} \alpha \, d\alpha + \int_{-\infty}^{\infty} 0.5\alpha\delta(\alpha - 1) \, d\alpha = \tfrac{1}{8} + \tfrac{1}{2} = \tfrac{5}{8}$$

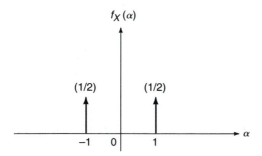

Fig. 3.8.9.   The pdf from Fig. 3.2.3.

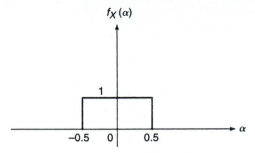

Fig. 3.8.10. The pdf from Fig. 3.7.1.

EXAMPLE 3.8.9.   Find the expected value of $g(X) = [X - E(X)]^2$ for each random variable in Examples 3.8.5 through 3.8.8.

SOLUTION: In Example 3.8.5, $g(X) = [X - E(X)]^2 = X^2$ since $E(X) = 0$. Equation 3.4.2b gives

$$E(X^2) = (-1)^2(\tfrac{1}{2}) + (1)^2(\tfrac{1}{2}) = 1$$

In Example 3.8.6 $g(X) = X^2$ again because $E(X) = 0$. Therefore Eq. 3.4.2a gives

$$E(X^2) = \int_{-0.5}^{0.5} \alpha \, d\alpha = \tfrac{1}{12}$$

(There are a few things you should memorize as you study. This value of $\tfrac{1}{12}$ for the mean square value of the distribution in Fig. 3.8.10 is one of them, for we will find it useful later.)

In Example 3.8.7 $g(X) = (X - 0.5)^2$, so Eq. 3.4.2b gives

$$E[(X - 0.5)^2] = \left(\frac{1}{6}\right)[(-2 - 0.5)^2 + (-1 - 0.5)^2 + (-0.5)^2$$

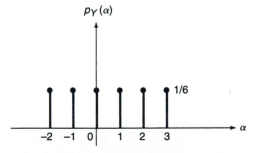

Fig. 3.8.11. The pmf from Fig. 3.2.4.

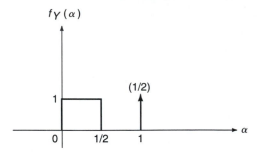

Fig. 3.8.12.   The pdf from Fig. 3.8.3.

$$+ (1 - 0.5)^2 + (2 - 0.5)^2 + (3 - 0.5)^2]$$
$$= \frac{17.5}{6} = 2.91667$$

For the pdf in Example 3.8.8 $g(X) = (X - 0.625)^2$, so Eq. 3.4.2a gives

$$E[(X - 0.625)^2] = \int_0^{0.5} (\alpha - 0.625)^2 \, d\alpha + \int_{-\infty}^{\infty} (\alpha - 0.625)^2(\tfrac{1}{2})\delta(\alpha - 1) \, d\alpha$$
$$= 0.3932 + 0.1406 = 0.5338$$

EXAMPLE 3.8.10.   Find the third moment for each random variable in Examples 3.8.5 through 3.8.8.

SOLUTION: In Example 3.8.5 $g(X) = X^3$, so Eq. 3.4.2 gives

$$m_3 = E(X^3) = (-1)^3(\tfrac{1}{2}) + (1)^3(\tfrac{1}{2}) = 0$$

In Example 3.8.6, $m_3 = 0$. The function $f_X(\alpha)$ is an even function of α. Upon multiplying by α^3, the product is odd, so the integral from -0.5 to 0.5 of this odd function is zero.
    In Example 3.8.7 we get

$$m_3 = (\tfrac{1}{6})[(-2)^3 + (-1)^3 + (0)^3 + (1)^3 + (2)^3 + (3)^3] = \tfrac{9}{2}$$

In Example 3.8.8 we get

$$m_3 = \int_0^{0.5} \alpha^3 \, d\alpha + \int_{-\infty}^{\infty} \alpha^3(\tfrac{1}{2})\delta(\alpha - 1) \, d\alpha = \tfrac{1}{64} + \tfrac{1}{2} = \tfrac{33}{64}$$

EXAMPLE 3.8.11.   Find the mean, the variance, and the mean square value for the waveform in Fig. 3.8.4.

SOLUTION: The pdf is shown in Fig. 3.8.6. The average value is given by

$$m_1 = \sum_{\alpha} \alpha p_V(\alpha) = 0(\tfrac{1}{6}) + 2(\tfrac{1}{3}) + 4(\tfrac{1}{3}) + 6(\tfrac{1}{6}) = 3$$

The average power is given by

$$m_2 = \sum_\alpha \alpha^2 p_V(\alpha) = 0^2(\tfrac{1}{6}) + 2^2(\tfrac{1}{3}) + 4^2(\tfrac{1}{3}) + 6^2(\tfrac{1}{6}) = \tfrac{38}{3}$$

The variance or ac power is

$$\sigma^2 = m_2 - m_1^2 = \tfrac{38}{3} - 9 = \tfrac{11}{3}$$

These same quantities can be found directly from the waveform by averaging over one period. Thus the dc or average value is (see Fig. 3.8.4)

$$m_1 = \frac{1}{6}\sum_{n=1}^{6} v(n) = \frac{2+4+6+4+2+0}{6} = 3$$

Likewise, the average power is given by

$$m_2 = \frac{1}{6}\sum_{n=1}^{6} v^2(n) = \frac{2^2 + 4^2 + 6^2 + 4^2 + 2^2 + 0^2}{6} = \frac{38}{3}$$

So the ac power is $m_2 - m_1^2 = \tfrac{38}{3} - 9 = \tfrac{11}{3}$.

3.9   Problems

3.1. Define the term random variable.

3.2. A box contains 3 red, 4 white, and 3 black marbles. The experiment is to draw one marble from the box. A random variable X is assigned values according to the following rules:

$X = 1$ if a red marble is drawn.
$X = -1$ if a white marble is drawn.
$X = 1$ if a black marble is drawn.

(a) Find and plot the cdf and the pdf for X.
(b) Find the variance of X.

3.3. Repeat Problem 3.2 for the random variable Y defined by

$Y = 1$ if a red marble is drawn.
$Y = 0$ if a white marble is drawn.
$Y = -1$ if a black marble is drawn.

3.4. (a) Find and plot the cdf and the pdf of the random variable X defined in the following table, where f_i stands for the ith die face in a die-toss experiment.

(b) Find the mean, the variance, and the mean square value of the random variable X.

X	-3	0	1	2	1	1
ζ	f_1	f_2	f_3	f_4	f_5	f_6

3.5. According to Definition 3.1.1, each value of α in $\{X \le \alpha\}$ defines an event. In Problem 3.4, which event corresponds to
(a) $\{X \le 1.5\}$
(b) $\{X \le 3\}$
(c) $\{X \le -3\}$

3.6. Find and plot the pdf $f_i(\alpha)$ corresponding to each cdf shown in Fig. 3.9.1.

3.7. Find and plot the cdf $F_i(\alpha)$ corresponding to each pdf shown in Fig. 3.9.2. (Enough information is given in the graphs for you to complete the problem.)

3.8. Find and plot the cdf and the pdf for the continuous-time waveforms in Fig. 3.9.3. The experiment is to choose a time at random in the interval $0 \le t < T$, where T is the period of the waveform. This value of time determines a value for $v(t)$, and it is the distribution of these values we seek.

3.9. Find and plot the cdf and the pdf for the discrete-time waveforms in Fig. 3.9.4. The experiment is to choose a time at random in the interval $0 \le n < N$, where N is the period of the waveform. This value of time determines a value for $v(n)$, and it is the distribution of these values we seek.

3.10. Calculate the dc value, dc power, ac power, and total power in each waveform in Problem 3.8. Do this in two ways: first by using the pdf, then by conventional waveform analysis techniques.

3.11. Calculate the dc value, dc power, ac power, and total power in each discrete-time waveform in Problem 3.9. Do this both ways, using the pdf and then by conventional waveform analysis. For discrete time, the conventional analysis is done by the following formulas:

$$\text{dc value} = \frac{1}{N} \sum_{n=0}^{N-1} v(n) \qquad (3.9.1)$$

$$\text{total power} = \frac{1}{N} \sum_{n=0}^{N-1} v^2(n) \qquad (3.9.2)$$

where N is the period of the waveform.

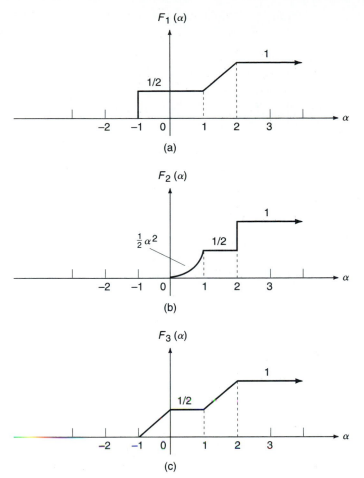

Fig. 3.9.1.   Three cdf's.

3.12.   The characteristic function for a random variable is given by

$$\phi_X(v) = \cos(v)$$

(a) Find the first two moments m_1 and m_2 for the random variable X.
(b) Find and plot the probability density function $f_X(\alpha)$. (In order to solve this problem, you have to know that the transform of two delta functions is a cosine waveform.)

3.13.   A random waveform is generated as follows: Two coins are tossed once each second, and the value of the discrete-time signal is set equal to the number of heads. Thus the waveform can assume values of 0, 1, or 2, with the value of 1 most likely. A typical sequence is shown in Fig. 3.9.5.

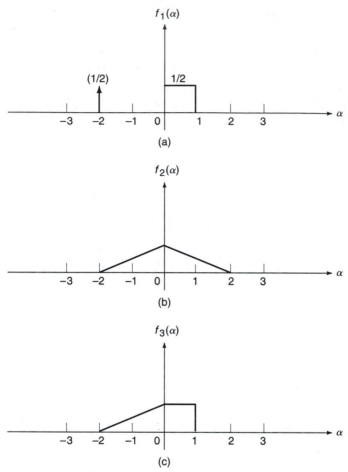

Fig. 3.9.2.   Three pdf's.

(a) Find and plot the cdf and the pmf for this waveform.
(b) Find the ac power, dc power, and total power in this waveform. (Assume that the waveform exists for all time.)
(c) Find and plot the characteristic function for this waveform distribution.

3.14.   For the sawtooth voltage waveform shown in Fig. 3.9.6.
(a) Find and plot the cdf and the pdf for values of the waveform.
(b) Find the total power in the waveform.
(c) Find and plot the characteristic function for this waveform distribution.

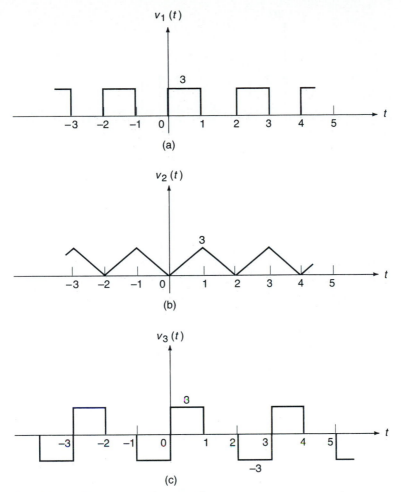

Fig. 3.9.3. Three periodic waveforms.

3.15. The waveform in Fig. 3.9.6 is passed through the ideal diode circuit in Fig. 3.9.7.
(a) Plot the output waveform.
(b) Plot the cdf and the pdf for the output waveform.
(c) Find the total power in the output waveform.

3.16. Calculate the entropy in (a) the experiment, and (b) the random variable in Problem 3.2. Recall that a random variable determines a partition of the sample space, so you should get two different answers for the two parts of this problem.

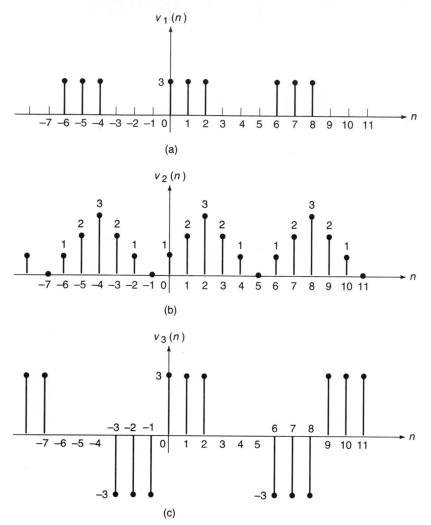

Fig. 3.9.4.   Three discrete-time waveforms.

3.17. Calculate the entropy in the random variable in Problem 3.3. Which is more random, the random variable in Problem 3.2 or the one in Problem 3.3?

3.18. Find the characteristic function for each random variable in Problem 3.6.

3.19. Find and plot the pdf for a binomial random variable Y given by

$$Y = \sum_{i=1}^{N} X_i$$

where $N = 4$ and $\Pr[X_i = 1] = 0.2$, $\Pr[X_i = 0] = 0.8$.

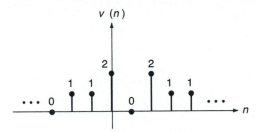

Fig. 3.9.5.   The random waveform $v(n)$.

3.20. The voltage at the output of an amplifier is normally distributed with 0 mean and variance equal to 4.
   (a) If the voltage is measured at a particular time t_1, what is the probability that this voltage is less than 2.4 V?
   (b) A 3-V battery is placed in series with the output so that the dc level is raised to $+3$ V. Now what is the probability that the voltage is less than 2.4 V?

3.21. The characteristic function for a random voltage sampled at a particular time is given by

$$\phi_V(v) = \frac{\sin(v)}{v}$$

   (a) Find the total power in the waveform.
   (b) Find the dc power.
   (c) Find the pdf.

3.22. Write down the formula for $(p + q)^5$ using Pascal's triangle.

3.23. A Gaussian random variable X has mean 2 and variance 4. What is the probability that $0 < X < 4$?

3.24. The cdf is a probability $F_X(\alpha) = P\{X \le \alpha\}$. Given a set of samples

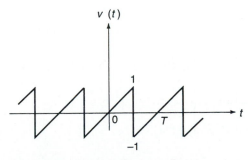

Fig. 3.9.6.   A periodic waveform.

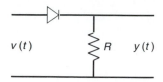

Fig. 3.9.7.   A diode rectifier.

from an unknown distribution, we an estimate the cdf by the ratio

$$\hat{F}_X(\alpha) = \frac{\text{number of samples} \le \alpha}{\text{total number of samples}} \qquad (3.9.1)$$

Given N samples $\{\alpha_1, \alpha_2, \ldots, \alpha_N\}$ from an unknown distribution, we can estimate the cdf in two different ways: (1) Use Eq. 3.9.1 directly; or (2) use predefined bins in the same manner used to plot the histogram in Problem 1.1. If we use method (1), the empirical pdf will have as many jumps as samples, and these steps will occur at values of α determined by the data. If we use method (2), the empirical cdf will have no more steps than bins, and these steps will occur at regular intervals determined by the bin widths. Use methods (1) and (2) for samples from an unknown distribution given by $\{0.03, 0.46, -0.45, 0.21, -0.36, -0.06, -0.49, 0.26, 0.12, -0.18, 0.34\}$. Find and plot the empirical cdf's for this set using each method.

3.25.   Repeat Problem 3.24 for samples given by $\{1.1, -0.45, -0.05, 1.30, 1.42, -0.10, -0.39, 1.48, 1.39, -0.40\}$. What do you think the pdf's look like here?

3.26.   Let's play a gambling game using the experiment in Fig. 3.9.8. I will toss a die, note the outcome (hidden from you), and tell you the value of Y. You must guess the value of X. We'll bet \$1.00 on each trial.
(a) Suppose that $Y = 0$. What is your guess for X?
(b) If $Y = 1$, what is your guess for X? (You get only one guess.)
(c) Would you be willing to play the game for a long time? To answer this, calculate your expected gain per trial. If this number is positive, you should play.

Y	4	1	0	1	4	9
X	−2	−1	0	1	2	3
ζ	⚀	⚁	⚂	⚃	⚄	⚅

Fig. 3.9.8

3.27. Sketch a continuous-time waveform that has the given distribution for each pdf shown in Fig. 3.9.9.

3.28. Suppose that X is normal $(0, 1)$. Find the distribution of Y for each relationship of the form $y = g(x)$ given below.
(a) $y = 2x$
(b) $y = 2x + 2$
(c) $y = 0.5x - 1$
(d) $y = x$ $x \geq 0$
    $y = 0$ otherwise

3.29. Repeat Problem 3.28 if X is uniform $(-\frac{1}{2}, \frac{1}{2})$.

3.30. Suppose that X normal $(0, 1)$ is supplied to a square law device $y = x^2$. Find the distribution of Y.

3.31. Suppose that X normal $(0, 1)$ is supplied to the limiter in Fig. 3.7.12. Find the distribution of Y.

3.32. The following statements are either correct, partially correct, or wrong. Change each statement as needed to make it correct.
(a) Probability and random variables are functions with the same domain.
(b) Experiments and only experiments have randomness.
(c) The characteristic function is the Fourier transform of the probability density function.
(d) The possible values of a Gaussian random variable extend from $-\infty$ to ∞.

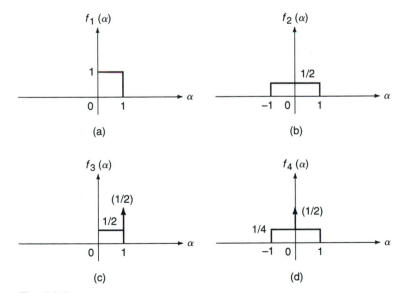

Fig. 3.9.9

(e) The average value of a random variable can be found by evaluating its characteristic function $\phi_X(\nu)$ at $\nu = 1$.

(f) If a waveform has mean 2 and total power 9, then its ac power is 3.

(g) If a random variable is normally distributed with mean 1 and variance 4, then the mean square value is 5.

(h) If a random variable is uniformly distributed between 0 and 1, then its mean square value is 1.

(i) The cdf for values of a continuous-time waveform is estimated by finding the ratio of time intervals.

(j) For a random variable X, the set $\{X \leq \alpha\}$ defines an event for every α.

Further Reading

1. ATHANASIOS PAPOULIS, *Probability, Random Variables, and Stochastic Processes*, 3rd ed., McGraw-Hill, New York, 1991.

2. CHARLES W. THERRIEN, *Discrete Random Signals and Statistical Signal Processing*, Prentice Hall, Englewood Cliffs, NJ, 1992.

3. MICHAEL O'FLYNN, *Probabilities, Random Variables, and Random Processes*, Harper & Row, New York, 1982.

CHAPTER 4 _____

Random Vectors

4.1 Joint Random Variables _____

Preview

There are two ways to get two numbers from an experiment: Perform the experiment twice; or perform the experiment once and let the outcome determine two numbers. For example, toss a coin and let $X = 1$ if heads occurs and $X = 0$ if tails occurs. If we toss the coin twice, this will generate a pair of numbers. But if one toss determines two numbers, we must define another random variable on the sample space. Let $Y = 1$ if heads occurs and $Y = -1$ if tails occurs, which gives us two random variables, X and Y. Now we could get $(1, 1)$ or $(0, -1)$ as the possible pair of values for the ordered pair (X, Y). An outcome of heads gives the first pair, and tails gives the second pair.

The case where two random variables X and Y are defined on the sample space is called a *two-dimensional random variable*. (See Fig. 4.1.1.) If the experimental outcome is ζ, then two numbers are determined: $X(\zeta)$ and $Y(\zeta)$. The ordered pair (X, Y) is called a two-dimensional random variable or a *random vector*. Random vectors have cdf and pdf functions, and can be dependent or independent. A measure of their dependence is their correlation or covariance, but these are topics for the next section. Here we introduce the concept of random vectors, define their cdf and pdf, and state the conditions for independence.

There is no randomness about a two-dimensional random variable, just as there is no randomness about a one-dimensional random variable. A two-dimensional random variable is a function. The randomness occurs in the experiment. The experiment is performed once and an ordered pair of numbers is generated by (X, Y). This is quite different from performing the experiment twice and using a single random variable to generate the ordered pair $(X(\zeta_1), X(\zeta_2))$.

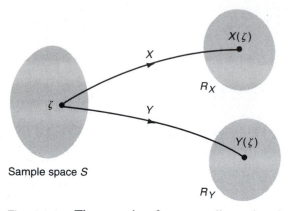

Fig. 4.1.1.   The mapping for a two-dimensional random variable.

The term *joint random variable* or *jointly distributed random variable* applies to a two-dimensional, three-dimensional, or higher-dimensional random variable. A three-dimensional random variable consists of the 3-tuple generated by one experimental outcome—say, $(X(\zeta), Y(\zeta), Z(\zeta))$. Thus an illustration similar to Fig. 4.1.1 for the three-dimensional random variable would consist of three range spaces, R_X, R_Y, and R_Z, and one sample space S.

Figure 4.1.2 illustrates a two-dimensional random variable defined on the sample space of die-toss outcomes. The experiment is performed, an outcome occurs, and we obtain the ordered pair (X, Y). For example, if the five-spot turns up, we obtain the ordered pair $(2, 4)$.

Let us apply this concept to sample values of a waveform taken at times $t_1, t_2, \ldots, t_k$. Consider any continuous-time signal. Suppose that we take k samples spaced 1 s apart. Thus we have k values $v(t_1), v(t_2), \ldots, v(t_k)$, and if the time of the first sample is arbitrary, we have a k-dimensional random variable. The experiment is to select the first sample time.

In this example the experimental outcome ζ determines the sampling time t_1 (and therefore all other sampling times). The domain or sample space for our random variable is a set of experimental outcomes (about which we have said nothing), and one element of the range is the k sample values $v(t_1), v(t_2), \ldots, v(t_k)$. Here we have chosen to let the experimental outcome

Y	4	1	0	1	4	9
X	−2	−1	0	1	2	3
ζ	⚀	⚁	⚂	⚃	⚄	⚅

Fig. 4.1.2.   A two-dimensional random variable (X, Y).

determine the sampling times. An alternative procedure for choosing a random vector is to fix the sampling times and let the experimental outcome determine different voltage waveforms. We will discuss this procedure in the chapter on stochastic processes.

Our ability to do anything useful with joint random variables depends on our ability to describe them mathematically. This mathematical description began, for one-dimensional random variables, with the cdf and the pdf. These descriptions also apply to vector random variables. We begin with the cdf.

DEFINITION 4.1.1.   The probability of the event $\{X \leq \alpha, Y \leq \beta\}$, denoted by

$$F_{XY}(\alpha, \beta) = P\{X \leq \alpha, Y \leq \beta\} \qquad (4.1.1)$$

is called the cumulative distribution function (cdf) of the random variable (X, Y).

This is a joint probability. A more accurate way to write it is $P(\{X \leq a\} \cap \{Y \leq b\})$. A comma is commonly used throughout the literature in place of the intersection symbol. The subscripts on F_{XY} distinguish this particular function from other (different) functions that might be labeled F. The domain of F_{XY} is the set of all ordered pairs (α, β), where $\alpha \in R_X$ and $\beta \in R_Y$ (see Fig. 4.1.1). The range of F_{XY} is the set of numbers $[0, 1]$.

$F_{XY}(\alpha, \beta)$ is the probability of the event $\{X \leq \alpha, Y \leq \beta\}$. There must be an equivalent event in S with probability equal to that of $F_{XY}(\alpha, \beta)$. We make use of this equivalent event to determine values of $F_{XY}(\alpha, \beta)$. This concept readily extends to higher dimensions (Eq. 4.1.1 defines the two-dimensional cdf). For a three-dimensional random variable the joint cdf is

$$F_{XYZ}(\alpha, \beta, \gamma) = P\{X \leq \alpha, Y \leq \beta, Z \leq \gamma\}$$

DEFINITION 4.1.2.   The second partial derivative of the joint cdf, given by

$$f_{XY}(\alpha, \beta) = \frac{\partial^2}{\partial \alpha \, \partial \beta} F_{XY}(\alpha, \beta) \qquad (4.1.2)$$

is called the joint probability density function (pdf).

For discrete random variables the joint pdf will have impulses in the two-dimensional plane. For continuous random variables the joint pdf will be a smooth surface. As before, the pdf is a density function and hence we must integrate it to find probability. For the two-dimensional random variable we must integrate the pdf twice to obtain probability. For example, the probability that (X, Y) is in the rectangle $(\alpha_1 \leq X \leq \alpha_2)$ and $(\beta_1 \leq Y \leq \beta_2)$ is given by

$$P(\alpha_1 \leq X \leq \alpha_2, \beta_1 \leq Y \leq \beta_2) = \int_{\alpha_1}^{\alpha_2} \int_{\beta_1}^{\beta_2} f_{XY}(\alpha, \beta) \, d\beta \, d\alpha \qquad (4.1.3)$$

Of course, this concept applies to higher dimensions also.

EXAMPLE 4.1.1.   Suppose that both X and Y are discrete, and our experiment consists of tossing two coins. The experimental outcome is the number of heads that turn up when the two coins are tossed. The three possible outcomes are

ζ_1 = no heads turn up.
ζ_2 = one head turns up.
ζ_3 = two heads turn up.

The joint random variable (X, Y) is defined in Fig. 4.1.3a. Figures 4.1.3b and 4.1.3c illustrate the pdf and cdf, respectively. The two-dimensional delta functions that describe the pdf are given by

$$f_{XY}(\alpha, \beta) = (\tfrac{1}{4})\,\delta(\alpha + 1)\,\delta(\beta - 1) + (\tfrac{1}{2})\,\delta(\alpha - 1)\,\delta(\beta - 1) \\ + (\tfrac{1}{4})\,\delta(\alpha - 2)\,\delta(\beta - 4)$$

Some authors write two-dimensional delta functions with two arguments, e.g., $\delta(\alpha + 1)\,\delta(\beta - 1) = \delta(\alpha + 1, \beta - 1)$. Both have the same meaning, so either way is appropriate.

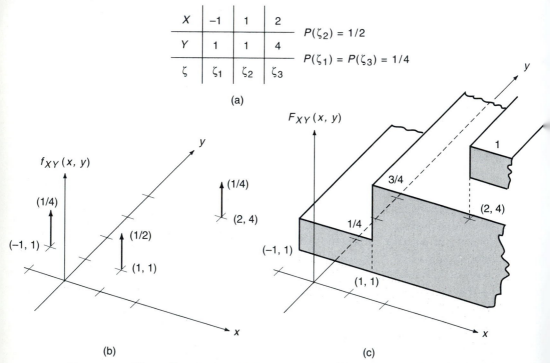

Fig. 4.1.3. The pdf and the cdf for a two-dimensional random variable.

From our discussion of the statistical independence of two events in Section 2.3, recall that events A and B are independent if and only if $P(A, B) = P(A) P(B)$. Since, by definition, $\{X \leq \alpha\}$ and $\{Y \leq \beta\}$ are events, we can apply this definition directly to obtain the following.

DEFINITION 4.1.3.   Two random variables X and Y are said to be independent if the events $\{X \leq \alpha\}$ and $\{Y \leq \beta\}$ are statistically independent for all values of α and β.

The independence of X and Y is equivalent to each of the following:

1. $F_{XY}(\alpha, \beta) = F_X(\alpha) \cdot F_Y(\beta)$ for all α, β.
2. $f_{XY}(\alpha, \beta) = f_X(\alpha) \cdot f_Y(\beta)$ for all α, β.

EXAMPLE 4.1.2.   Are the random variables in Fig. 4.1.2 independent?

SOLUTION: If we can find one pair of values for α, β such that $F_{XY}(\alpha, \beta) \neq F_X(\alpha) \cdot F_Y(\beta)$, this shows dependence. For $\alpha = 0$ and $\beta = 2$ we have

$$F_X(0) = P[X \leq 0] = \tfrac{1}{2}$$
$$F_Y(2) = P[Y \leq 2] = \tfrac{1}{2}$$

but

$$F_{XY}(0, 2) = P[X \leq 0, Y \leq 2] = \tfrac{1}{3}$$

which is not equal to the product $(\tfrac{1}{2})(\tfrac{1}{2})$, showing that X and Y are dependent.

In dealing with two or more jointly distributed random variables, we call the distribution of each random variable the *marginal distribution*. This marginal distribution is related to the joint distribution as follows. First, the marginal cdf is given by

$$F_X(\alpha) = F_{XY}(\alpha, \infty) \tag{4.1.4}$$

Similarly,

$$F_Y(\beta) = F_{XY}(\infty, \beta) \tag{4.1.5}$$

To see that Eq. 4.1.4 defines the distribution of X, note that the events $\{X \leq \alpha, Y \leq \infty\}$ and $\{X \leq \alpha\}$ are equivalent, since it is always true that $\{Y \leq \infty\}$. Hence their probabilities are equal, leading to Eq. 4.1.4.

Second, the marginal density functions can be derived from Eq. 4.1.4 (or 4.1.5) as follows:

$$F_X(\alpha) = F_{XY}(\alpha, \infty) = \int_{-\infty}^{\infty} \int_{-\infty}^{\alpha} f_{XY}(\gamma, \beta) \, d\gamma \, d\beta$$

Differentiating both sides with respect to α gives

$$f_X(\alpha) = \int_{-\infty}^{\infty} f_{XY}(\alpha, \beta) \, d\beta \tag{4.1.6}$$

Similarly,

$$f_Y(\beta) = \int_{-\infty}^{\infty} f_{XY}(\alpha, \beta) \, d\alpha \qquad (4.1.7)$$

Given the joint distribution (density) function, we can find the marginal distribution (density) functions. We cannot, in general, find the joint function if the marginal functions are known. An exception occurs when X and Y are independent. Then the joint function is the product of the marginal functions.

EXAMPLE 4.1.3.   Find and plot the two-dimensional pdf for the joint random variables (X, Y) in Fig. 4.1.2. Write an equation for this pdf in terms of delta functions.

SOLUTION: Figure 4.1.4 displays the six delta functions in the pdf. There is one delta function for each possible outcome, each occurring with probability $\frac{1}{6}$. The equation for these delta functions is

$$\begin{aligned} f_{XY}(\alpha, \beta) = (\tfrac{1}{6})[&\delta(\alpha + 2)\,\delta(\beta - 4) + \delta(\alpha + 1)\,\delta(\beta - 1) \\ &+ \delta(\alpha)\,\delta(\beta) + \delta(\alpha - 1)\,\delta(\beta - 1) \\ &+ \delta(\alpha - 2)\,\delta(\beta - 4) + \delta(\alpha - 3)\,\delta(\beta - 9)] \end{aligned}$$

Applications to Waveforms

In Section 3.3 we discussed the connection between one-dimensional random variables and waveform analysis. Our reasoning went something like this: A random variable is a function, so we can think of it as a process whereby

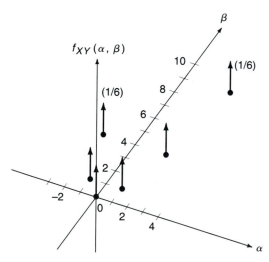

Fig. 4.1.4.   Solution to Example 4.1.3.

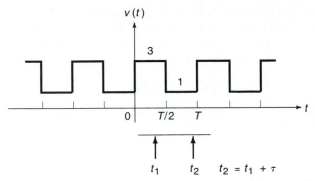

Fig. 4.1.5.   The periodic waveform $v(t)$ with two selected time values, t_1 and t_2.

an experiment is performed, the outcome occurs, and this determines a value for the random variable. Now, in our application to waveform analysis we imagine that a fictitious experiment is performed, the outcome of which determines a value of time. This in turn gives a value for $v(t)$ that we liken to the value of a random variable.

We wish to apply the idea of two-dimensional random variables to waveform analysis. To do this we choose two time values, t_1 and t_2, which then produce two values, $v(t_1)$ and $v(t_2)$. The distribution of these two values is related to important parameters of the waveform, just as the distribution of the univariate values is related to the dc power, ac power, and total power in waveforms. To introduce this topic let us return to the waveform $v(t)$ in Fig. 3.3.1. This is reproduced as Fig. 4.1.5, where we show two time values t_1 and t_2 separated by a fixed increment τ. As t_1 changes value throughout one period, $0 < t_1 < T$, the value of t_2 ranges over the interval $\tau < t_2 < T + \tau$. Since τ is fixed, t_1 and t_2 move together.

The two-dimensional cdf given by

$$F_V(\alpha_1, \alpha_2) = P\{v(t_1) \le \alpha_1 \cap v(t_2) \le \alpha_2\} \tag{4.1.8}$$

depends on both the separation τ and the values of α_1 and α_2. To illustrate, suppose that $\tau = T/2$. Then for different values of α_1 and α_2 the following table gives values for the cdf in Fig. 4.1.5.

α_1	α_2	cdf
2	2.5	0
4	2.5	0.5
1.8	1.7	0
1.8	4.7	0.5

You should be sure to understand where these numbers come from. To illustrate, consider the second entry and look at Fig. 4.1.6. This shows t_1 in the interval $0 \le t_1 < T/2$. Therefore $v(t_1) = 3$, and $v(t_2) = 1$. Since the probability is 0.5 that t_1 is in a half-interval where $v(t_1) = 3$, we have

$$P\{v(t_1) \le 4, v(t_2) \le 2.5\} = 0.5$$

For another example, let $\tau = 3T/4$. Now, for these same values of α_1 and α_2, we get the following values for the cdf:

α_1	α_2	cdf
2	2.5	0.25
4	2.5	0.5
1.8	1.7	0.25
1.8	4.7	0.5

When $\alpha_1 = 1.8$, $\alpha_2 = 1.7$, and $\tau = 3T/4$, both conditions $v(t_1) \le \alpha_1$ and $v(t_2) \le \alpha_2$ are satisfied for one-fourth of the period as t_1 (and therefore t_2) range over one period. The condition is met while t_1 is in the interval $3T/4 \le t_1 < T$. During that time interval, $v(t_1) = 1$ and $v(t_2) = 1$. This gives the value 0.25 for the cdf.

EXAMPLE 4.1.4.   The periodic sawtooth waveform in Fig. 4.1.7 ranges over values between -1 and $+1$ with a period of 2. In the following we evaluate Eq. 4.1.8 for the given combinations of α_1, α_2, and τ.

(a) $\alpha_1 = 0$, $\alpha_2 = 0$, $\tau = 1$. $F_v(\alpha_1, \alpha_2) = 0$

(b) $\alpha_1 = 0$, $\alpha_2 = 1$, $\tau = 1$. $F_v(\alpha_1, \alpha_2) = 0.5$

(c) $\alpha_1 = 0$, $\alpha_2 = 0$, $\tau = 2$. $F_v(\alpha_1, \alpha_2) = 0.5$

(d) $\alpha_1 = 0$, $\alpha_2 = 1$, $\tau = 2$. $F_v(\alpha_1, \alpha_2) = 0.5$

(e) $\alpha_1 = 0$, $\alpha_2 = 0$, $\tau = 1.5$. $F_v(\alpha_1, \alpha_2) = 0.25$

(f) $\alpha_1 = 0$, $\alpha_2 = 1$, $\tau = 1.5$. $F_v(\alpha_1, \alpha_2) = 0.5$

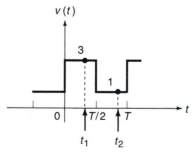

Fig. 4.1.6

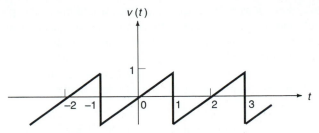

Fig. 4.1.7. The sawtooth waveform $v(t)$.

Parts (b), (d), and (f) all have $\alpha_2 = 1$. Since the waveform always has a value less than or equal to 1, the probability F_V depends only on the value of $v(t)$ at $t = t_1$. But $v(t_1) \leq 0$ for half the period, so the probability is 0.5 in all three parts.

Parts (a), (c), and (e), on the other hand, give three different values for the probability. In part (a), $v(t)$ is never less than 0 at the two times. If $v(t_1) \leq 0$, then $v(t_2)$ is not. If $v(t_2) \leq 0$, then $v(t_1)$ is not. Thus the probability is 0 for part (a). In part (c) we always have $v(t_1) = v(t_2)$ because the two times are separated by one period. This gives 0.5 for the probability. In part (e), $v(t_1)$ and $v(t_2)$ are both less than or equal to 0 when t_1 is in the interval $1.5 \leq t_1 < 2$. This is one-fourth of the period, so the probability is 0.25.

Review

Two or more random variables defined on the same sample space constitute a random vector. The cdf is a probability, and the pdf is the derivative of the cdf, as defined in Eqs. 4.1.1 and 4.1.2. Two random variables X and Y are independent if the events $\{X \leq \alpha\}$ and $\{Y \leq \beta\}$ are independent for all α, β. The distribution of each random variable in a vector is called the marginal distribution. We can always find the marginal distribution from the given joint distribution by Eqs. 4.1.4 through 4.1.7. If X and Y are independent, the joint pdf is the product of the marginal pdf's. If X and Y are dependent, we must know how they are dependent before we can find the joint distribution from the marginal distributions.

A natural extension of the concepts described in Section 3.3 gives us a two-dimensional probability description of waveforms. We did not explain how this will be helpful in real-world problems, perhaps because the truth is that we do not use it much. We are simply preparing you for some of the ideas we will encounter when we discuss stochastic processes.

4.2 Correlation and Covariance

Preview

We have been talking about correlation all along. In Chapter 1 we introduced the idea of a correlated signal when a coin-tossing experiment was used to generate a random signal. We talked about signals x_0, x_6, x_9, and x_{11}, where the subscript matched the number of coin tosses in common between sample values. The correlation between samples was related to the number of common coin tosses. In this section we give a more precise meaning to correlation. We will have to wait for the chapter on stochastic processes to realize the full potential of this topic, but we begin our study of this important concept with the introduction of correlation between two random variables.

DEFINITION 4.2.1.   The expected value of the product $X^i Y^j$ (X raised to the ith power times Y raised to the jth power) is said to be the joint moment of order $i + j$, given by

$$m_{ij} = E(X^i Y^j) = \int_{-\infty}^{\infty} \int_{-\infty}^{\infty} \alpha^i \beta^j f_{XY}(\alpha, \beta) \, d\alpha \, d\beta \qquad (4.2.1)$$

The order of the moment is $i + j$. There are three second-order moments, m_{20}, m_{02}, and m_{11}, and they all have special significance:

$$m_{20} = E(X^2)$$
$$m_{02} = E(Y^2)$$
$$m_{11} = E(XY)$$

The first two terms are the mean square values of X and Y, respectively. The last term, $m_{11} = E(XY)$, is called the correlation of X and Y.

EXAMPLE 4.2.1.   Find the correlation between X and Y in Fig. 4.1.2.

SOLUTION: This is the die-toss experiment with Y the square of X. The correlation is given by

$$E(XY) = \sum_{i=1}^{n} \alpha_i \beta_i P(X = \alpha_i, Y = \beta_i)$$
$$= (-2)(4)(\tfrac{1}{6}) + (-1)(1)(\tfrac{1}{6}) + (0)(0)(\tfrac{1}{6})$$
$$+ (1)(1)(\tfrac{1}{6}) + (2)(4)(\tfrac{1}{6}) + (3)(9)(\tfrac{1}{6}) = \tfrac{9}{2}$$

DEFINITION 4.2.2.   If $E(XY) = E(X) \, E(Y)$, then X and Y are said to be uncorrelated.

If X and Y are independent (see Definition 4.1.3), then they are uncorrelated. If X and Y are uncorrelated, they may or may not be independent.

EXAMPLE 4.2.2.   Are X and Y in the example above independent or uncorrelated?

SOLUTION: For independence we must have $P(X = \alpha, \ Y = \beta) = P(X = \alpha) \, P(Y = \beta)$ for every possible value of α and β. Let $\alpha = -2$ and $\beta = 9$. Then $P(X = -2, Y = 9) = 0$ (it is impossible for X to equal -2 and Y to equal 9 on the same toss of the die). But $P(X = -2) = \tfrac{1}{6}$, and $P(Y = 9) = \frac{1}{6}$, giving the product $\frac{1}{36}$. This does not equal 0, so we have found at least one case where $P(X, Y)$ does not equal $P(X) P(Y)$. Therefore X and Y are dependent.

If they are uncorrelated we must have $E(X, Y) = E(X) E(Y)$. The value of $E(X, Y)$ is $\frac{9}{2}$, from Example 4.2.1, and $E(X)$ and $E(Y)$ are given by

$$E(X) = \sum_{i=1}^{6} \alpha_i P(X = \alpha_i) = (\tfrac{1}{6})(-2 - 1 + 0 + 1 + 2 + 3) = \tfrac{1}{2}$$

$$E(Y) = \sum_{i=1}^{6} \alpha_i P(Y = \alpha_i) = (\tfrac{1}{6})(4 + 1 + 0 + 1 + 4 + 9) = \tfrac{19}{6}$$

Therefore $E(X) E(Y) = \frac{19}{12}$, which is not $\frac{9}{2}$. We conclude that X and Y are correlated.

DEFINITION 4.2.3.   The second central moment μ_{11} is called the covariance of X and Y. It is given by

$$\mu_{11} = E[(X - m_X)(Y - m_Y)] = E(XY) - E(X) E(Y) = m_{11} - m_X m_Y$$
$$(4.2.2)$$

Note: If $\mu_{11} = 0$, then X and Y are uncorrelated. This is rather confusing, because when the correlation $E(XY) = 0$ it does not necessarily mean that the random variables are uncorrelated. However, when the covariance $\mu_{11} = 0$, it does mean that the random variables are uncorrelated. Only when the means m_X and m_Y are 0 does zero correlation imply uncorrelated random variables.

The normalized covariance is called the correlation coefficient. It is given by

$$\rho_{XY} = \frac{\mu_{11}}{\sigma_X \sigma_Y} \qquad (4.2.3)$$

The correlation coefficient has the important property that

$$-1 \le \rho_{XY} \le 1 \qquad (4.2.4)$$

Thus the correlation coefficient is a normalized measure of the correlation between X and Y. If $\rho_{XY} = 0$, the random variables X and Y are uncorrelated.

Once again, these terms can be confusing. The correlation coefficient ρ_{XY} has a closer relation to the covariance than to the correlation. The correlation coefficient is 0 if and only if the covariance is 0, and this implies that the random variables are uncorrelated. It would be much better if we could start over and rename these terms, but this is impossible. We are stuck with the present system of confusing terms.

Here is a list of related terms and their meanings:

1. Independent: $f_{XY}(\alpha, \beta) = f_X(\alpha) \cdot f_Y(\beta)$ for all α, β.
2. Uncorrelated: $E(XY) = E(X) E(Y)$

3. Correlation: $m_{11} = E(XY)$
4. Covariance: $\mu_{11} = E(XY) - E(X) \, E(Y)$
5. Correlation coefficient: $\rho_{XY} = \mu_{11}/\sigma_X \sigma_Y$.

EXAMPLE 4.2.3.   Refer to Example 4.1.3, with the random variables X and Y as defined in Fig. 4.1.3. Calculate the correlation m_{11}, the covariance μ_{11}, and the correlation coefficient ρ_{XY}.

SOLUTION:

$$m_{11} = E(XY) = \sum_{\alpha, \beta} \alpha\beta f_{XY}(\alpha, \beta)$$

$$= (-1)(1)\left(\frac{1}{4}\right) + (1)(1)\left(\frac{1}{2}\right) + (2)(4)\left(\frac{1}{4}\right) = \frac{9}{4}$$

$$\mu_{11} = E[(X - m_X)(Y - m_Y)] = \sum_{\alpha, \beta} (\alpha - m_X)(\beta - m_Y) f_{XY}(\alpha, \beta)$$

$$= \left(-1 - \frac{3}{4}\right)\left(1 - \frac{7}{4}\right)\left(\frac{1}{4}\right) + \left(1 - \frac{3}{4}\right)\left(1 - \frac{7}{4}\right)\left(\frac{1}{2}\right)$$

$$+ \left(2 - \frac{3}{4}\right)\left(4 - \frac{7}{4}\right)\left(\frac{1}{4}\right) = \frac{15}{16}$$

$$\rho_{XY} = \frac{\mu_{11}}{\sigma_X \sigma_Y} = \frac{\frac{15}{16}}{\sqrt{\left(\frac{19}{16}\right)\left(\frac{27}{16}\right)}} = 0.663$$

Now we are in a position to examine the coin-tossing experiment from Chapter 1, where we tossed the coin 12 times and counted the number of heads to obtain a digital signal. Figure 1.2.6 displayed what we called the correlation between two samples of the signal $x_{11}(n)$. This is actually the correlation coefficient ρ between samples. We called it correlation in Chapter 1 because it would have been confusing to use all the correct terms (it still is confusing).

We moved the window one toss at a time to obtain values of $x_{11}(n)$ given by

$$X_1 = \sum_{i=1}^{12} Z_i, \qquad X_2 = \sum_{i=2}^{13} Z_i, \qquad \ldots, \qquad X_n = \sum_{i=n}^{n+11} Z_i, \qquad \ldots \quad (4.2.5)$$

where the Z_i are random variables defined by

$$Z_i = \begin{cases} 1 & \text{if heads on the } i\text{th toss} \\ 0 & \text{if tails on the } i\text{th toss} \end{cases}$$

Let us derive the correlation, the covariance, and the correlation coefficient between samples for this signal. This gives us a wealth of problems to explore, one for each pair of values, and this for three calculations.

First, the correlation between two samples is given by $E(X_iX_j)$. Since each X is the sum of 12 Z's, the correlation can be expressed in terms of the Z variables. For $i = j = 1$ we have

$$
\begin{aligned}
E(X_1^2) &= E[(Z_1 + Z_2 + \cdots + Z_{12})(Z_1 + Z_2 + \cdots + Z_{12})] \\
&= E(Z_1^2 + Z_1Z_2 + Z_1Z_3 + \cdots \\
&\quad + Z_1Z_{12} + Z_2Z_1 + Z_2^2 + \cdots + Z_{12})
\end{aligned}
\tag{4.2.6}
$$

This has two types of terms, $E(Z_i^2)$ and $E(Z_iZ_j)$, where $i \neq j$. These expected values are given by

$$
E(Z_i^2) = \sum_\alpha \alpha^2 P(Z_i = \alpha) = 0^2(\tfrac{1}{2}) + 1^2(\tfrac{1}{2}) = \tfrac{1}{2}
\tag{4.2.7}
$$

$$
\begin{aligned}
E(Z_iZ_j) &= \sum_{\alpha, \beta} \alpha\beta P(Z_i = \alpha, Z_j = \beta) \\
&= 0 \cdot 0 \cdot (\tfrac{1}{4}) + 0 \cdot 1 \cdot (\tfrac{1}{4}) + 1 \cdot 0 \cdot (\tfrac{1}{4}) + 1 \cdot 1 \cdot (\tfrac{1}{4}) = \tfrac{1}{4} \qquad \text{for } i \neq j
\end{aligned}
\tag{4.2.8}
$$

There are 12 $E(Z_i^2)$ terms and $144 - 12 = 132$ $E(Z_iZ_j)$ terms in the expression for $E(X_1^2)$, giving

$$
E(X_1^2) = 12(\tfrac{1}{2}) + 132(\tfrac{1}{4}) = 39
$$

There is no reason to favor X_1 over any other value. That is, $E(X_2^2)$ should also equal 39, as should the mean square value at any other time. This signal is said to be stationary, because statistical values at one time are the same at any other time. (We will give a more formal definition of stationarity in Chapter 6.)

Now let us calculate $E(X_iX_j)$ for $j = i + 1$, i.e., for adjacent values of the signal. Since the signal is stationary we can calculate $E(X_1X_2)$, and this will equal $E(X_iX_j)$ for any i and j with $j = i + 1$.

$$
\begin{aligned}
E(X_1X_2) &= E[(Z_1 + Z_2 + \cdots + Z_{12})(Z_2 + Z_3 + \cdots + Z_{13})] \\
&= E(Z_1Z_2 + Z_1Z_3 + \cdots + Z_1Z_{13} + Z_2^2 + Z_2Z_3 \\
&\quad + \cdots + Z_{12}Z_{13})
\end{aligned}
\tag{4.2.9}
$$

Out of 144 terms in this expression, 11 are $E(Z_i^2)$ and 133 are $E(Z_iZ_j)$ with $i \neq j$. Each $E(Z_i^2)$ term equals $\tfrac{1}{2}$, and each $E(Z_iZ_j)$ term equals $\tfrac{1}{4}$, giving

$$
E(X_1X_2) = 11(\tfrac{1}{2}) + 133(\tfrac{1}{4}) = 38.75
$$

Of course this is the correlation between any X_iX_j with $j = i + 1$.

We can see the trend now by comparing Eqs. 4.2.6 and 4.2.9. The next expression, for $E(X_iX_j)$ with $i = j + 2$, will contain 10 $E(Z_i^2)$ terms with a value of $\tfrac{1}{2}$ each, and 134 $E(Z_iZ_j)$ terms with a value of $\tfrac{1}{4}$ each. Therefore

$$
E(X_1X_3) = 10(\tfrac{1}{2}) + 134(\tfrac{1}{4}) = 38.5
$$

This trend continues, with the correlation reduced by $\frac{1}{4}$ each step, until the samples are separated by 12, when all correlation values for X are equal to 36, which is just the dc (mean) value squared. The autocorrelation function for $X_{11}(n)$ is plotted in Fig. 4.2.1.

The covariance defined in Eq. 4.2.2 is the correlation with the product of the means subtracted. Here, each mean is 6, so if we subtract 36 from the autocorrelation function in Fig. 4.2.1 we get the autocovariance function. This function has a maximum value of 3 at zero shift, and declines by $\frac{1}{4}$ until the samples are separated by 12. Therefore the graph of this function looks just like Fig. 4.2.1, but with the abscissa moved up to 36.

The correlation coefficient is just the autocovariance function normalized to have a maximum value of 1. Therefore if we divide each value of the autocovariance function by 3, we should get the graph in Section 1.2, Fig. 1.2.6. To check that this agrees with Eq. 4.2.3 for the correlation coefficient, we should calculate σ_X and σ_Y. These terms are the standard deviation for the waveform at two different times, so they are equal. We have already made the calculations necessary to find the variance, given by

$$\sigma_x^2 = E[(X - m_1)^2] = E(X^2) - m_1^2 = 39 - 36 = 3$$

Therefore the standard deviation is the square root of 3. Using this in Eq. 4.2.3 for both σ_X and σ_Y gives the graph in Fig. 1.2.6.

Review

The correlation, the covariance, and the correlation coefficient are related to each other by the sequence:

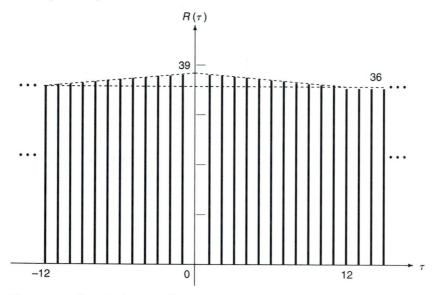

Fig. 4.2.1.   Correlation coefficients for X_{11}.

Correlation: $m_{11} = E(XY)$.
Covariance: $\mu_{11} = E(XY) - E(X)\,E(Y)$.
Correlation coefficient: $\rho_{XY} = \mu_{11}/\sigma_X \sigma_Y$.

If X and Y are independent, they are uncorrelated. If X and Y are uncorrelated, they may or may not be independent. One big problem with the term "correlation" occurs when we test for correlation. The correlation m_{11} need not be 0 for the random variables to be uncorrelated. It is the covariance that must be 0 for the random variables to be uncorrelated.

4.3 Conditional Random Variables _____

Preview

In Section 2.2 we introduced the concept of conditional probability and stressed the point that a reduction in the sample space results from the conditioning. The expression $P(A|B)$ meant that B had occurred, and therefore the sample space was reduced from the original space S to the given event B. This changed the probability of the event A from its original value to something different, depending on the relationship between A and B. We now extend this concept to random variables and consider the meaning of expressions such as $f_X(\alpha|Y = \beta)$ and $F_X(\alpha|Y = \beta)$. We will do this by drawing an analogy between conditional cdf or pdf and conditional probability.

Let us start with the simplest case. The expression $F_X(\alpha|A)$ is the cdf for the random variable X, given that the event A has occurred. Since A is a subset of the sample space (by definition), the possible values of X are restricted to those related to the experimental outcomes in A. Figure 4.3.1 is supposed to illustrate this concept, the idea being that the possible values of X are no longer all those associated with the sample space, but are restricted to those associated with the experimental outcomes in A. An example should help.

Let X be the random variable that counts the number of spots on a die face. That is, if a 1 is rolled, then $X = 1$; if a 2 is rolled, then $X = 2$; etc. The cdf $F_X(\alpha)$ is plotted in Fig. 4.3.2a. (This is the unconditional cdf.)

Let A be the event "X is greater than 3." Figure 4.3.2b shows the conditional cdf $F_X(\alpha|A)$. This plot is determined from the following reasoning: Recall that conditional probability is related to the joint and the marginal probability by

$$P(B|A) = \frac{P(B \cap A)}{P(A)}$$

Since F_X is a probability, we can write

$$F_X(\alpha|A) = P\{(X \leq \alpha)|A\} = \frac{P[(X \leq \alpha) \cap A]}{P(A)} \qquad (4.3.1)$$

The numerators in this expression for several values of α are given by

$$P\{(X \leq 3.5) \cap A\} = 0$$
$$P\{(X \leq 4.5) \cap A\} = \tfrac{1}{6}$$
$$P\{(X \leq 5.8) \cap A\} = \tfrac{2}{6}$$
$$P\{(X \leq 6.2) \cap A\} = \tfrac{3}{6}$$

The event A has probability $\tfrac{1}{2}$, so Eq. 4.3.1 gives

$$P\{(X \leq 3.5)|A\} = 0$$
$$P\{(X \leq 4.5)|A\} = \tfrac{1}{3}$$
$$P\{(X \leq 5.8)|A\} = \tfrac{2}{3}$$
$$P\{(X \leq 6.2)|A\} = 1$$

in agreement with Fig. 4.3.2b.

Incidentally, many authors use the symbol $F_{X|A}$ for the function in Eq. 4.3.1 instead of our notation F_X. They are correct, because the two functions are different, as you can see from Fig. 4.3.2. Our use of the same notation for both functions can be confusing, so we will use the more complex notation when necessary.

Now let us consider the next simplest case. Suppose that there are two random variables X and Y defined on the same sample space, as shown in Fig. 4.3.3. The experiment is to throw a die, and the experimental outcome determines values for X and Y. (This is the same example we used before, and we will see it again several times.) By the expression $F_X(\alpha|\beta) =$

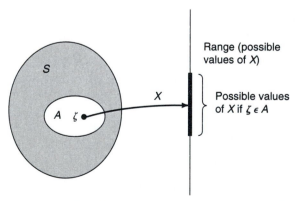

Fig. 4.3.1. Possible values of a conditional random variable.

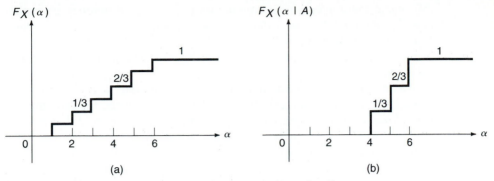

Fig. 4.3.2.   The unconditional cdf and the conditional cdf.

$F_X(\alpha|Y = \beta)$ we mean the probability that X is less than or equal to α, given that $Y = \beta$.

$$F_X(\alpha|\beta) = P\{X \le \alpha|Y = \beta\} = \frac{P[X \le \alpha|Y = \beta]}{P[Y = \beta]} \qquad (4.3.2)$$

For example, given $Y = 4$ in Fig. 4.3.3, the cdf for X is given by

$$F_X(\alpha|Y = 4) = P\{X \le \alpha|Y = 4\} = \begin{cases} 0, & \alpha \le -2 \\ \frac{1}{2}, & -2 \le \alpha < 2 \\ 1, & 2 \le \alpha \end{cases}$$

This is shown in Fig. 4.3.4. Once again we have been lax in our notation. The more correct symbol for the cdf in Eq. 4.3.2 is $F_{X|Y}$.

The pdf associated with each cdf is just the derivative. For example,

$$f_X(\alpha|\beta) = \frac{d}{d\alpha} F_X(\alpha|\beta) \qquad (4.3.3)$$

We have discussed two situations: first, $F_X(\alpha|A)$, where A is an event; and second, $F_X(\alpha|\beta)$, where β is the value of the random variable Y that is somehow related to X. For discrete random variables there is virtually no

Y	4	1	0	1	4	9
X	-2	-1	0	1	2	3
ζ	⚀	⚁	⚂	⚃	⚄	⚅

Fig. 4.3.3

difference, but for continuous random variables the difference is important. Equation 4.3.2 is not valid if Y is continuous at $Y = \beta$, because then
$P\{Y = \beta\} = 0$. After all, if Y is continuous then there are an infinite number
of possible values for Y. The probability that Y will equal any one of these
values must be 0.

We are on the horns of a dilemma. If Y is continuous at $Y = \beta$, then the
conditional cdf can no longer be defined by Eq. 4.3.2 because the denominator
is 0. However, if the experiment is performed and we are told that $Y = \beta$,
then the expressions $f_X(\alpha|\beta)$ and $F_X(\alpha|\beta)$ must have some meaning (and
value), and these values should be different, in general, from the unconditional functions.

It is more convenient simply to define the functions $f_X(\alpha|\beta)$ and $F_X(\alpha|\beta)$
if Y is continuous at β, rather than relate their meaning to previous definitions
as in Eq. 4.3.2. In doing this we should make our definition consistent with
the probability laws and our previous definitions. For example, the total area
under $f_X(\alpha|\beta)$ must always be 1, just as the area under $f_X(\alpha)$ must be 1.
Also, our definition for $F_X(\alpha|\beta)$ should reduce to Eq. 4.3.2 if Y is discrete
at $Y = \beta$. This is accomplished by the following definitions.

DEFINITION 4.3.1.   The conditional pdf of the random variable X,
given that $Y = \beta$, is

$$f_X(\alpha|\beta) = \frac{f_{XY}(\alpha, \beta)}{f_Y(\beta)}, \qquad f_Y(\beta) > 0 \qquad (4.3.4)$$

Now the denominator is a number. For a given β, $f_Y(\beta)$ is just the
value of the probability density function evaluated at β. The area under
$f_X(\alpha|\beta)$ is 1 because

$$\int_{-\infty}^{\infty} f_X(\alpha|\beta) \, d\alpha = \int_{-\infty}^{\infty} \frac{f_{XY}(\alpha, \beta)}{f_Y(\beta)} \, d\alpha$$

$$= \frac{1}{f_Y(\beta)} \int_{-\infty}^{\infty} f_{XY}(\alpha, \beta) \, d\alpha = \frac{f_Y(\beta)}{f_Y(\beta)} = 1$$

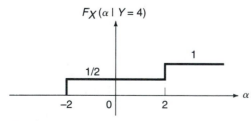

Fig. 4.3.4

DEFINITION 4.3.2.   The conditional cdf is the integral of the pdf, given by

$$F_X(\alpha|\beta) = \int_{-\infty}^{\alpha} f_X(\gamma|\beta)\, d\gamma \qquad (4.3.5)$$

This function satisfies the same conditions as the unconditional cdf.

$$\lim_{\alpha \to -\infty} F_X(\alpha|\beta) = 0 \qquad \text{and} \qquad \lim_{\alpha \to \infty} F_X(\alpha|\beta) = 1$$

If Y is discrete at $Y = \beta$, then $f_X(\beta) = P(Y = \beta)$, and Eq. 4.3.5 gives the same result as Eq. 4.3.2.

EXAMPLE 4.3.1.   The negative exponential distribution models the time between successive telephone calls, the time between customers at a bank teller's window, and the time between wars in our society. Suppose that the random variable X represents the time between successive telephone calls and

$$f_X(\alpha|\lambda) = \lambda e^{-\lambda\alpha}, \qquad 0 \le \alpha < \infty$$

where λ is a value of the random variable Λ, which depends on the time of day, the weather, and day of the week. Suppose that Λ has a density function given by

$$f_\Lambda(\lambda) = \begin{cases} \lambda e^{-\lambda}, & 0 \le \lambda < \infty \\ 0 & \text{otherwise} \end{cases}$$

So in this problem we are given the conditional pdf for X and the marginal pdf for Λ. Find the marginal density for X, the time between telephone calls.

SOLUTION: From Eq. 4.3.4 the two-dimensional pdf is given by

$$f_{X\Lambda}(\alpha, \lambda) = f_X(\alpha|\lambda) \cdot f_\Lambda(\lambda) = \lambda^2 e^{-\lambda(\alpha+1)} \qquad \text{for } \alpha \text{ and } \lambda \ge 0$$

Now use Eq. 4.1.6 to obtain the marginal pdf.

$$f_X(\alpha) = \int_{-\infty}^{\infty} f_{X\Lambda}(\alpha, \lambda)\, d\lambda = \int_{0}^{\infty} \lambda^2 e^{-\lambda(\alpha+1)}\, d\lambda$$

$$= \frac{2}{(\alpha+1)^3}, \qquad 0 \le \alpha < \infty$$

Figure 4.3.5 shows the conditional and marginal pdf's for X.

Review

Conditional distribution is similar to conditional probability. The conditioning changes the distribution for a random variable, just as the conditioning changes the probability for an event. This change is described by Eq. 4.3.4. Notice in

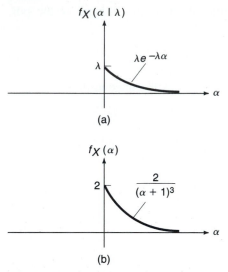

$f_X(\alpha \mid \lambda)$

$\lambda e^{-\lambda \alpha}$

λ

α

(a)

$f_X(\alpha)$

$\dfrac{2}{(\alpha + 1)^3}$

2

α

(b)

Fig. 4.3.5. (a) The conditional pdf and (b) the marginal pdf, both from Example 4.3.1.

Example 4.3.1 how the form of the pdf for X differs between the conditional and the marginal pdf.

4.4   The Gaussian Random Variable _____

Preview

In Section 3.6 we introduced the one-dimensional Gaussian random variable, along with several other important distributions. Here we will spend some time on the important two-dimensional Gaussian random variable and then extend these concepts to random variables of arbitrary dimensions.

The two-dimensional Gaussian pdf with arbitrary means, variances, and covariance takes some room to write down. Higher dimensions become unmanageable without matrix notation. An important part of this section explains this matrix notation. We also explore marginal and conditional distributions and list some important properties of Gaussian random variables.

Let X_1 and X_2 be two Gaussian random variables with zero mean and unit variance. If they are defined on the same sample space, then they form a two-dimensional random variable (X_1, X_2). In that case, the two-dimensional pdf is given by

$$f_{X_1 X_2}(\alpha_1 \alpha_2) = \frac{1}{2\pi\sqrt{1-\rho^2}} \exp\left[-\frac{\alpha_1^2 - 2\rho\alpha_1\alpha_2 + \alpha_2^2}{2(1-\rho^2)}\right] \qquad (4.4.1)$$

where α_1 is the value of X_1 and α_2 is the value of X_2. The correlation coefficient ρ relates X_1 and X_2 as defined in Eq. 4.2.3. Figure 4.4.1 shows the effect that different values of ρ have on the two-dimensional pdf. The shape of the two-dimensional pdf changes with ρ, although each marginal random variable has zero mean and unit variance. When $\rho = 0$, the pdf is a symmetrical hill with each level contour forming a circle, as shown in Fig. 4.4.1a. For $\rho < 0$, the mound representing the pdf stretches in the direction of a line with $-45°$ slope in the X_1, X_2 plane, as shown in Fig. 4.4.1b. When $\rho > 0$, the mound stretches the other way, as shown in Fig. 4.4.1c. But regardless of the value of ρ (so long as $-1 \le \rho \le 1$), the marginal density of each random variable remains $N(0, 1)$.

Recall that the term "density" implies that we must integrate to find probability. The volume under the $f_{X_1 X_2}$ curve over a particular region of the $x_1 x_2$ plane defines the probability of X_1 and X_2 being in that region. The entire volume is 1.

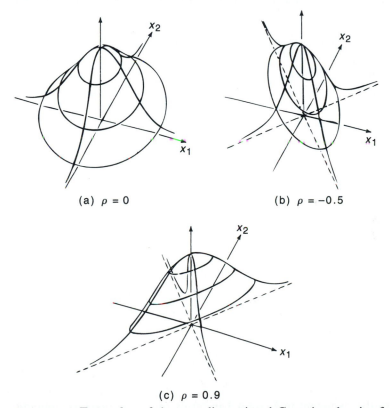

(a) $\rho = 0$

(b) $\rho = -0.5$

(c) $\rho = 0.9$

Fig. 4.4.1. Examples of the two-dimensional Gaussian density function. (Adapted from Wozencraft J.M. and Jacobs I.M., *Principles of Communication Engineering,* John Wiley, New York, 1965. With permission.)

EXAMPLE 4.4.1.   Let X_1, X_2 be a two-dimensional Gaussian random variable with each marginal density $N(0, 1)$. If $\rho = 0$, find the probability that $X_1 > 1$.

SOLUTION: We seek the volume under the curve in Fig. 4.4.1a to the right of $X_1 = 1$. This volume is given by

$$P(X_1 > 1) = \int_{-\infty}^{\infty} \int_{1}^{\infty} f_{X_1 X_2}(\alpha_1, \alpha_2)\, d\alpha_1\, d\alpha_2$$

With $\rho = 0$, the two-dimensional pdf in Eq. 4.4.1 becomes

$$f_{X_1 X_2}(\alpha_1, \alpha_2) = \frac{1}{2\pi} \exp\left(-\frac{\alpha_1^2 + \alpha_2^2}{2}\right)$$
$$= \frac{1}{\sqrt{2\pi}} \exp\left(-\frac{\alpha_1^2}{2}\right) \frac{1}{\sqrt{2\pi}} \exp\left(-\frac{\alpha_2^2}{2}\right)$$

(4.4.2)

Therefore the double integral may be separated into the product of two integrals:

$$P(X_1 \geq 1) = \int_{1}^{\infty} f_{X_1}(\alpha_1)\, d\alpha_1 \int_{-\infty}^{\infty} f_{X_2}(\alpha_2)\, d\alpha_2 = 0.1587$$

The first integral finds the area under the normal pdf from 1 to ∞, and the second finds the area under the pdf from $-\infty$ to ∞. The first integral has value $1 - F_X(1)$, and the second has value 1. From the table of normal areas (Appendix A), $1 - F_X(1) = 1 - 0.8413 = 0.1587$.

The Marginal Density

Recall from Section 4.1 that we integrate over all values of X_2 to find the marginal density for X_1. We will carry out the manipulations for the Gaussian pdf to show you how and to illustrate our remark that each marginal density is $N(0, 1)$.

$$f_{X_1}(\alpha_1) = \int_{-\infty}^{\infty} f_{X_1 X_2}(\alpha_1, \alpha_2)\, d\alpha_2$$
$$= \int_{-\infty}^{\infty} \frac{1}{2\pi\sqrt{1 - \rho^2}} \exp\left[-\frac{\alpha_1^2 - 2\rho\alpha_1\alpha_2 + \alpha_2^2}{2(1 - \rho^2)}\right] d\beta$$

(4.4.3)

Now change the variable of integration. Let

$$\lambda = \frac{\alpha_2 - \rho\alpha_1}{\sqrt{1 - \rho^2}}$$

(4.4.4)

We express the exponent in Eq. 4.4.3 in terms of $(\alpha_2 - \rho\alpha_1)$ by noticing that

$$(\alpha_2 - \rho\alpha_1)^2 + \alpha_1^2(1 - \rho^2) = \alpha_1^2 - 2\rho\alpha_1\alpha_2 + \alpha_2^2$$

(4.4.5)

Therefore, Eq. 4.4.3 becomes

$$f_{X_1}(\alpha_1) = \frac{e^{-\alpha_1^2/2}}{\sqrt{2\pi}} \int_{-\infty}^{\infty} \frac{1}{\sqrt{2\pi(1-\rho^2)}} \exp\left[-\frac{(\alpha_2 - \rho\alpha_1)^2}{2(1-\rho^2)}\right] d\alpha_2$$

Substituting λ from Eq. 4.4.4 now gives the desired result:

$$f_{X_1}(\alpha_1) = \frac{e^{-\alpha_1^2/2}}{\sqrt{2\pi}} \int_{-\infty}^{\infty} \frac{1}{\sqrt{2\pi}} e^{-\lambda^2/2} d\lambda = \frac{1}{\sqrt{2\pi}} e^{-\alpha_1^2/2} \qquad (4.4.6)$$

and we recognize this as the one-dimensional Gaussian form with zero mean and unit variance.

EXAMPLE 4.4.2.   Let X_1, X_2 be a two-dimensional Gaussian random variable with each marginal density $N(0, 1)$. If $\rho = -0.5$, find the probability that $X_1 \geq 1$.

SOLUTION: Here we repeat Example 4.4.1 but with a different correlation coefficient. As you can see from Fig. 4.4.1b, we still seek the volume under the pdf to the right of $\alpha = 1$, but with different pdf. We can either repeat the steps in Example 4.4.1, or we can use our new-found formula 4.4.6 to find

$$P(X_1 \geq 1) = \int_1^{\infty} f_{X_1}(\alpha_1)\, d\alpha_1 = 0.1587$$

The Conditional Density

We defined the conditional pdf in Section 4.3 (Definition 4.3.1) for general random variables. Here we amplify that discussion and apply the definition to Gaussian random variables. Suppose that we observe one of the random variables—say, X_2—and the observed value of X_2 is v. This additional information should change the marginal pdf of X_1. What we are saying is that the conditional pdf $f_{X_1}(\alpha_1|X_2 = v)$ should be different from the unconditional (marginal) pdf $f_{X_1}(\alpha_1)$. This statement may not be obvious, so we will present an intuitive argument based on Fig. 4.4.1b before proceeding to derive the conditional pdf.

Suppose that $X_2 = -0.5$. Now imagine that the two-dimensional pdf is cut vertically along a line parallel to the x_1 axis and passing through $x_2 = -0.5$. Figure 4.4.2, which is a redrawn version of Fig. 4.4.1b, shows this silhouette. The area under the silhouette is not equal to 1. If it were, this would be the conditional pdf $f_{X_1}(\alpha_1|X_2 = -0.5)$. We use a normalizing constant to adjust the area to 1. Note that the mean is not 0.

From Definition 4.3.1, given that $X_2 = v$, the conditional pdf of X_1 is

$$f_{X_1}(\alpha_1|X_2 = v) = \frac{f_{X_1 X_2}(\alpha_1, v)}{f_{X_2}(v)} \qquad (4.4.7)$$

The denominator normalizes the area under $f_{X_1}(\alpha_1|X_2 = v)$ to 1. The numerator forms the silhouette (such as in Fig. 4.4.2) as X_1 is varied and X_2 is held constant at the value of v. The area under this silhouette is not 1, but division by $f_{X_2}(v)$ makes it so.

Now we shall derive the conditional pdf of the Gaussian random variable. Suppose that $f_{X_1 X_2}$ is given by Eq. 4.4.1. Then the conditional pdf is

$$
\begin{aligned}
f_{X_1}(\alpha_1|X_2 = v) &= \frac{f_{X_1 X_2}(\alpha_1, v)}{f_{X_2}(v)} \\
&= \frac{(1/2\pi\sqrt{1 - \rho^2}) \exp[-(\alpha_1^2 - 2\rho\alpha_1 v + v^2)/2(1 - \rho^2)]}{(1/\sqrt{2\pi}) \exp(-v^2/2)} \\
&= \frac{1}{\sqrt{2\pi(1 - \rho^2)}} \exp\left[-\frac{\alpha_1^2 - 2\rho\alpha_1 v + v^2}{2(1 - \rho^2)} + \frac{v^2}{2}\right] \\
&= \frac{1}{\sqrt{2\pi(1 - \rho^2)}} \exp\left[-\frac{(\alpha_1 - \rho v)^2}{2(1 - \rho^2)}\right]
\end{aligned}
\tag{4.4.8}
$$

Compare this to Eq. 3.6.8 (the one-dimensional Gaussian pdf). Here the mean is ρv and the variance is $(1 - \rho^2)$.

EXAMPLE 4.4.3.   Let X_1, X_2 be a two-dimensional Gaussian random variable with each marginal density $N(0, 1)$. If $\rho = -0.5$, find and plot the conditional pdf for X_1 given that $X_2 = -0.5$.

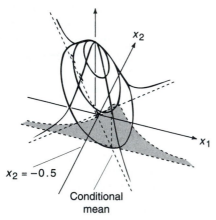

Fig. 4.4.2

SOLUTION: From Eq. 4.4.8,

$$f_{X_1}(\alpha_1|X_2 = -0.5) = \frac{1}{\sqrt{2\pi(1 - 0.25)}} \exp\left[-\frac{(\alpha_1 - 0.25)^2}{2(1 - 0.25)} \right]$$

$$= \frac{1}{\sqrt{3\pi/2}} \exp\left[-\frac{(\alpha_1 - 0.25)^2}{3/2} \right]$$

This function has a mean of 0.25 and a variance 0.75, and is plotted in Fig. 4.4.3.

The Two-Dimensional Gaussian pdf with Arbitrary Mean and Variance

So far we have restricted our attention to the special case where each marginal pdf is normal with zero mean and unit variance. We now remove these restrictions one at a time. First, suppose that the means are each 0 and the variances are equal but arbitrary. With each variance equal to σ^2, we have

$$f_{X_1 X_2}(\alpha_1, \alpha_2) = \frac{1}{2\pi\sigma^2\sqrt{1 - \rho^2}} \exp\left[-\frac{\alpha_1^2 - 2\rho\alpha_1\alpha_2 + \alpha_2^2}{2\sigma^2(1 - \rho^2)} \right] \qquad (4.4.9)$$

Next, suppose that the variances are unequal, given by σ_1^2 and σ_2^2. Then the joint pdf is given by

$$f_{X_1 X_2}(\alpha_1, \alpha_2) = \frac{1}{2\pi\sigma_1\sigma_2\sqrt{1 - \rho^2}} \exp\left[-\frac{1}{2(1 - \rho^2)} \left(\frac{\alpha_1^2}{\sigma_1^2} - \frac{2\rho\alpha_1\alpha_2}{\sigma_1\sigma_2} + \frac{\alpha_2^2}{\sigma_2^2} \right) \right]$$

$$(4.4.10)$$

Finally, if the means are m_1 and m_2, we replace α_1 by $(\alpha_1 - m_1)$ and α_2 by $(\alpha_2 - m_2)$ in Eq. 4.4.10 to yield the general form. This gives

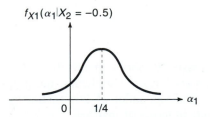

$f_{X1}(\alpha_1|X_2 = -0.5)$

0   1/4 α_1

Fig. 4.4.3

$$f_{X_1 X_2}(\alpha_1, \alpha_2) = \frac{1}{2\pi\sigma_1\sigma_2\sqrt{1-\rho^2}} \exp\left\{-\frac{1}{2(1-\rho^2)}\left[\frac{(\alpha_1 - m_1)^2}{\sigma_1^2}\right.\right.$$
$$\left.\left. -\frac{2\rho(\alpha_1 - m_1)(\alpha_2 - m_2)}{\sigma_1\sigma_2} + \frac{(\alpha_2 - m_2)^2}{\sigma_2^2}\right]\right\} \tag{4.4.11}$$

As we said, this takes some room to write down.

Properties of the Gaussian Random Variable

1. The Gaussian distribution depends only on the first- and second-order moments of the random variables, that is, only on the means, the variances, and the covariances.

2. If X_1 and X_2 are jointly Gaussian, they are individually Gaussian. Both the one-dimensional marginal and the conditional distributions are Gaussian.

3. Uncorrelated Gaussian random variables are statistically independent. The converse is true for any distribution, but the Gaussian is one of only a few for which being uncorrelated implies independence. See Eq. 4.4.2.

4. Linear and affine transformations on Gaussian random variables result in Gaussian random variables. A linear transformation is of the form

$$Y = aX$$

where a is a constant. An affine transformation is of the form

$$Y = aX + b$$

where a and b are constants. If X is Gaussian, then so is Y. Another example of a linear transformation is given by

$$Y = aX_1 + bX_2$$

If X_1 and X_2 are Gaussian, then Y is a one-dimensional Gaussian random variable.

Matrix Representations

Note that the formulas have grown progressively more complex from Eq. 4.4.1 to Eq. 4.4.11. Adding more dimensions further complicates the pdf, unless we use matrix notation. Therefore the remainder of this section is devoted to developing formulas for the pdf in terms of matrices.

The general form for the multidimensional Gaussian pdf is given in matrix form by

$$f_X(\alpha) = \frac{1}{(2\pi)^{n/2}|\Lambda_x|^{1/2}} \exp\left[-\frac{1}{2}(\alpha - m)^t \Lambda_X^{-1}(\alpha - m) \right] \quad (4.4.12)$$

where α is the n-dimensional vector

$$\alpha = \begin{bmatrix} \alpha_1 \\ \alpha_2 \\ \vdots \\ \alpha_n \end{bmatrix} \quad (4.4.13)$$

m is the vector of means, α^t is the transpose, and Λ_X is the matrix

$$\Lambda_X = \begin{bmatrix} \sigma_1^2 & \rho_{12}\sigma_1\sigma_2 & \cdots & \rho_{1n}\sigma_1\sigma_n \\ \rho_{21}\sigma_2\sigma_1 & \sigma_2^2 & \cdots & \rho_{2n}\sigma_2\sigma_n \\ \vdots & \vdots & \vdots & \vdots \\ \rho_{n1}\sigma_n\sigma_1 & & \cdots & \sigma_n^2 \end{bmatrix} \quad (4.4.14)$$

You can see that the two-dimensional pdf of Eq. 4.4.11 is equivalent to the form in Eq. 4.4.12 if

$$\Lambda_X = \begin{bmatrix} \sigma_1^2 & \rho\sigma_1\sigma_2 \\ \rho\sigma_1\sigma_2 & \sigma_2^2 \end{bmatrix}$$

so that $|\Lambda_X| = \sigma_1^2\sigma_2^2(1 - \rho^2)$, giving the inverse

$$\Lambda_X^{-1} = \frac{1}{|\Lambda_X|} \begin{bmatrix} \sigma_2 & -\rho\sigma_1\sigma_2 \\ -\rho\sigma_1\sigma_2 & \sigma_1^2 \end{bmatrix} = \frac{1}{1 - \rho^2} \begin{bmatrix} \dfrac{1}{\sigma_1^2} & -\dfrac{\rho}{\sigma_1\sigma_2} \\ -\dfrac{\rho}{\sigma_1\sigma_2} & \dfrac{1}{\sigma_2^2} \end{bmatrix}$$

After carrying out the indicated matrix multiplication in Eq. 4.4.12, we see that the expansion yields Eq. 4.4.11.

When the variables are statistically independent, the covariance matrix has the diagonal form

$$\Lambda_X = \begin{bmatrix} \sigma_1^2 & 0 & \cdots & 0 \\ 0 & \sigma_2^2 & \cdots & 0 \\ \vdots & \vdots & \vdots & \vdots \\ 0 & 0 & \cdots & \sigma_n^2 \end{bmatrix} \quad (4.4.15)$$

The quadratic form $Q = (\alpha - m)^t \Lambda_X^{-1}(\alpha - m)$ in Eq. 4.4.12 is just a number. That is, for given α, m, and Λ_X, Q is just a number. The matrix

Λ_X is called the *covariance matrix* and consists of the terms $\lambda_{ij} = E[(X_i - m_i)(X_j - m_j)] = E(X_iX_j) - m_im_j$. The term λ_{ij} is the covariance when $i \neq j$, and the variance when $i = j$. In general, the covariance matrix is not diagonal, so the off-diagonal terms are nonzero. However, when the variables are mutually independent (or uncorrelated), the covariance matrix takes the form of Eq. 4.4.15.

EXAMPLE 4.4.4.   The three-dimensional Gaussian density function for nonzero means and correlation coefficients has the form

$$f_X(\alpha) = \frac{1}{(2\pi)^{3/2}|\Lambda_X|^{1/2}} e^{-Q/2}$$

where

$$Q = (\alpha - m)^t \Lambda_X^{-1}(\alpha - m), \qquad \alpha = \begin{bmatrix} \alpha_1 \\ \alpha_2 \\ \alpha_3 \end{bmatrix}, \qquad m = \begin{bmatrix} m_1 \\ m_2 \\ m_3 \end{bmatrix}$$

$$\Lambda_X = \begin{bmatrix} \lambda_{11} & \lambda_{12} & \lambda_{13} \\ \lambda_{21} & \lambda_{22} & \lambda_{23} \\ \lambda_{31} & \lambda_{32} & \lambda_{33} \end{bmatrix}, \qquad \text{where } \lambda_{ij} = E[(\alpha_i - m_i)(\alpha_j - m_j)]$$

EXAMPLE 4.4.5.   Let the three-dimensional Gaussian random variable have statistically independent components with means and standard deviations given by $m_1 = 1$, $m_2 = 0$, $m_3 = -2$, $\sigma_1 = 1$, $\sigma_2 = 2$, and $\sigma_3 = 1$. Find the value of this function at the point $(0, 0, -2)$.

SOLUTION: From the example above,

$$Q = (\alpha - m)^t \Lambda_X^{-1}(\alpha - m) = \begin{bmatrix} 1 & 0 & 0 \end{bmatrix} \begin{bmatrix} 1 & 0 & 0 \\ 0 & \frac{1}{4} & 0 \\ 0 & 0 & 1 \end{bmatrix} \begin{bmatrix} 1 \\ 0 \\ 0 \end{bmatrix} = 1$$

giving

$$f_X(\alpha) = \frac{1}{(2\pi)^{3/2}|\Lambda_X|^{1/2}} e^{-Q/2} = \frac{1}{(2\pi)^{3/2}(2)} e^{-1/2} = 0.01926$$

Concentration Ellipses

The location and spread of Gaussian random variables is determined by the mean and covariance matrix. Let us assume zero mean for convenience, and write the pdf from Eq. 4.4.12 for a two-dimensional case as

$$f_X(\alpha) = \frac{1}{2\pi|\Lambda_X|^{1/2}} \exp\left(-\frac{1}{2}\alpha^t \Lambda_X^{-1}\alpha\right) \tag{4.4.16}$$

Figure 4.4.1 shows the equal-height contours as ellipses. From Eq. 4.4.16 we see that these equal-height ellipses are defined by the equation

$$\alpha^t \Lambda_X^{-1} \alpha = d^2 \tag{4.4.17}$$

Compare Eq. 4.4.17 to the component form in Eq. 4.4.10 to get

$$\frac{1}{(1-\rho^2)}\left(\frac{\alpha_1^2}{\sigma_1^2} - \frac{2\rho\alpha_1\alpha_2}{\sigma_1\sigma_2} + \frac{\alpha_2^2}{\sigma_2^2}\right) = d^2 \tag{4.4.18}$$

In general, the equation for an ellipse is given by

$$a\alpha_1^2 + 2k\alpha_1\alpha_2 + b\alpha_2^2 = c$$

provided $k^2 < ab$. From Eq. 4.4.18 we see that $a = 1/\sigma_1^2$, $b = 1/\sigma_2^2$, and $k^2 = (\rho/\sigma_1\sigma_2)^2$. Since $\rho^2 < 1$, the condition $k^2 < ab$ is satisfied, so Eq. 4.4.17 represents the equation of an ellipse.

These ellipses become larger with increasing d. The parameter d is called the *Mahalanobis distance,* and it finds wide use in statistics. The probability of being inside the ellipse for a given Mahalanobis distance depends only on the value of d. In fact, with a little algebra we could show that this probability is given by $1 - \exp(-d^2/2)$.

> **EXAMPLE 4.4.6.**   Let X be a two-dimensional random variable with zero mean and covariance matrix given by
>
> $$\Lambda_X = \begin{bmatrix} 1 & \rho \\ \rho & 1 \end{bmatrix}$$
>
> Sketch the contours of Mahalanobis distance one from the origin if $\rho = 0$, 0.7, and -0.7.
>
> SOLUTION: Since σ_1 and $\sigma_2 = 1$, the concentration ellipse for $d = 1$ is given by
>
> $$\frac{\alpha_1^2 - 2\rho\alpha_1\alpha_2 + \alpha_2^2}{1 - \rho^2} = 1$$
>
> For $\rho = 1$, 0.7, and -0.7, we obtain ellipses similar to those in Fig. 4.4.4.

We will introduce the concept of eigenvalues and eigenvectors for matrices, and in particular covariance matrices, in Section 9.1. We will discover that these eigenvectors form the principal axes of the concentration ellipses, and that the eigenvalues are equal to the variances of the random variables in the direction of the eigenvectors. Hence Gaussian random variables can be described by the eigenvectors and eigenvalues of the covariance matrix if we know the mean vector. This is not true for non-Gaussian random variables, but the covariance matrix may provide valuable information about these random variables in the absence of their pdf.

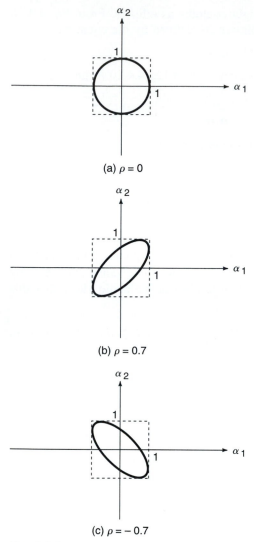

(a) $\rho = 0$

(b) $\rho = 0.7$

(c) $\rho = -0.7$

Fig. 4.4.4

Review

We have derived the marginal and conditional density functions from the two-dimensional pdf to show you how this is done. The Gaussian pdf has the general form in Eq. 4.4.12. This simplifying matrix notation is convenient for arbitrary dimensions. The concepts related to the concentration ellipses allow us to visualize operations related to any random variable, particularly the Gaussian random variable. But non-Gaussian random vectors have covariance matrices, and although their covariance matrices may not completely describe the shape of the pdf as they do for the Gaussian case, they still provide valuable information about the distribution.

The Gaussian distribution is important because of its prevalence in nature, its useful properties, and the ease with which we can manipulate the pdf mathematically. It may not look easy to you, but its exponential form makes it so.

4.5   Transformation of Variables _____

Preview

In this section we continue our discussion of transformation of variables begun in Section 3.7. There we were concerned with one-dimensional random variables, where $Y = g(X)$ formed a cascade of maps with the random variable X. Since X is a map from the sample space S to the real numbers R, and the function g operates on real numbers, then gX forms a composite map from S to R. The problem addressed there was to determine the distribution of $Y = gX$ from the known distribution of X.

Now we address a similar problem. Given a random vector $X = (X_1, X_2, \ldots, X_n)$ with known distribution, and a function $y = g(x_1, x_2, \ldots, x_n)$, determine the distribution of $Y = gX$. We will concentrate on two-dimensional random variables X_1, X_2, but our discussion is general and also applies to higher-dimensional random variables.

We will need the concept of set cross product to deal with n-dimensional random variables. Given two sets $A = \{a_1, a_2, \ldots, a_m\}$ and $B = \{b_1, b_2, \ldots, b_n\}$, the cross (or Cartesian) product $A \times B$ is the set of all ordered pairs (a_i, b_j), where a_i is from A and b_j is from B. For three sets A, B, and C the cross product is the set of all ordered triplets (a_i, b_j, c_k), where $a_i \in A$, $b_j \in B$, and $c_k \in C$. Similar definitions hold for higher dimensions.

EXAMPLE 4.5.1.   Let $A = \{f_2, f_4, f_6\}$ be the set of even die faces, and let $B = \{H, T\}$ be the set of possible coin-toss outcomes. Find the cross product $A \times B$.

SOLUTION:

$$A \times B = \{(f_2, H), (f_2, T), (f_4, H), (f_4, T), (f_6, H), (f_6, T)\}$$

Notice two things in Example 4.5.1. First, the cross product contains *all* ordered pairs; None is missing. Thus the set $\{(f_2, H), (f_4, H), (f_4, T), (f_6, H), (f_6, T)\}$ is not the cross product because (f_2, T) is missing. Second, the order is important: The ordered pair (H, f_2) is not in $A \times B$ because f_2 must precede H in the pair. Thus $A \times B \neq B \times A$.

Cartesian products are important for several reasons. For one thing, a binary relation is a subset of the cross product. We use binary relations extensively, especially since a function is a special type of binary relation. For another, the concept of cross product allows us to use the concept of function to describe multiple-input–multiple-output systems. And of course we will use them in our present discussion of transformation of variables.

Let $X = (X_1, X_2)$ be a two-dimensional random variable, and let $Y = g(X) = g(X_1, X_2)$ be a function of X. Then we have a situation similar to the one in Section 3.7. The random variable X is a function with domain S, the sample space, and range $R_{X_1} \times R_{X_2}$, where R_{X_1} is the range of X_1 and R_{X_2} is the range of X_2. Now we place in cascade the functions g with X to get the composite function $Y = g(X)$. Any function g with domain $D_g \supset R_{X_1} \times R_{X_2}$ can be a function of the two-dimensional random variable X, and therefore can be correctly labeled a random variable. Note that Y need not be a two-dimensional random variable, although it can be. In this section we will consider only the one-dimensional case for Y.

Let A be a subset of the sample space, $A \subset S$. Then A is an event. Suppose that an experiment is performed and the outcome ζ is an element of A, $\zeta \in A$. The point ζ is mapped to the pair $[X_1(\zeta), X_2(\zeta)]$ by the random variable X, and if there is a relation of the form $Y = g(X)$, then the pair $[X_1(\zeta), X_2(\zeta)]$ is mapped into $g[X_1(\zeta), X_2(\zeta)]$. See Fig. 4.5.1. Each element ζ of A is mapped into some element $[X_1(\zeta), X_2(\zeta)]$ of B, where B is a subset of the cross product, by the random variable X, which is in turn mapped into some element $g[X_1(\zeta), X_2(\zeta)]$ of C by the function g. Since sets A, B, and C all occur together, they must have the same probability of occurrence. This leads to the use once again of equivalent events.

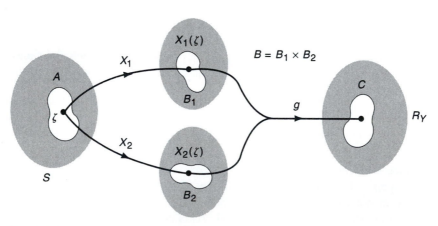

Fig. 4.5.1

Finding the cdf of $Y = g(X) = g(X_1, X_2)$

We assume that the cdf for X and the relationship $Y = g(X)$ are known, and we wish to find the cdf for the random variable Y. For a particular value y_1 the function F_Y defines an event in R_Y, namely, $\{Y \le y\}$. There is an equivalent event in $R_{X_1} \times R_{X_2}$; call it B. Therefore

$$F_Y(y) = P\{Y \le y\} = P\{X \in B\} \tag{4.5.1}$$

and we can express the probability $P\{X \in B\}$ in terms of F_X or f_X. In particular,

$$P\{Y \le y\} = P\{X \in B\} = \iint_B f_X(\alpha_1 \alpha_2)\, d\alpha_1 \alpha_2 \tag{4.5.2}$$

One of the most important functions to consider is the sum $Y = X_1 + X_2$. The cdf for Y is given by

$$F_Y(y) = P(Y \le y) = P(X_1 + X_2 \le y) = \iint_B f_X(\alpha_1, \alpha_2)\, d\alpha_1\, d\alpha_2 \tag{4.5.3}$$

where

$$B = \{(\alpha_1, \alpha_2): \alpha_1 + \alpha_2 \le y\}$$

This situation is pictured in Fig. 4.5.2. Therefore

$$F_Y(y) = \int_{-\infty}^{\infty} \int_{-\infty}^{y-\alpha_1} f_X(\alpha_1, \alpha_2)\, d\alpha_2\, d\alpha_1$$

Now change the variable. Let $\alpha_2 = \lambda - \alpha_1$ in the inner integral, and change the order of integration to get

$$F_Y(y) = \int_{-\infty}^{y} \int_{-\infty}^{\infty} f_X(\alpha_1, \lambda - \alpha_1)\, d\alpha_1\, d\lambda \tag{4.5.4}$$

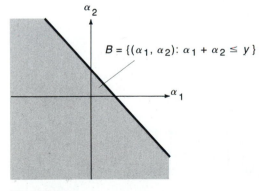

Fig. 4.5.2

This formula can be simplified if X_1 and X_2 are independent, $f_X(\alpha_1, \alpha_2) = f_{X_1}(\alpha_1) f_{X_2}(\alpha_2)$. In this case Eq. 4.5.4 becomes

$$F_Y(y) = \int_{-\infty}^{y} \left[\int_{-\infty}^{\infty} f_{X_1}(\alpha_1) f_{X_2}(\lambda - \alpha_1) d\alpha_1 \right] d\lambda \qquad (4.5.5)$$

Differentiating with respect to y gives the pdf:

$$f_Y(y) = \int_{-\infty}^{\infty} f_{X_1}(\alpha_1) f_{X_2}(\lambda - \alpha_1) d\alpha_1 \qquad (4.5.6)$$

which is the convolution integral. Therefore, if we wish to find the distribution of the sum of two independent random variables, we convolve their pdf's.

EXAMPLE 4.5.2.   In the game of craps, two dice are tossed and the only outcome of interest is the sum of the spots that turn up on the two dice. Assume fair dice and find the probability mass function (pmf) for the sum.

SOLUTION: Implicit in this problem is the definition of two random variables, X_1 and X_2, where X_1 is the number of spots turned up on the first die and X_2 is the number of spots on the second die. These are independent, and for fair dice they have equally likely outcomes for all six die faces. Figure 4.5.3 displays their pmf's. Let Y be their sum,

$$Y = X_1 + X_2$$

Then

$$p_Y(y) = \sum_{\alpha=-\infty}^{\infty} p_1(l) p_2(\alpha - l)$$

Figure 4.5.4 shows the result of this convolution. These are the probabilities of the possible outcomes in the game of craps.

EXAMPLE 4.5.3.   The input $x(n)$ to a discrete-time signal processing system is the sum of the signal plus independent noise,

$$x(n) = s(n) + w(n)$$

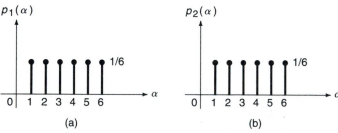

(a)   (b)

Fig. 4.5.3   The pmf's of two random variables: (a) X_1; (b) X_2.

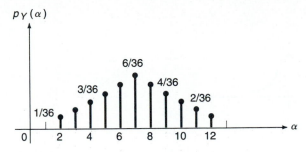

$p_Y(\alpha)$

6/36

4/36

3/36

2/36

1/36

0   2   4   6   8   10   12 α

Fig. 4.5.4

where the signal $s(n)$ is binary with equally likely $+1$ and -1 values, and the noise $w(n)$ is Gaussian with zero mean and variance 0.25. Find the pdf for the input signal $x(n)$.

SOLUTION: The pdf's for signal and noise are shown in Fig. 4.5.5. Since signal and noise are independent, we can convolve their pdf's to obtain the distribution of $x(n)$ shown in Fig. 4.5.6. Since $f_W(\alpha)$ is given by

$$f_W(\alpha) = \frac{2}{\sqrt{2\pi}} \exp(-2\alpha^2)$$

the pdf for x is the sum of the convolution of this function with each delta function. This gives

$$f_X(\alpha) = \frac{1}{\sqrt{2\pi}} \exp[-2(\alpha + 1)^2] + \frac{1}{\sqrt{2\pi}} \exp[-2(\alpha - 1)^2]$$

We often wish to determine the minimum or maximum value of a set of random variables. Suppose that $Y = \max(X_1, X_2)$. Then for a given number y, the random variable Y is less than or equal to y if and only if X_1 and X_2 are both less than or equal to y:

$$Y \leq y \qquad \text{if and only if } X_1 \leq y \text{ and } X_2 \leq y$$

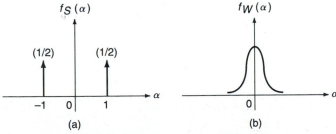

$f_S(\alpha)$

(1/2)   (1/2)

-1 0   1 α

(a)

$f_W(\alpha)$

0   α

(b)

Fig. 4.5.5   The pdf's for (a) signal and (b) noise.

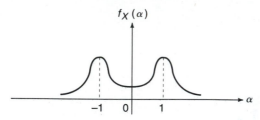

Fig. 4.5.6

Thus the set B in Fig. 4.5.1 is given by

$$B = \{(\alpha_1, \alpha_2): \alpha_1 \le y \text{ and } \alpha_2 \le y\}$$

Therefore

$$F_Y(y) = P(Y \le y) = P(X_1 \le y \text{ and } X_2 \le y) = F_X(y, y) \quad (4.5.7)$$

EXAMPLE 4.5.4.   Suppose that we toss 12 coins and count the number of heads that occur. Let X_1 be the number of heads in one trial of this experiment, and let X_2 be the number of heads on a second trial. What is the probability that the number of heads will not exceed 6?

SOLUTION: We derived the pdf for X_i in Example 3.6.3. The value of the cdf for $y = 6$ is the sum of the probabilities for outcomes less than or equal to 6. This sum is approximately 0.61337. Since the trials are independent, this gives

$$F_Y(y) = F_X(y, y) = F_{X_1}(y)F_{X_2}(y)$$

or $(0.61337)^2 = 0.377623$.

If X_1 and X_2 are independent, zero-mean, Gaussian random variables with equal variance, then the function $y = (x_1^2 + x_2^2)^{1/2}$ has Rayleigh distribution:

$$f_Y(y) = \begin{cases} \dfrac{y}{\sigma^2} \exp\left(-\dfrac{y^2}{2\sigma^2}\right), & y > 0 \\ 0 & \text{otherwise} \end{cases} \quad (4.5.8)$$

If $y > 0$, then the set B in Fig. 4.5.1 is that portion of the α_1, α_2 plane such that $\alpha_1^2 + \alpha_2^2 \le y^2$ is a circle with radius y. Hence $F_Y(y)$ equals the probability mass in this circle.

If $y > 0$, then

$$f_Y(y) = \iint\limits_{\alpha_1^2 + \alpha_2^2 \le y^2} f_X(\alpha_1, \alpha_2) \, d\alpha_1 \, d\alpha_2 \quad (4.5.9)$$

If $y < 0$, then $F_Y(y) = 0$. Now let X be a two-dimensional Gaussian random variable with pdf given by

$$f_X(\alpha_1, \alpha_2) = \frac{1}{2\pi\sigma^2} \exp\left(-\frac{\alpha_1^2 + \alpha_2^2}{2\sigma^2}\right)$$

Substitute this into Eq. 4.5.9 and change variables. Let $\alpha_1 = r \cos \theta$ and $\alpha_2 = r \sin \theta$. This gives

$$F_Y(y) = \frac{1}{2\pi\sigma^2}\int_0^y 2\pi r \exp\left(-\frac{r^2}{2\sigma^2}\right) dr = 1 - \exp\left(-\frac{y^2}{2\sigma^2}\right), \qquad y > 0$$

Differentiating, we get Eq. 4.5.8.

> **EXAMPLE 4.5.5.**   In shooting at a circular target, we assume that the points of impact in both the horizontal and vertical directions have independent Gaussian distribution $N(0, \sigma^2)$, where the bullseye is centered at the origin. Suppose for a particular shooter that the variance is $\sigma^2 = 4$ in^2. Find his average distance from the bullseye.
>
> SOLUTION: The radial distance from the target has Rayleigh distribution with parameter $\sigma^2 = 4$. The average value of the Rayleigh distribution is
>
> $$E(Y) = \int_0^\infty \frac{y^2}{\sigma^2} \exp\left(-\frac{y^2}{2\sigma^2}\right) dy = \sigma \sqrt{\frac{\pi}{2}}$$
>
> Thus the average error is $2\sqrt{\pi/2} = \sqrt{2\pi} = 2.5$ in.

Review

The procedure for finding the distribution of a function of a multivariate random variable depends on the concept of equivalent events. We first described this concept in Section 3.7, and we extended it to n-dimensional random variables with the help of the set cross product in this section. Figure 4.5.1 pictures these equivalent events for a two-dimensional random variable, where $A \subset S$, B is the cross product $B_1 \times B_2$ which is a subset of the range of X, and $C \subset R_Y$. Let $C = \{Y \leq y\}$, then the equivalent event is $\{X \in B\}$, giving

$$F_Y(y) = P\{Y \leq y\} = P\{X \in B\}$$

and we can express the probability $P\{X \in B\}$ in terms of f_X:

$$P\{Y \leq y\} = P\{X \in B\} = \iint_B f_X(\alpha_1\alpha_2)\, d\alpha_1\alpha_2$$

The set B depends on the function g. We described three important applications, the sum, the maximum, and the Euclidean distance. The sum led to the convolution operation for independent random variables, the maximum led to the simple formula in Eq. 4.5.7, and the Euclidean distance led to the Rayleigh distribution for independent Gaussian random variables.

4.6   Chebyshev's Inequality ————————————————

Preview

We now introduce Chebyshev's inequality and apply it to the weak law of large numbers. So far, the only link between theory and practice has been the reasonable solutions that our examples and exercises have provided. That is, probability theory seems to provide reasonable answers to complicated problems, but there has been no guarantee that our solutions have been correct. The inequalities and limit theorems in this section provide such a link, for they relate long sequences of experimental outcomes to the probabilities in the experiments. We begin with Chebyshev's inequality.

CHEBYSHEV'S INEQUALITY.   Let X be a random variable with $E(X) = 0$ and variance σ_X^2. Then for any positive number ε we have

$$P\{|X| \geq \varepsilon\} \leq \frac{\sigma_X^2}{\varepsilon^2} \qquad (4.6.1)$$

or, equivalently,

$$P\{|X| < \varepsilon\} \geq 1 - \frac{\sigma_X^2}{\varepsilon^2} \qquad (4.6.2)$$

The first form, Eq. 4.6.1, sets a bound on the likelihood that X is large, and relates this bound to the variance σ_X^2. If σ_X^2 is large, then X is more likely to deviate from its mean of 0 by a large amount. If σ_X^2 is small, then X is less likely to have a large magnitude.

The proof of this powerful theorem begins with the mean square value of X:

$$E(X^2) = \int_{-\infty}^{\infty} \alpha^2 f_X(\alpha) \, d\alpha \qquad (4.6.3)$$

Refer to Fig. 4.6.1, which shows a typical pdf with the values ε and $-\varepsilon$ marked on the graph. The shaded area is the probability $P\{|X| \geq \varepsilon\}$. Since the integrand of Eq. 4.6.3 is positive, we can write

$$E(X^2) \geq \int_{|\alpha| \geq \varepsilon} \alpha^2 f_X(\alpha) \, d\alpha$$

That is, the mean square value of X is

$$E(X^2) = \int_{-\infty}^{-\varepsilon} \alpha^2 f_X(\alpha) \, d\alpha + \int_{-\varepsilon}^{+\varepsilon} \alpha^2 f_X(\alpha) \, d\alpha + \int_{+\varepsilon}^{\infty} \alpha^2 f_X(\alpha) \, d\alpha$$

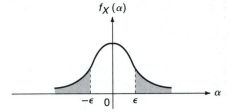

Fig. 4.6.1.   A typical pdf. Shaded portion represents $P[|X| \geq \varepsilon]$.

and we are simply leaving out the middle term. Now replace α^2 by its smallest value ε^2, making this inequality even stronger:

$$E(X^2) \geq \varepsilon^2 \int_{|\alpha| \geq \varepsilon} f_X(\alpha) \, d\alpha = \varepsilon^2 P\{|X| \geq \varepsilon\}$$

Since $m_X = 0$, Eq. 4.6.1 follows immediately.

This is an extremely useful and powerful theorem, for look what it says! If you know only the mean and the variance of some random variable Y, you can make a statement about its probable value. Simply subtract the mean from Y to obtain $X = Y - m_Y$ for use in either Eq. 4.6.1 or 4.6.2. Both inequalities say that the value of Y is likely to be near its mean.

Do not infer that this inequality is good only for the Gaussian distribution, as you might from looking at Fig. 4.6.1. This inequality applies to any random variable, continuous or discrete.

Consider the degenerate case $\sigma_X = 0$. Then Eq. 4.6.2 implies that $P\{|X| < \varepsilon\} = 1$, since a probability can never be greater than 1. This implies that for any experimental outcome the random variable X will assume its mean value, and hence the width of the pdf is 0. All the probability is concentrated at the mean, as in Fig. 4.6.2.

EXAMPLE 4.6.1.   Let X be a random variable with mean $m_X = 0$ and variance $\sigma_X^2 = 2$. Find the largest probability that $|X| \geq 2$.

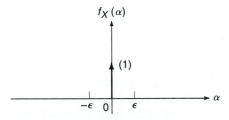

Fig. 4.6.2.   A pdf with zero variance.

SOLUTION: From Eq. 4.6.1 this is

$$P\{|X| \geq 2\} \leq \tfrac{2}{4} = \tfrac{1}{2}$$

EXAMPLE 4.6.2.   Let Y be a random variable with mean $m_Y = 2$ and mean square value $E(Y^2) = 5$. Find the smallest probability that Y is between -1 and 5.

SOLUTION: Subtract the mean from Y to obtain

$$X = Y - m_Y = Y - 2$$

Then the events $\{-1 \leq Y \leq 5\}$ and $\{-3 \leq X \leq 3\}$ are equivalent. The variance of Y (and X) is

$$\sigma_Y^2 = \sigma_X^2 = E(Y^2) - m_Y^2 = 1$$

Applying Chebyshev's inequality, Eq. 4.6.2, we have

$$P\{|X| < 3\} \geq 1 - \left(\frac{\sigma_X^2}{3^2}\right) = 1 - \frac{1}{9} = \frac{8}{9}$$

or, from the equivalence of events,

$$P\{-1 \leq Y \leq 5\} \geq \frac{8}{9}$$

These results may be interpreted (loosely) as follows: If the experiment is performed a large number of times, then the value of X in Example 4.6.1 will be outside the interval $(-2, 2)$ somewhat less than half the time. In Example 4.6.2 the value of Y will be between -1 and 5 most ($\tfrac{8}{9}$) of the time.

If we knew the distribution of these random variables, rather than just their mean and variance, we could determine the probabilities exactly. For example, if X is Gaussian $N(0, 2)$ in Example 4.6.1, then $P\{|X| \geq 2\} = 0.3174$. This agrees with our result in Example 4.6.1, where we have $P\{|X| \geq 2\} \leq 0.5$.

The Weak Law of Large Numbers

In Chapter 1 we tossed a coin 12 times to generate a random signal. Although the expected number of heads is 6, the actual number of heads was often 4, 5, 7, 8, or sometimes other numbers in the range from 0 to 12. Now consider what might happen if the number of tosses is increased from 12 to 24, or 50, or 100. If you toss a coin 100 times, the number of heads will likely be close to 50, although any number from 0 to 100 is possible. Typically the number of heads will be between 40 and 60. One feels intuitively that heads should turn up roughly half the time, especially if the number of tosses is increased. If h is the number of heads in N tosses, then the ratio h/N should be close to $\frac{1}{2}$ for large N; furthermore, this ratio should get closer to $\frac{1}{2}$ as N increases. This is precisely the statement of the law of large numbers.

The weak law of large numbers follows directly from Chebyshev's inequality. Define the random variable Z as

$$Z = \frac{1}{N} \sum_{i=1}^{N} X_i \tag{4.6.4}$$

where the $\{X_i\}$ are identically distributed random variables with mean m_X and variance σ_X^2. Let X_i be the success or failure random variable:

$$X_i = \begin{cases} 1 & \text{if success on the } i\text{th trial} \\ 0 & \text{otherwise} \end{cases}$$

With reference to our coin-toss experiment, $X_i = 1$ if heads and $X_i = 0$ if tails. The number of heads in N tosses is

$$h = \sum_{i=1}^{N} X_i \quad \text{and} \quad Z = \frac{h}{N}$$

Now we want to apply Chebyshev's inequality to Z. For this we need the mean and the variance of Z. The mean is

$$E(Z) = E\left(\frac{1}{N} \sum_{i=1}^{N} X_i\right) = \frac{1}{N} \sum_{i=1}^{N} E(X_i) = m_X \tag{4.6.5}$$

and the variance is

$$\sigma_Z^2 = \frac{\sigma_X^2}{N} \tag{4.6.6}$$

We derive the variance of Z as follows: First we recognize that the variance of the sum of statistically independent random variables is the sum of the variances. That is,

$$\text{var}\left(\sum_{i=1}^{N} X_i\right) = \sum_{i=1}^{N} \sigma_X^2$$

Next, the variance of a constant times a random variable is

$$\text{var}(aX) = a^2 \, \text{var}(X)$$

Therefore we can write

$$\sigma_Z^2 = \frac{1}{N^2} \sum_{i=1}^{N} \sigma_X^2 = \frac{N\sigma_X^2}{N^2} = \frac{\sigma_X^2}{N}$$

as stated above.

Now to use Chebyshev's inequality, Eq. 4.6.1, on the random variable Z, we define Y as

$$Y = Z - m_Z = Z - m_X$$

The mean of Y is 0, and the variance is

$$\sigma_Y^2 = \sigma_Z^2 = \frac{\sigma_X^2}{N}$$

Therefore Eq. 4.6.1 gives

$$P\{|Y| \geq \varepsilon\} \leq \frac{\sigma_Y^2}{\varepsilon^2}$$

or

$$P\{|Z - m_X| \geq \varepsilon\} \leq \frac{\sigma_X^2}{N\varepsilon^2} \tag{4.6.7}$$

Equation 4.6.7 is called the *weak law of large numbers.*

EXAMPLE 4.6.3.   Let $\varepsilon = 0.25$ and compare the probabilities obtained from the law of large numbers for tossing a coin 12, 50, and 100 times.

SOLUTION: Let

$$Z = \frac{1}{N}\sum_{i=1}^{N} X_i$$

where

$$X_i = \begin{cases} 1 & \text{if heads on the } i\text{th toss} \\ 0 & \text{otherwise} \end{cases}$$

Then $\sigma_X^2 = E(X^2) - [E(X)]^2 = \frac{1}{2} - \frac{1}{4} = \frac{1}{4}$. Now apply Eq. 4.6.7. ·

For $N = 12$, $P\{|Z - 0.5| \geq 0.25\} \leq \dfrac{\sigma_X^2}{N\varepsilon^2} = \dfrac{\frac{1}{4}}{12(\frac{1}{16})} = \dfrac{1}{3}$

For $N = 50$, $P\{|Z - 0.5| \geq 0.25\} \leq \dfrac{\frac{1}{4}}{50(\frac{1}{16})} = \dfrac{2}{25}$

For $N = 100$, $P\{|Z - 0.5| \geq 0.25\} \leq \dfrac{\frac{1}{4}}{100(\frac{1}{16})} = \dfrac{1}{25}$

The weak law of large numbers is a statement about the probable value of the sum of N identically distributed random variables. Specifically, if this sum is divided by N, then Eq. 4.6.7 states that the random variable Z is very likely to be near the mean value of X if N is large. This law provides justification for our intuitive feeling that if an experiment is performed a large number of times, the relative frequency of a particular outcome should be near the probability of that outcome.

The weak law of large numbers is opposed to the *strong law of large numbers,* due to Borel. The weak law is concerned with statements about the probability of events such as $\{|Z - m_X| < \varepsilon\}$. Proof of these statements follows from Chebyshev's inequality. In contrast, the strong law allows us to make statements such as

$$P\{Z \to m_X\} = 1 \qquad \text{as } N \to \infty$$

and is simply a stronger statement of the weak law.

> **EXAMPLE 4.6.4.**   Here is one more example of the application of the weak law of large numbers. In tossing a die a large number of times, the one-spot should turn up about one-sixth of the time. Let n be the number of times the one-spot turns up in N tosses of the die. How many times must the die be thrown for the probability that n/N will be between 0 and $\frac{1}{3}$ to be greater than 0.5?
>
> SOLUTION: We wish to find N so that
>
> $$P\left\{0 < \frac{n}{N} < \frac{1}{3}\right\} \geq 0.5$$
>
> Define the random variable Z as
>
> $$Z = \frac{1}{N} \sum_{i=1}^{N} X_i$$
>
> where
>
> $$X_i = \begin{cases} 1 & \text{if the one-spot turns up on the } i\text{th toss} \\ 0 & \text{otherwise} \end{cases}$$
>
> Then from Eqs. 4.6.5 and 4.6.6 we have
>
> $$E(Z) = m_X = \frac{1}{6} \qquad \text{and} \qquad \sigma_Z^2 = \frac{\sigma_X^2}{N} = \frac{5}{36N}$$
>
> and ε is $\frac{1}{6}$ because if $0 < (n/N) < \frac{1}{3}$, then $-\frac{1}{6} < (Z - m_Z) < \frac{1}{6}$. The problem can be restated as follows: Find N so that
>
> $$P\left\{|Z - m_Z| < \frac{1}{6}\right\} \geq 0.5$$
>
> Applying Chebyshev's inequality in the form of Eq. 4.6.2, we have
>
> $$P\left\{|Z - m_Z| < \frac{1}{6}\right\} \geq 1 - \frac{\frac{5}{36}}{N(\frac{1}{36})} = 1 - \frac{5}{N}$$

If we set

$$\left(1 - \frac{5}{N}\right) \geq 0.5 \tag{4.6.8}$$

we are assured that $P\{|Z - m_Z| < \frac{1}{6}\} \geq 0.5$. Manipulating the inequality 4.6.8, we finally obtain

$$N \geq 10$$

This is a conservative estimate, for we used only the mean and the variance of Z in arriving at this bound on N. If we had used more information (e.g., the distribution of Z), we would have found a more accurate estimate.

As a further check on the validity of the law of large numbers, suppose that the probability bound is changed from 0.5 to 0.8. That is, if we require $P\{0 < n/N < \frac{1}{3}\} \geq 0.8$, then the minimum N should increase. Using this, our inequality becomes

$$\left(1 - \frac{5}{N}\right) \geq 0.8$$

so $N \geq 25$.

Review

We have seen a direct link between theory and practice in this section. Chebyshev's inequality makes a statement about a random variable's probable value based only on the mean and the variance. The weak law of large numbers uses this to establish a probable value for a long sequence of random variables. If we are willing to toss a coin several thousand times, we can check to see if it is fair.

4.7   The Central Limit Theorem _____

Preview

The Gaussian distribution is prevalent in nature. If you have a random variable about which you know nothing, your best guess is to say that it has Gaussian distribution. You might be wrong, but you would be correct with this guess more often than with any other. The central limit theorem serves to explain this phenomenon, for it states that the cumulative effect of many small independent contributions leads to the Gaussian distribution. Small but frequent disturbances in the transmission medium add Gaussian noise to the signal. Individual electrons move about in a resistor due to thermal agitation, and their cumulative effect produces a small but measurable Gaussian noise across the terminals. We model these effects by the Gaussian distribution because of the central limit theorem.

The central limit theorem is one of the most remarkable theorems in science. It is a property of convolution. That we are able to link convolution to probability by this theorem is most remarkable.

When discussing the binomial distribution, we were concerned with the sum of independent random variables $X_1 + X_2 + \cdots + X_N$. The variance of this sum is $\sigma_1^2 + \sigma_2^2 + \cdots + \sigma_N^2$ and is equal to $N\sigma_X^2$ if all the N random variables have the same variance. In our discussion of the law of large numbers above, we divided this sum by N, and found that the variance of

$$Z = \frac{1}{N} \sum_{i=1}^{N} X_i$$

approaches zero as $N \to \infty$.

Now we wish to discuss the behavior of the sum $\sum_{i=1}^{N} X_i$ as $N \to \infty$, particularly the tendency of this sum to approach the Gaussian distribution. For this discussion the variance should become neither infinite nor zero, and this is accomplished by dividing by $\sqrt{N}$. Therefore let us define the random variable Y by

$$Y = \frac{1}{\sqrt{N}} \sum_{i=1}^{N} X_i \tag{4.7.1}$$

If the variance of each X_i is the same, then the variance of Y is equal to σ_X^2 for any N.

CENTRAL LIMIT THEOREM.   Suppose that the X_i of Eq. 4.7.1 are statistically independent random variables with zero mean and finite variance, and that all have identical distribution. Then, for any t,

$$\lim_{N \to \infty} F_Y(t) = \int_{-\infty}^{t} \frac{1}{\sqrt{2\pi}\,\sigma_X} e^{-\lambda^2/2\sigma_X^2} \, d\lambda \tag{4.7.2}$$

The theorem states that the area under the pdf f_Y is equal to the area under an appropriate Gaussian pdf. Note specifically that the theorem does not say that the pdf f_Y is equal to a Gaussian pdf in the limit. As a counterexample, suppose each X_i has binary distribution. Then the pdf of Y consists of impulses and can never approach the smooth Gaussian pdf.

Note the assumptions made in the statement of the theorem: namely, that all X_i must be identically distributed and statistically independent. There are several degrees of complexity for the central limit theorem, depending on the assumptions made. For example, the distribution of all X_i need not be identical. We have chosen the form above to simplify the following discussion.

The central limit theorem is really a property of convolution, where we are convolving a large number of functions. Consider the sum of two statistically independent random variables, say, $X = X_1 + X_2$. Then the pdf

f_X is the convolution of the pdf's for X_1 and X_2. This is shown by finding the characteristic function for X as follows:

$$\phi_X(v) = E(e^{jvX}) = E[e^{jv(X_1+X_2)}]$$
$$= E(e^{jvX_1})E(e^{jvX_2}) = \phi_{X_1}(v)\phi_{X_2}(v) \tag{4.7.3}$$

Since the characteristic function of X is the product of the characteristic functions of X_1 and X_2, the pdf f_X is the convolution of the two pdf's f_{X_1} and f_{X_2}. (This follows from the convolution property of Fourier transforms.) Now the random variable Y in the theorem is related to the sum of N statistically independent random variables. We can expect the pdf f_Y to be related somehow to the convolution of the N pdf's f_{X_i}, and the characteristic function ϕ_Y should be related to the product of the N characteristic functions ϕ_{X_i}.

To proceed, let us derive the relationship between ϕ_Y and ϕ_X for the variables in Eq. 4.7.1.

$$\phi_Y(v) = E(e^{jvY}) = E\left[\exp\left(jv\frac{1}{\sqrt{N}}\sum_{i=1}^{N}X_i\right)\right]$$
$$= E\left[\prod_{i=1}^{N}\exp\left(jv\frac{X_i}{\sqrt{N}}\right)\right] \tag{4.7.4}$$
$$= \prod_{i=1}^{N}E\left[\exp\left(jv\frac{X_i}{\sqrt{N}}\right)\right] = \prod_{i=1}^{N}\left[\phi_X\left(\frac{v}{\sqrt{N}}\right)\right]$$

The remainder of our discussion is straightforward, though lengthy. We wish to show that as $N \to \infty$ the characteristic function ϕ_Y is of the form of the Gaussian characteristic function. The Maclaurin series of e^{jvX} is

$$e^{jvX} = 1 + jvX + \frac{(jv)^2}{2!}X^2 + \frac{(jv)^3}{3!}X^3 + \cdots$$

Assume that all moments of X are finite. Then

$$E(e^{jvX}) = 1 + jvm_X + \frac{(jv)^2}{2!}E(X^2) + \frac{(jv)^3}{3!}E(X^3) + \cdots = \phi_X(v)$$

But $m_X = 0$, so $E(X^2) = \sigma_X^2$. Therefore $\phi_X(v) = 1 - (v^2/2!)\sigma_X^2 +$ terms involving v^3 and higher powers of v. For simplicity we will denote these terms involving v^3 and higher powers of v as $v^3f(v)$. Then

$$\phi_X(v) = 1 - \frac{v^2}{2!}\sigma_X^2 + v^3f(v) \tag{4.7.5}$$

Now let us return to the relation between ϕ_Y and ϕ_X, Eq. 4.7.4. Taking logarithms and substituting Eq. 4.7.5, we have

$$\ln \phi_Y(v) = N \ln \phi_X \left[\frac{v}{\sqrt{N}} \right]$$

$$= N \ln \left[1 - \frac{v^2}{N} \frac{\sigma_X^2}{2} + \left(\frac{v}{\sqrt{N}} \right)^3 f \left(\frac{v}{\sqrt{N}} \right) \right]$$

(4.7.6)

Let us digress to point out that an expansion for $\ln(1 + \alpha)$ is

$$\ln(1 + \alpha) = \alpha - \frac{\alpha^2}{2} + \frac{\alpha^3}{3} - \frac{\alpha^4}{4} + \cdots \qquad |\alpha| < 1$$

Using this we can write Eq. 4.7.6 as

$$N \ln \phi_X \left(\frac{v}{\sqrt{N}} \right) = N \left[-\frac{v^2}{N} \frac{\sigma_X^2}{2} + \left(\frac{v}{\sqrt{N}} \right)^3 f \left(\frac{v}{\sqrt{N}} \right) + \text{terms of } N^{-2}, N^{-3}, \ldots \right]$$

(4.7.7)

The condition $|\alpha| < 1$ is satisfied by choosing N large enough. After all, we are going to use this expression in the limit as $N \rightarrow \infty$.

To complete our discussion, we take the limit as $N \rightarrow \infty$ in Eq. 4.7.7:

$$\lim_{N \to \infty} \ln \phi_Y(v) = \lim_{N \to \infty} N \ln \phi_X \left(\frac{v}{\sqrt{N}} \right) = -\frac{v^2 \sigma_X^2}{2}$$

or, on raising both sides to the exponential power,

$$\lim_{N \to \infty} \phi_Y(v) = e^{-v^2 \sigma_X^2 / 2}$$

(4.7.8)

which we recognize as the characteristic function of a Gaussian random variable. See Eq. 3.6.12.

The above derivation is not very precise. We took several questionable steps without justification, and although we could justify these steps, we must not call this a proof. The above should be regarded as an outline or explanation.

EXAMPLE 4.7.1.   If a coin is tossed 10,000 times, estimate $P\{4900 < n < 5100\}$, where n is the number of heads.

(a) Use Chebyshev's inequality to obtain a bound on this probability.
(b) Use the central limit theorem to estimate this probability.

SOLUTION: (a) Let $X_i = 1$ if heads and $X_i = 0$ if tails on the ith toss. Then

$$h = \sum_{i=1}^{N} X_i$$

will be the number of heads in N tosses of the coin ($N = 10,000$). Let

$$Y = \frac{h}{N} - 0.5$$

If $0.49 < h/N < 0.51$, then $-0.01 < Y < 0.01$. Applying Chebyshev's inequality gives

$$P\{|Y| < 0.01\} \geq 1 - \frac{\sigma_X^2}{N\varepsilon^2} = 1 - \frac{\frac{1}{4}}{1} = 0.75$$

(b) Define X_i, N, and h as above. Now if

$$4900 < n < 5100$$

then

$$49 < \frac{n}{\sqrt{N}} < 51$$

and

$$-1 < \left(\frac{n}{\sqrt{N}} - 50\right) < 1$$

Let $Y = n/\sqrt{N} - 50$. Then $m_Y = 0$ and $\sigma_Y^2 = \sigma_X^2 = \frac{1}{4}$. Therefore the probability that the number of heads will be between 4900 and 5100 is (approximately) equal to the probability that a Gaussian random variable Y with 0 mean and variance $\frac{1}{4}$ will be between -1 and 1:

$$P\{-1 < Y < 1\} = \int_{-1}^{1} \frac{1}{\sqrt{2\pi(\frac{1}{4})}} e^{-2\alpha^2} \, d\alpha$$

Change the variable. Let $Z = 2Y$ (or $\lambda = 2\alpha$). Then if $\{-1 < Y < 1\}$, we have $\{-2 < Z < 2\}$. This gives

$$P\{-1 < Y < 1\} = \int_{-2}^{2} \frac{1}{\sqrt{2\pi}} e^{-\lambda^2/2} \, d\lambda = 0.9544$$

where 0.9544 is the area under the standard normal curve from $\lambda = -2$ to $\lambda = 2$. Comparing the two estimates, we see that they agree. Chebyshev's inequality gives a bound of 0.75, while the Gaussian approximation gives an estimate of about 0.95. Since $N = 10,000$ is a very large number of trials, we can have confidence in these results.

EXAMPLE 4.7.2.   A die is tossed 36 times. Use the central limit theorem to estimate $P\{4 < n < 8\}$, where n is the number of times a one-spot is rolled.

SOLUTION: First define the success or failure random variable. Let

$$X = \begin{cases} 1 & \text{if a one-spot is rolled on the } i\text{th toss} \\ 0 & \text{otherwise} \end{cases}$$

Then

$$n = \sum_{i=1}^{N} X_i \qquad \text{where } N = 36$$

If $4 < n < 8$, then

$$\frac{4}{6} < \frac{n}{\sqrt{N}} < \frac{8}{6}$$

and

$$-\frac{1}{3} < Y < \frac{1}{3} \qquad \text{where } Y = \frac{n}{\sqrt{N}} - 1$$

Also, we have $m_Y = 0$ and $\sigma_Y^2 = \sigma_X^2 = \frac{5}{36}$. Therefore

$$P\{4 < n < 8\} \approx \int_{-1/3}^{1/3} \frac{1}{\sqrt{2\pi(\frac{5}{36})}} e^{-36\alpha^2/10} \, d\alpha$$

Change the variable; let $\alpha = \lambda\sqrt{5/36}$. Then

$$P\{4 < n < 8\} \approx \int_{-2/\sqrt{5}}^{2/\sqrt{5}} \frac{1}{\sqrt{2\pi}} e^{-\lambda^2/2} \, d\lambda = 0.629$$

EXAMPLE 4.7.3. A pointer is spun on a dial marked from 0 to 1. This experiment is repeated 100 times. Use the central limit theorem to estimate $P\{40 < S < 60\}$, where S is the sum of the 100 numbers obtained in the trials.

SOLUTION: Let $S = \Sigma_{i=1}^N X_i$, where the pdf for each X_i is uniform $(0, 1)$, and $N = 100$. If $40 < S < 60$, then

$$4 < \frac{S}{\sqrt{N}} < 6$$

and

$$-1 < Y < 1 \qquad \text{where } Y = \frac{S}{\sqrt{N}} - 5$$

Also, we have $m_Y = 0$ and $\sigma_Y^2 = \sigma_X^2 = \int_0^1 (\alpha - \frac{1}{2})^2 \, d\alpha = \frac{1}{12}$. Therefore

$$P\{40 < S < 60\} \approx \int_{-1}^{1} \frac{1}{\sqrt{2\pi(\frac{1}{12})}} e^{-6\alpha^2} \, d\alpha$$

$$= \int_{-\sqrt{12}}^{\sqrt{12}} \frac{1}{\sqrt{2\pi}} e^{-\lambda^2/2} \, d\lambda = 0.999$$

Let us return to a statement made at the beginning of this section. We said that the central limit theorem is really a property of convolution, and we now wish to demonstrate the meaning of this statement with three examples.

EXAMPLE 4.7.4. Two independent random variables X_1 and X_2 are added together. The distribution of each is uniform $(-1, 1)$. Find and plot the pdf of the sum.

SOLUTION: Let $Y = X_1 + X_2$. Then, according to the discussion concerning Eq. 4.7.3, we should convolve the pdf's of X_1 and X_2 to obtain the pdf for Y.

The two pdf's to be convolved are shown in Fig. 4.7.1a and b. The convolution formula is

$$f_Y(\alpha) = \int_{-\infty}^{\infty} f_{X_1}(\lambda) f_{X_2}(\alpha - \lambda)\, d\lambda$$

Figures 4.7.1c through 4.7.1f show $f_{X_2}(\alpha - \lambda)$ displaced to various values of α along with the corresponding limits on the integral. Then Fig. 4.7.1g

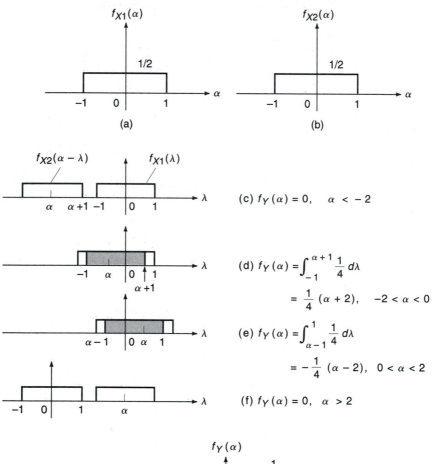

(a)

(b)

(c) $f_Y(\alpha) = 0$, $\alpha < -2$

(d) $f_Y(\alpha) = \int_{-1}^{\alpha+1} \dfrac{1}{4}\, d\lambda$

$\qquad = \dfrac{1}{4}(\alpha + 2)$, $-2 < \alpha < 0$

(e) $f_Y(\alpha) = \int_{\alpha-1}^{1} \dfrac{1}{4}\, d\lambda$

$\qquad = -\dfrac{1}{4}(\alpha - 2)$, $0 < \alpha < 2$

(f) $f_Y(\alpha) = 0$, $\alpha > 2$

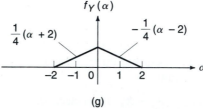

(g)

Fig. 4.7.1

shows the resulting pdf for Y. This does not look very "Gaussian," but we are just getting started.

EXAMPLE 4.7.5.   Repeat Example 4.7.4 for three random variables X_1, X_2, and X_3.

SOLUTION: Let $Z = X_1 + X_2 + X_3 = Y + X_3$. Therefore, to find the pdf for Z we need only convolve f_Y with f_{X_3}.

$$f_Z(\alpha) = \int_{-\infty}^{\infty} f_Y(\lambda) f_{X_3}(\alpha - \lambda) \, d\lambda$$

The steps are shown in Figs. 4.7.2a through 4.7.2d, with the result in Fig. 4.7.2e. Now we are beginning to get somewhere. Notice that $f_Z(\alpha)$ resembles the Gaussian pdf.

EXAMPLE 4.7.6.   In formulating the derivation of the central limit theorem, we divided the sum by $\sqrt{N}$ because we wanted to maintain a constant variance as the number of terms in the sum increased. To see the effect of this, repeat Example 4.7.5 for the sum divided by $\sqrt{3}$.

SOLUTION: Let $V_i = X_i/\sqrt{3}$. Then the pdf for each V_i is uniform but with restricted limits from $-1/\sqrt{3}$ to $1/\sqrt{3}$, as shown in Fig. 4.7.3. Let

$$W = \frac{1}{\sqrt{3}} \sum_{i=1}^{3} X_i = \frac{1}{\sqrt{3}} (X_1 + X_2 + X_3)$$

The $f_W(\alpha)$ is given by the curve in Fig. 4.7.4. Notice that this is just $f_Z(\alpha)$ squeezed or reduced along the α axis by $\sqrt{3}$. Thus W has the same variance as X.

If these examples are extended to find the pdf for the sum of four X_i variables, the resulting curves will look even more Gaussian. You can see that with only a small number of terms, any initial distribution on X will rapidly converge to a Gaussian form for the sum.

Review

Take any pdf and convolve it with itself a number of times and the result begins to look Gaussian. Discrete pdf's remain discrete, but they soon take the bell-shaped appearance of the Gaussian pdf. Continuous distributions look Gaussian in only a few iterations of the convolution operation.

When we add independent random variables, we convolve their pdf's. That is because the characteristic function for the sum of independent random variables results in the product form. If we multiply characteristic functions, we convolve their transform, which is the pdf. This convolution of pdf's produces the Gaussian-like result in only a few operations.

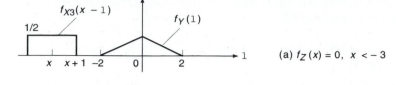

(a) $f_Z(x) = 0, \quad x < -3$

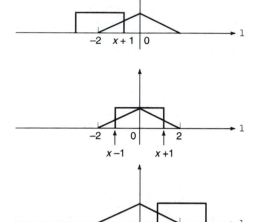

(b) $f_Z(x) = \int_{-2}^{x+1} \frac{1}{2} \cdot \frac{1}{4}(1+2)\,d1$

$\quad = \dfrac{x^2}{16} + \dfrac{3}{8}x + \dfrac{9}{16}, \quad -3 < X < -1$

(c) $f_Z(x) = \int_{x-1}^{0} \frac{1}{2} \cdot \frac{1}{4}(1+2)\,d1$

$\quad + \int_{0}^{x+1} \frac{1}{2}\left(-\frac{1}{4}\right)(1-2)\,d1$

$\quad = \dfrac{x^2}{8} + \dfrac{3}{8}, \quad -1 < x < +1$

(d) $f_Z(x) = \int_{x-1}^{2} -\frac{1}{8}(1-2)\,d1$

$\quad = \dfrac{x^2}{16} + \dfrac{3}{8}x + \dfrac{9}{16}, \quad 1 < x < 3$

$f_Z(x) = 0, \quad x > 3$

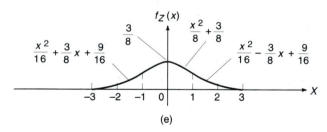

(e)

Fig. 4.7.2

4.8 Problems

4.1. (a) State what it means for two random variables X and Y to be statistically independent; that is, give a definition.

(b) What is the condition for statistical independence of X and Y in terms of their characteristic functions?

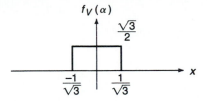

Fig. 4.7.3

4.2. Let $X = (X_1, X_2)$ be the two-dimensional random variable in Fig. 4.8.1.
(a) Find and plot the two-dimensional pdf $f_X(x_1, x_2)$.
(b) Find and plot the two-dimensional cdf $F_X(x_1, x_2)$.
(c) Are X_1 and X_2 independent? Uncorrelated? State why.

4.3. Find the marginal pdf $f_{X_1}(x_1)$ in Fig. 4.8.1. Do this two ways, directly without regard for X_2 and by using the two-dimensional distribution.

4.4. What is the event (i.e., the subset of the sample space) corresponding to the following conditions on the cdf for the random variable $X = (X_1, X_2)$ in Fig. 4.8.1?
(a) $F_X(0, 0)$
(b) $F_X(1.5, 1.5)$
(c) $F_X(-0.5, -0.5)$
(d) $F_X(1.5, -0.5)$

4.5. Let $t_2 - t_1 = 0.5$ in Fig. 4.8.2. Thus $t_2 - t_1 = T/4$, where $T = 2$ is the period. Suppose that t_1 is chosen at random (i.e., with uniform probability) in the interval $0 \leq t_1 < 2$. What are the following probabilities?
(a) $F_v(0, 0)$
(b) $F_v(10, 0)$
(c) $F_v(0, -10)$
(d) $F_v(5, 5)$

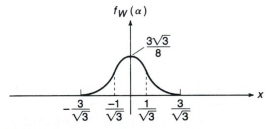

Fig. 4.7.4

X_1	1	−1
X_2	1	0
ζ	H	T

Fig. 4.8.1

4.6. Find and plot the two-dimensional pdf for Problem 4.5.

4.7. Repeat Problem 4.5 if $t_2 - t_1 = 0.25$ (or $T/8$).

4.8. Let $t_2 = t_1 + T/2$ in Fig. 4.8.3. For $v_1(t)$, evaluate the two-dimensional cdf $F_V(\alpha_1, \alpha_2) = P\{v(t_1) \leq \alpha_1, v(t_2) \leq \alpha_2\}$ for the following values:
(a) $\alpha_1 = 5, \alpha_2 = 5$
(b) $\alpha_1 = 5, \alpha_2 = -20$
(c) $\alpha_1 = 15, \alpha_2 = 5$

4.9. Repeat Problem 4.8 for $v_2(t)$.

4.10. Repeat Problem 4.8 for $v_1(t)$ with $t_2 = t_1 + 3T/4$.

4.11. Repeat Problem 4.10 for $v_2(t)$.

4.12. The function $v(t)$ shown in Fig. 4.8.4 is periodic with period $T = 3$. Suppose we sample this function at time t_1 chosen at random in the interval $0 \leq t_1 < 3$, and at time $t_2 = t_1 + 2$. Notice therefore that t_2 can take on values $2 \leq t_2 < 5$. Finding the two-dimensional pdf is equivalent to finding four probabilities.
(a) Supply the values for the four probabilities:

$$P\{v(t_1) = 5, v(t_2) = 5\}$$
$$P\{v(t_1) = 5, v(t_2) = 1\}$$
$$P\{v(t_1) = 1, v(t_2) = 5\}$$
$$P\{v(t_1) = 1, v(t_2) = 1\}$$

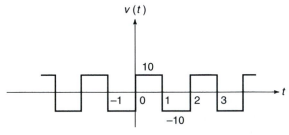

Fig. 4.8.2

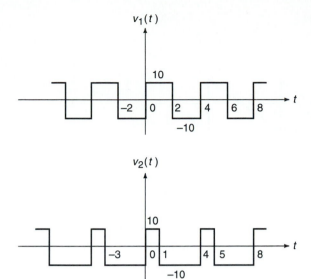

Fig. 4.8.3

(b) Find the correlation $E[V(t_1)V(t_2)]$ for $t_2 = t_1 + 2$.

4.13. Suppose that the signal $v(n) = n$ for $n = 0, 1, 2,$ and 3 is periodic with period $N = 4$.
(a) Plot about three periods of this signal.
(b) Find and plot the cdf $F_v(\alpha)$ and the pdf $f_v(\alpha)$.
(c) Evaluate $P\{v(n) \le 2, v(n + 1) \le 2\}$.
(d) Evaluate $P\{v(n) \le 2.5, v(n + 1) \le 3.2\}$.

4.14. Evaluate $F_V(\alpha_1, \alpha_2) = P\{v(t_1) \le \alpha_1, v(t_2) \le \alpha_2\}$ for the sawtooth waveform in Fig. 4.8.5 if $\alpha_1 = 1$, $\alpha_2 = 1$ for the following values:
(a) $t_2 = t_1 + 1$
(b) $t_2 = t_1 + 1.5$
(c) $t_2 = t_1 + 2$

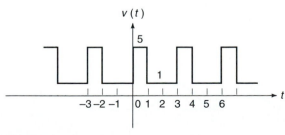

Fig. 4.8.4

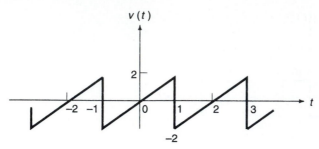

Fig. 4.8.5

4.15.   Repeat Problem 4.14 for the waveform in Fig. 4.8.6.

4.16.   Two random variables X and Y have joint pdf

$$f_{XY}(x, y) = \begin{cases} K, & 0 < x < 1, 0 < y < 1 \\ 0 & \text{elsewhere.} \end{cases}$$

(a) Sketch $f_{XY}(x, y)$.
(b) What is the value of K?
(c) What is the probability that $X \le \frac{1}{2}$ and $Y \le \frac{1}{2}$?
(d) Are X and Y statistically independent? Why?

4.17.   Suppose that X and Y are statistically independent, zero-mean random variables.
(a) What is the characteristic function of $Z = aX + bY$?
(b) What is the correlation coefficient between Z and $W = cX + dY$? That is, find ρ_{ZW}.

4.18.   For the two random variables in Fig. 4.8.7, find and plot
(a) The marginal pdf for X
(b) The conditional pdf $f_X(x|Y = 4)$
(c) The conditional pdf $f_X(x|Y = 0)$

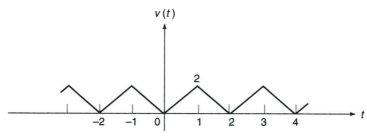

Fig. 4.8.6

4.19. For the random variables in Fig. 4.8.7, calculate
(a) The covariance μ_{11}
(b) The correlation coefficient ρ_{XY}

4.20. The random variable X is uniformly distributed from $(-1, 1)$:

$$f(x) = \begin{cases} 0.5, & -1 < x < 1 \\ 0 & \text{otherwise} \end{cases}$$

Find and plot the conditional density $f_X(x|B)$ if B is the event
(a) $B = \{X \le 0\}$
(b) $B = \{X = 0\}$

4.21. Two independent random variables X and Y are jointly Gaussian with 0 means and variances given by $\sigma_X^2 = 1$, $\sigma_Y^2 = 4$. Find the probability that $-1 < X < 1$ and $-1 < Y < 2$.

4.22. Prove Chebyshev's inequality.

4.23. A boy shooting baskets from the free-throw line has probability 0.4 of success. He shoots 100 times. What are his chances of making between 30 and 50 baskets?
(a) Use Chebyshev's inequality to obtain a bound on this probability.
(b) Use the central limit theorem to estimate this probability.

4.24. If a coin is tossed 100 times, estimate the chances of the number of heads falling between 40 and 60.
(a) Use Chebyshev's inequality to obtain a bound on this probability.
(b) Use the central limit theorem to estimate this probability.

4.25. Let $X = (X_1, X_2)$ be a two-dimensional Gaussian random variable with 0 mean and covariance matrix

$$\Lambda_X = \begin{bmatrix} 2 & 1 \\ 1 & 4 \end{bmatrix}$$

(a) Write an expression for the two-dimensional pdf.
(b) Write an expression for the conditional density $f_X(x_1|X_2 = 2)$.

4.26. Repeat Problem 4.25 given that the means are $E(X_1) = 1$, $E(X_2) = -3$.

Y	4	1	0	1	4	9
X	-2	-1	0	1	2	3
ζ	f_1	f_2	f_3	f_4	f_5	f_6

Fig. 4.8.7

4.27. Determine the probability that the sum of the random variables in Problem 4.25 is greater than 0.

4.28. State (with words and equations) the central limit theorem.

4.29. For a two-dimensional random variable X with

$$\Lambda_X = \begin{bmatrix} 9 & 1 \\ 1 & 4 \end{bmatrix} \quad \text{and} \quad m_X = \begin{bmatrix} 0 \\ 0 \end{bmatrix}$$

(a) Write the two-dimensional pdf in matrix form.
(b) Write the conditional pdf for X_1 given that $X_2 = 0$.
(c) Set up an expression for $P\{-1 < X_1 < 1\}$.

4.30. Given n vectors $x_1, x_2, \ldots, x_n$, we can estimate the mean vector, the correlation matrix, and the covariance matrix from this data by the formulas

$$\hat{m}_x = \frac{1}{n}(x_1 + x_2 + \cdots + x_n)$$

$$\hat{R}_x = \frac{1}{n}(x_1 x_1^t + x_2 x_2^t + \cdots + x_n x_n^t)$$

$$\hat{\Lambda}_x = \hat{R}_x - \hat{m}_x \hat{m}_x^t$$

Suppose that you are given four vectors:

$$x_1 = \begin{bmatrix} 3 \\ 1 \end{bmatrix} \quad x_2 = \begin{bmatrix} 5 \\ 1 \end{bmatrix} \quad x_3 = \begin{bmatrix} 3 \\ 3 \end{bmatrix} \quad x_4 = \begin{bmatrix} 5 \\ 3 \end{bmatrix}$$

(a) Plot these points on the plane and note their positions.
(b) Estimate the mean, the correlation, and the covariance matrices.
(c) Center these points about the origin by subtracting the mean of the data from each vector. (The data mean is $[4 \quad 2]^t$.) Repeat parts (a) and (b). Can you account for the differences and similarities?

4.31. Given that X_1 and X_2 are independent random variables with identical uniform $(0, 1)$ distributions, find the distribution of $Y = X_1 + X_2$.

4.32. Repeat Problem 4.31 if X_1 is uniform $(0, 1)$ but X_2 is exponential with distribution $f_2(\alpha) = e^{-\alpha}$, $\alpha > 0$. $f_2(\alpha) = 0$ for $\alpha < 0$.

4.33. Suppose that $Y = X_1 + X_2$, where X_1 and X_2 are independent. If X_1 has known uniform $(0, 1)$ distribution and Y has the following distribution, find the distribution of X_2.

$$f_Y(\alpha) = \begin{cases} 0, & \alpha < 0.5 \\ 2(\alpha - 0.5), & 0.5 < \alpha < 1 \\ 1, & 1 < \alpha < 1.5 \\ -2(\alpha - 2), & 1.5 < \alpha < 2 \\ 0, & 2 < \alpha \end{cases}$$

4.34. If two random variables X_1 and X_2 are independent and normal $(0, \sigma^2)$, then their ratio $Y = X_1/X_2$ has a Cauchy density function given by

$$f_Y(\alpha) = \frac{1/\pi}{1 + \alpha^2}$$

Find the probability that $X_1 \geq 2X_2$.

4.35. An overall system S will fail if either subsystem S_1 or S_2 fails. Let X_1 and X_2 be the times to failure for each system, and suppose that X_1 and X_2 are identically distributed with exponential density $f(\alpha) = e^{-\alpha}$, $\alpha > 0$. Find the probability that the system S will last more than one time unit.

4.36. Repeat Problem 4.35 if S will fail only if both subsystems fail.

4.37. The following statements are either correct, partially correct, or wrong. Change each statement as needed to make it correct.
(a) If events A and B are independent, then $P(A \cap B) = 0$.
(b) A random variable is Gaussian if it is the sum of a large number of random variables.
(c) Random variables X and Y are uncorrelated if $E(XY) = 0$.
(d) The conditional cdf $F(x|y)$ is the derivative of the conditional pdf $f(x|y)$ with respect to x.
(e) The pdf for the n-dimensional Gaussian random variable can be written down if we know the means, the variances, and the covariances.
(f) The central limit theorem says that the sum of a large number of random variables is Gaussian, even though none of the random variables may be Gaussian.
(g) In order to apply Chebyshev's inequality to a random variable X, we need to know the mean and variance of X.
(h) The weak law of large numbers applies to the sum of independent random variables X_i.

Further Reading

1. ATHANASIOS PAPOULIS, *Probability, Random Variables, and Stochastic Processes,* 3rd ed., McGraw-Hill, New York, 1991.
2. CHARLES W. THERRIEN, *Discrete Random Signals and Statistical Signal Processing,* Prentice Hall, Englewood Cliffs, NJ, 1992.
3. MICHAEL O'FLYNN, *Probabilities, Random Variables, and Random Processes,* Harper & Row, New York, 1982.

CHAPTER 5 _____

Signal Analysis Techniques

5.1 Power and Energy Signals _____

Preview

The title of this chapter is "Signal Analysis Techniques." We will spend this entire chapter discussing the relationship between time and frequency, leading to a demonstration that the autocorrelation function and spectral density function form a Fourier transform pair for deterministic signals. This is important because wide-sense stationary random signals (which we will discuss in the next chapter) have this same property. What this really refers to is the use of second-order statistics to describe the correlation function and the frequency content of signals.

There are several types of correlation functions and spectral density functions, and the type depends on the signal, specifically on whether it is an energy or a power signal. Therefore this section is devoted to the classification of signals. We can classify signals as

Discrete-time or continuous-time signals
Power, energy, and "monster" signals

We will ignore the monster category, because those signals increase with time. We don't know how to handle them mathematically, but we don't need to because we seldom use them in practice. This gives us four different signal types:

1. Continuous-time energy signals
2. Continuous-time power signals
3. Discrete-time energy signals
4. Discrete-time power signals

The purpose of this section is to describe how to classify signals into one of these four categories.

A resistor dissipates power as heat. One of the more memorable experiments in any sophomore circuits laboratory demonstrates this fact with a candle leaning against a hot resistor. We heat the resistor by impressing a steady voltage of moderate value across it. If the power dissipated, the candle size, and the power rating of the resistor are about right, then the melting of the candle, which rests against the resistor, will be proportional to the power dissipated. This experiment demonstrates not only that power is dissipated as heat, but also that power is dissipated over time. If the voltage is not too large, the resistor warms to a constant temperature and remains there. In other words, it reaches a steady state. Thus we speak of "average power" dissipated in the resistor. This implies the following:

1. The heat produced is proportional to average power. Note the emphasis on "average" power, not instantaneous power.
2. This implies a steady-state signal. Ideally, a steady-state signal exists forever, starting before the world was formed, and lasting until long after the world has burned to a crisp. For the laboratory experiment, however, the signal must last only for the duration of the experiment. Thus the idealization to infinite time is one of mathematical convenience, as we will see in the following definition for power.

For a continuous-time signal voltage $v(t)$ impressed across a resistor, the average power is defined by

$$P = \lim_{T \to \infty} \frac{1}{2T} \int_{-T}^{T} \frac{v^2(t)}{R}\, dt \qquad (5.1.1)$$

To make this quantity depend only on the signal, we set the resistor equal to 1 Ω. The total power, average power, or mean square value of the signal is thus defined to be

$$P = \lim_{T \to \infty} \frac{1}{2T} \int_{-T}^{T} v^2(t)\, dt \qquad (5.1.2)$$

The square root of P is the root-mean-square (rms) value. Notice that just as easily we could have defined the power in terms of the current through the resistor. Our definition of power in a signal applies to any signal, whether it is voltage, current, or some nonelectrical function such as pressure or velocity.

This serves to introduce the classification of signals into one of two categories: energy signals or power signals. A signal $v(t)$ is called a *power signal* if the expression in Eq. 5.1.2 is greater than 0 and less than ∞, that is, $0 < P < \infty$. A signal $v(t)$ is called an *energy signal* if the energy is greater than 0 and less than ∞, where the energy E is given by

$$E = \int_{-\infty}^{\infty} v^2(t)\, dt \qquad (5.1.3)$$

Ch. 5 / Signal Analysis Techniques

This is the total energy, not the instantaneous energy. We will concern ourselves only with those signals that can be classified into one of the two categories, but you should understand that there are other signals. A ramp signal, or an increasing exponential signal, has infinite power, and therefore is neither a power nor an energy signal. We will have no need for such signals, so as far as we are concerned, every signal is either an energy signal or a power signal. Notice that any signal lasting for only a finite duration is an energy signal. A signal that lasts for an infinite duration can be either a power or an energy signal, depending on the value of the expressions in Eqs. 5.1.2 and 5.1.3. Here are some examples.

EXAMPLE 5.1.1.   The square pulse in Fig. 5.1.1 has zero power. To see this, apply Eq. 5.1.2 with increasing values of T as follows:

For $T = 5$, $$P = \frac{1}{10} \int_0^1 2^2 \, dt = \frac{4}{10}$$

For $T = 10$, $$P = \frac{1}{20} \int_0^1 2^2 \, dt = \frac{4}{20}$$

For $T = 100$, $$P = \frac{1}{200} \int_0^1 2^2 \, dt = \frac{4}{200}$$

As T gets larger and larger, the power gets smaller and smaller, until in the limit it is 0.

The energy in the pulse is finite, however, because Eq. 5.1.3 gives

$$E = \int_0^1 2^2 \, dt = 4$$

Since the power is 0 and the energy is finite, this is an energy signal.

EXAMPLE 5.1.2.   The periodic square wave in Figure 5.1.2 has finite power and infinite energy. If we apply Eq. 5.1.2 gradually, as we did above, we get

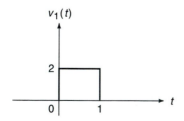

Fig. 5.1.1.   A square pulse with zero power.

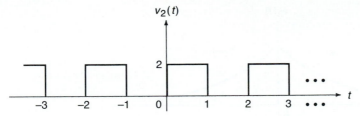

Fig. 5.1.2.   A periodic square wave with finite power and infinite energy.

For $T = 5$, $\qquad P = \dfrac{1}{10} \displaystyle\int_{-5}^{5} v_2^2(t)\, dt = \dfrac{20}{10}$

For $T = 10$, $\qquad P = \dfrac{1}{20} \displaystyle\int_{-10}^{10} v_2^2(t)\, dt = \dfrac{40}{20}$

For $T = 100$, $\qquad P = \dfrac{1}{200} \displaystyle\int_{-100}^{100} v_2^2(t)\, dt = \dfrac{400}{200}$

As this continues with larger and larger values of T we get the same value, 2. Notice that power in a periodic signal of period T is given by the simplified formula

$$P = \frac{1}{T} \int_0^T v^2(t)\, dt$$

which should be familiar to you from circuit analysis courses. The rms value is the square root of the power. Therefore the rms value of the signal in this example is $\sqrt{2}$.

The energy in the signal is infinite because

$$E = \int_{-\infty}^{\infty} v_2^2(t)\, dt = \infty$$

Since the energy is infinite and the power is finite, this is a power signal.

Any signal that we actually generate and use must be of finite duration, and therefore an energy signal. The classification above applies only to our mathematical model of the signal. We will often define a signal as lasting forever, although this is practically impossible. But just because a signal lasts forever does not mean it is a power signal. Here is an example.

EXAMPLE 5.1.3.   The decaying exponential in Fig. 5.1.3 lasts forever but has finite energy.

$$E = \int_{-\infty}^{\infty} v_3^2(t)\, dt = 2 \int_0^{\infty} 4e^{-2t}\, dt = 4$$

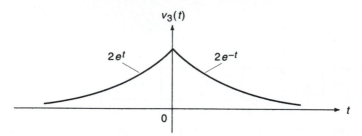

Fig. 5.1.3.   A decaying exponential with finite energy.

The power is 0, as we see from the following sequence of integrals:

For $T = 5$, $\qquad\qquad P = \dfrac{1}{10} \displaystyle\int_{-5}^{5} v_3^2(t)\, dt = 0.4$

For $T = 10$, $\qquad\qquad P = \dfrac{1}{20} \displaystyle\int_{-10}^{10} v_3^2(t)\, dt = 0.2$

For $T = 100$, $\qquad\qquad P = \dfrac{1}{200} \displaystyle\int_{-100}^{100} v_3^2(t)\, dt = 0.02$

In the limit, $P = 0$. Because the power is 0 and the energy is finite, this is an energy signal.

For discrete-time signals these definitions become

$$P = \lim_{N \to \infty} \frac{1}{2N + 1} \sum_{n=-N}^{N} v^2(n) \qquad\qquad (5.1.4)$$

and

$$E = \sum_{n=-\infty}^{\infty} v^2(n) \qquad\qquad (5.1.5)$$

In the formula for P, Eq. 5.1.4, the sum over the range $(-N, N)$ has $2N + 1$ terms in it, so we divide by this number of terms to find the average. This is also called the mean square value. Thus the mean square value is "average power," although the concept of power loses its meaning for discrete-time signals. The energy in Eq. 5.1.5 is the *total* energy, not the instantaneous energy.

EXAMPLE 5.1.4.   The signal $x(n)$ shown in Fig. 5.1.4 is an energy signal. To see this, apply Eq. 5.1.4 with increasing values of N to get

For $N = 5$, $\qquad\qquad P = \dfrac{1}{11} \displaystyle\sum_{n=-5}^{5} x^2(n) = \dfrac{12}{11}$

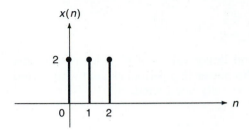

Fig. 5.1.4.   An energy signal $x(n)$.

For $N = 10$,

$$P = \frac{1}{21} \sum_{n=-10}^{10} x^2(n) = \frac{12}{21}$$

For $N = 100$,

$$P = \frac{1}{201} \sum_{n=-100}^{100} x^2(n) = \frac{12}{201}$$

As N gets larger and larger, the power gets smaller and smaller, until in the limit it is 0.

The energy in the pulse is finite, however, because Eq. 5.1.5 gives

$$E = \sum_{n=0}^{2} 2^2 = 12$$

Since the power is 0 and the energy is finite but greater than 0, this is an energy signal.

EXAMPLE 5.1.5.   The periodic digital square wave in Fig. 5.1.5 has finite power and infinite energy. Apply Eq. 5.1.4 gradually, as above, to get

For $N = 5$,

$$P = \frac{1}{11} \sum_{n=-5}^{5} x^2(n) = \frac{20}{11}$$

For $N = 10$,

$$P = \frac{1}{21} \sum_{n=-10}^{10} x^2(n) = \frac{40}{21}$$

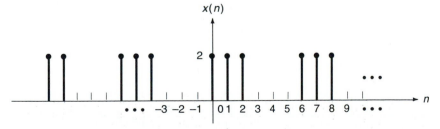

Fig. 5.1.5.   A periodic digital square wave with finite power and infinite energy.

For $N = 100$, $P = \dfrac{1}{201} \displaystyle\sum_{n=-100}^{100} x^2(n) = \dfrac{400}{201}$

As this continues with larger and larger values of N, P approaches the value 2. Notice that here, too, the power in a periodic signal with period N can be found by averaging over only one period.

$$P = \frac{1}{N} \sum_{n=0}^{N-1} x^2(n)$$

The rms value is again the square root of the power. If we calculate the energy we get

$$E = \sum_{n=-\infty}^{\infty} x^2(n) = \infty$$

Since the energy is infinite and the power is finite, this is a power signal.

As with continuous time signals, just because a signal lasts forever does not mean it is a power signal. Here is an example to illustrate that fact for discrete-time signals.

EXAMPLE 5.1.6.   The exponential signal in Fig. 5.1.6 lasts forever in both directions, but it is an energy signal. The energy is

$$E = \sum_{n=-\infty}^{\infty} x^2(n) = 2 \sum_{n=1}^{\infty} x^2(n) + x^2(0)$$

The terms on the right are derived by noticing that for the waveform in Fig. 5.1.6, the sum for $-\infty$ to -1 is the same as the sum for 1 to ∞. Then the energy is given by twice the sum from 1 to ∞ plus the term for $n = 0$.

The sum $\sum_{n=0}^{\infty} (0.81)^n$ is a geometric series. This familiar series has the value given by

$$\sum_{n=0}^{\infty} a^n = \frac{1}{1-a}, \qquad |a| < 1 \tag{5.1.6}$$

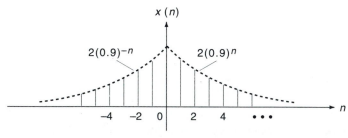

Fig. 5.1.6.   An exponential energy signal.

The sum we seek has a lower limit of 1 instead of 0, so we should calculate the sum using Eq. 5.1.6 and then subtract the term at $n = 0$. With $a = 0.81$ this gives

$$\sum_{n=1}^{\infty} x^2(n) = 4 \sum_{n=1}^{\infty} (0.81)^n = \frac{4(0.81)}{1 - 0.81} = 17.0526$$

Therefore the energy is double this number plus $x^2(0)$.

$$E = 2(17.0526) + 4 = 38.1052$$

The power is 0, as we see from the following sequence of sums:

For $N = 5$, $$P = \frac{1}{11} \sum_{n=-5}^{5} x^2(n) = 2.383$$

For $N = 10$, $$P = \frac{1}{21} \sum_{n=-10}^{10} x^2(n) = 0.5471$$

For $N = 100$, $$P = \frac{1}{201} \sum_{n=-100}^{100} x^2(n) = 0.0552$$

In the limit, $P = 0$. Because the power is 0 and the energy is greater than 0 but finite, this is an energy signal. (The formula for calculating finite sums is derived in the next section; see Eq. 5.2.6.)

We will use only those signals that can be classified into one of the two categories of energy or power signals, according to the value of P or E from these definitions. Notice once again that a power signal must last forever, so these definitions apply only to their mathematical models.

Review

For continuous-time signals Eqs. 5.1.2 and 5.1.3 define power and energy, respectively. Equations 5.1.4 and 5.1.5 define power and energy for discrete-time signals. You should understand how to apply these formulas to find power and energy in a signal. You should also understand that these concepts apply equally to deterministic or random signals.

5.2 Time Correlation Functions _____

Preview

Correlation is a binary operation. This means that we operate on two waveforms and produce a third waveform. Convolution does the same thing, and convolution and correlation are almost identical, differing only in the details of calculation and in their use. Convolution is a mathematical description of the operation of linear systems on input signals, while correlation has many practical applications.

Initial applications were to radar and pulse communication, where a pulse of known shape is either present or absent during each specified period. The receiver uses a correlation receiver to determine (guess) whether or not the signal is present during that time interval. Almost all data and long-distance voice communication is handled in this way.

Pattern recognition illustrates another application. Despite all the theoretical and experimental work done to improve automatic pattern-recognition systems, most systems use correlation to decide (guess) which pattern is present. One familiar and very successful system is the bank check readers now used by every bank in the United States. The numbers and symbols printed across the bottom of your checks are specially designed for a correlation detector. A scanning device produces a signal that is sampled nine times during the span of each symbol. Therefore each symbol is represented by a digital signal of nine samples. There are 14 different symbols used in this system, 10 numbers and 4 special characters, so there are 14 different matched filters used to detect each symbol by correlation.

Automatic page scanners use correlation. These devices read and enter a page full of text into a computer automatically, a great labor-saving device in many offices. Correlation decides which letter or number is present at each possible symbol location. Page scanners are reliable for one particular font and type size, but as you can imagine, they are of little value for reading arbitrary type styles and sizes without adjustment. This is an example of two-dimensional correlation. We match a template for each class to the unknown symbol and make the decision on the basis of the highest correlation value. This is called template matching in pattern recognition.

These examples illustrate only some of the many applications of correlation to technical problems. For a better understanding of this general technique we need to describe the conditions under which we may apply the various forms of correlation. In this chapter we look at the four forms of time correlation. In later chapters we look at statistical correlation. In this section we define correlation for each of the four types of signals: continuous-time power and energy signals, and discrete-time power and energy signals. After completing this section you should be able to find and plot the correlation function for any two given signals when both are the same type.

Correlation is a binary operation. The black box with two input lines and one output line in Fig. 5.2.1 shows a binary operation. The chief requirement for this box is that the same type of things that go in come out. If the two input terms are numbers, then the output must be a number. If the two input terms are functions, then the output must be a function. (I suppose if the two inputs are elephants, then the output must be an elephant, but we'll stick to mathematics.) Ordinary addition of two numbers is a binary operation. We add two numbers x and y to produce the sum $z = x + y$. Convolution is a binary operation, where the two input terms are functions and the output

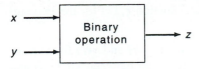

Fig. 5.2.1.   A black box.

term is a function. Both statistical and time correlations are binary operations, and as we will see, time correlation is closely related to convolution.

The four forms of the correlation operation correspond to the four signal types we deal with. These four signal types are

Type 1. Continuous-time energy signals
Type 2. Continuous-time power signals
Type 3. Discrete-time energy signals
Type 4. Discrete-time power signals

See Section 5.1 for a discussion of these signal types. There is a different correlation formula for each type. We discuss each formula in turn below.

Type 1. Continuous-Time Energy Signals

The correlation $r_{xy}(t)$ between two continuous-time energy signals $x(t)$ and $y(t)$ is given by

$$r_{xy}(\tau) = \int_{-\infty}^{\infty} x(t)\, y(t - \tau)\, dt \qquad (5.2.1a)$$

$$= \int_{-\infty}^{\infty} x(t + \tau)\, y(t)\, dt \qquad (5.2.1b)$$

If $x(t) = y(t)$, this is called the *autocorrelation* and is written $r_{xx}(\tau)$. If $x(t) \neq y(t)$, this is called the *cross correlation* between x and y. Notice that $r_{xy}(\tau) \neq r_{yx}(\tau)$. The two formulas in Eq. 5.2.1 give the same result because a change of variable relates the two integrals. Let $\lambda = t - \tau$ in the first form to obtain the second form. We will use lowercase r for time correlation, and uppercase R for statistical correlation.

Figure 5.2.2 shows correlation as a binary operation. The input terms are continuous-time signals x and y, so the output $r_{xy}(\tau)$ must also be a continuous-time signal. When the input terms are discrete-time signals, the correlation is a discrete-time signal. Here are some examples.

$$x(t) \longrightarrow \boxed{\int_{-\infty}^{\infty} x(t)\, y(t - \tau)\, dt} \longrightarrow r_{xy}(\tau)$$
$$y(t) \longrightarrow$$

Fig. 5.2.2

EXAMPLE 5.2.1.   Find the correlation function $r_{xy}(\tau)$ for the two signals in Figs. 5.2.3a and 5.2.3b.

SOLUTION: Figure 5.2.4 shows a plot of $x(t - \tau)$ and $y(t)$ for various values of shift τ. In Fig. 5.2.4a no overlap occurs between $x(t - \tau)$ and $y(t)$, so the correlation from Eq. 5.2.1a is 0. Notice that the abscissa is labeled t in the diagram, and the variable of integration in Eq. 5.2.1 is t. You should plot functions versus the variable of integration for both correlation and convolution so that they will provide a picture of the integration process. Figures 5.2.4b and 5.2.4c show the picture for $-1 < \tau < 0$, and for $0 < \tau < 1$. The picture for $1 < \tau$ is not shown, but there is no overlap so the correlation is 0 in that interval.

Figure 5.2.5 shows the correlation function $r_{xy}(\tau)$ plotted versus τ. Since $x(t) = y(t)$ this is also the autocorrelation function for signal $x(t)$. Notice that $r_{xx}(0)$ is the energy in the signal $x(t)$, which we calculated in Example 5.1.1.

EXAMPLE 5.2.2.   Find the correlation function $r_{xy}(\tau)$ for the two signals in Fig. 5.2.6a and 5.2.6b.

SOLUTION: Figure 5.2.7 shows the answer. You should be able to arrive at the answer by using the same techniques for evaluating the correlation integral as in the previous example.

Type 2. Continuous-Time Power Signals

The correlation $r_{xy}(\tau)$ between two continuous-time power signals $x(t)$ and $y(t)$ is given by

$$r_{xy}(\tau) = \lim_{T \to \infty} \frac{1}{2T} \int_{-T}^{T} x(t)\, y(t - \tau)\, dt \qquad (5.2.2a)$$

$$= \lim_{T \to \infty} \frac{1}{2T} \int_{-T}^{T} x(t + \tau)\, y(t)\, dt \qquad (5.2.2b)$$

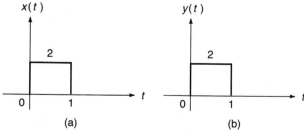

(a)

(b)

Fig. 5.2.3

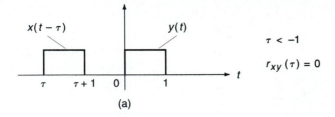

$\tau < -1$

$r_{xy}(\tau) = 0$

(a)

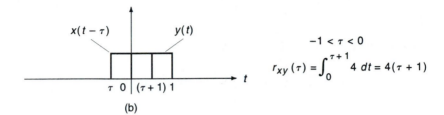

$-1 < \tau < 0$

$r_{xy}(\tau) = \int_0^{\tau+1} 4\, dt = 4(\tau + 1)$

(b)

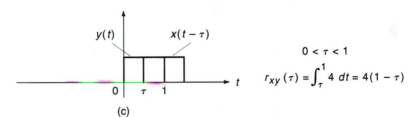

$0 < \tau < 1$

$r_{xy}(\tau) = \int_{\tau}^{1} 4\, dt = 4(1 - \tau)$

(c)

Fig. 5.2.4

This is similar to our definition for mean square value or power (see Eq. 5.1.2), because we have the limit as $T \to \infty$ in the formula.

If $x(t)$ and $y(t)$ are periodic with the same period T, we may replace Eq. 5.2.2 by the simpler form given by

$$r_{xy}(\tau) = \frac{1}{T}\int_0^T x(t)\, y(t - \tau)\, dt \qquad (5.2.3a)$$

$$= \frac{1}{T}\int_0^T x(t + \tau)\, y(t)\, dt \qquad (5.2.3b)$$

EXAMPLE 5.2.3.   Find the autocorrelation function $r_{xx}(\tau)$ for the signal in Fig. 5.2.8a.

SOLUTION: Figure 5.2.8b shows $x(t - \tau)$ for $0 < \tau < T/2$. Figure 5.2.8c shows the product $x(t) \times x(t - \tau)$. Integrate over one period and divide

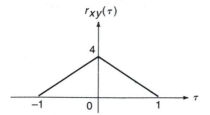

Fig. 5.2.5.   The correlation of $x(t)$ with $y(t)$.

by T according to Eq. 5.2.3a to get

$$r_{xx}(\tau) = \frac{1}{T} \int_{\tau}^{T/2} E^2 \, dt = \frac{E^2}{T} \left(\frac{T}{2} - \tau \right), \qquad 0 < \tau < \frac{T}{2}$$

Figure 5.2.9 is like Fig. 5.2.8, but for $T/2 < \tau < T$. Figure 5.2.9a shows the original function $x(t)$. Figure 5.2.9b shows $x(t - \tau)$ for τ approximately equal to $3T/2$, and Figure 5.2.9c shows the product. Now integrate over one period to get

$$r_{xx}(\tau) = \frac{1}{T} \int_{0}^{\tau - T/2} E^2 \, dt = \frac{E^2}{T} \left(\tau - \frac{T}{2} \right), \qquad \frac{T}{2} < \tau < T$$

If this process is duplicated for all values of τ, the result is the periodic autocorrelation function shown in Fig. 5.2.10.

This example illustrates that the result of correlating two periodic signals with the same period is a periodic correlation function with the same period. Do not, however, draw the incorrect conclusion that the result of correlating two power signals is a power correlation function. In this example, the correlation $r_{xx}(\tau)$ is periodic, and is therefore a power function. But that is only in this example. The correlation function for the random power signal in Section 1.2 is an energy function.

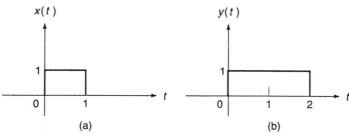

(a)   (b)

Fig. 5.2.6

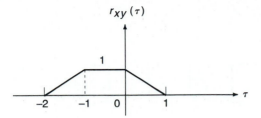

Fig. 5.2.7.   The solution to Example 5.2.2.

Type 3. Discrete-Time Energy Signals

The correlation $r_{xy}(t)$ between two discrete-time energy signals $x(n)$ and $y(n)$ is given by

$$r_{xy}(\tau) = \sum_{n=-\infty}^{\infty} x(n) \, y(n - \tau) \qquad (5.2.4a)$$

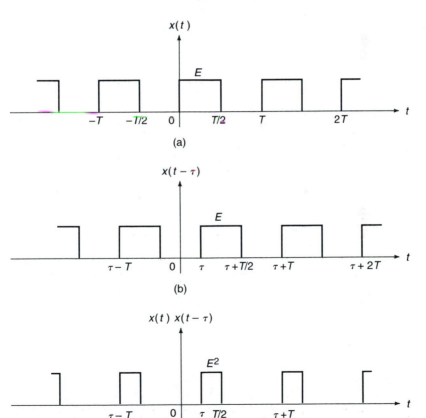

Fig. 5.2.8.   Calculating correlation for $0 < \tau < T/2$.

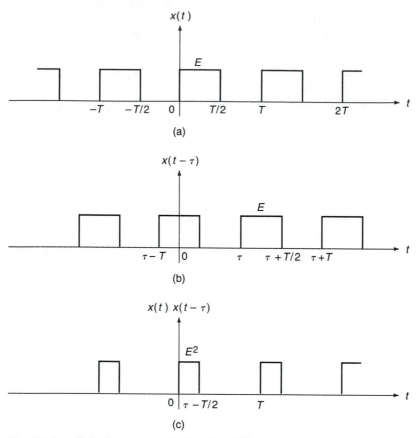

$x(t)$

E

$-T$ $-T/2$ 0 $T/2$ T $2T$ t

(a)

$x(t-\tau)$

E

$\tau-T$ 0 τ $\tau+T/2$ $\tau+T$ t

(b)

$x(t)\,x(t-\tau)$

E^2

0 $\tau-T/2$ T t

(c)

Fig. 5.2.9.   Calculating correlation for $T/2 < \tau < T$.

$$= \sum_{n=-\infty}^{\infty} x(n+\tau)\,y(n) \qquad (5.2.4b)$$

We will use τ for the shift parameter for both continuous- and discrete-time signals. You should be aware, however, that τ is a continuous variable in

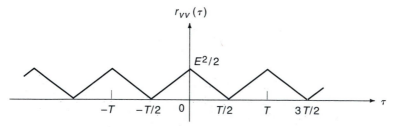

$r_{vv}(\tau)$

$E^2/2$

$-T$ $-T/2$ 0 $T/2$ T $3T/2$ τ

Fig. 5.2.10.   The autocorrelation function for Example 5.2.3.

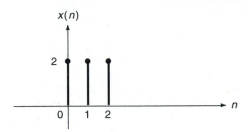

Fig. 5.2.11

Eqs. 5.2.1 through 5.2.3, and τ is a discrete variable here for digital signals.

EXAMPLE 5.2.4.   Find the autocorrelation function $r_{xx}(\tau)$ for the signal in Fig. 5.2.11.

SOLUTION: This is the energy signal from Example 5.1.4. Figure 5.2.12a shows $x(n - \tau)$ for $\tau < 0$, along with $x(n)$. You can see that as τ increases, overlap occurs at $\tau = -2$. The sum in Eq. 5.2.4 therefore has the values shown in Fig. 5.2.12b.

EXAMPLE 5.2.5.   Find the autocorrelation function $r_{xx}(\tau)$ for the exponential energy signal shown in Fig. 5.2.13.

SOLUTION: The problem here is to evaluate the sum

$$r_{xx}(\tau) = \sum_{n=-\infty}^{\infty} x(n - \tau) x(n)$$

for each value of τ. We begin with $\tau < 0$, as shown in Fig. 5.2.14a. After multiplying the functions $x(n - \tau)$ and $x(n)$, we have two regions with different terms in the product.

Region 1: $(n < 0)$, product $= 0$.
Region 2: $(n \geq 0)$, product $= (0.9)^{(n-\tau)}(0.9)^{n} = (0.9)^{-\tau}(0.9)^{2n}$.

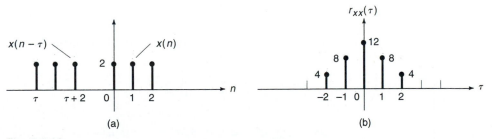

(a)   (b)

Fig. 5.2.12

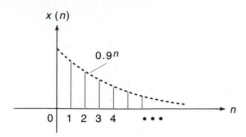

Fig. 5.2.13

The sum over these two regions gives the autocorrelation function for $\tau < 0$.

$$r_{xx}(\tau) = \sum_{n=-\infty}^{-1} 0 + \sum_{n=0}^{\infty} (0.9)^{-\tau}(0.9)^{2n} = (0.9)^{-\tau} \sum_{n=0}^{\infty} (0.81)^n$$

The sum $\sum_{n=0}^{\infty} (0.81)^n$ is a geometric series. This familiar series has the value given by

$$\sum_{n=0}^{\infty} a^n = \frac{1}{1-a}, \qquad |a| < 1 \tag{5.2.5}$$

To see this, divide $1 - a$ into 1 by long division. With $a = 0.81$ the autocorrelation function is

$$r_{xx}(\tau) = \frac{1}{0.19} (0.9)^{-\tau} \qquad \text{for } T < 0$$

Figure 5.2.14b shows the picture for $\tau > 0$. For this situation the autocorrelation sum is given by

$$r_{xx}(\tau) = \sum_{n=-\infty}^{\tau-1} 0 + \sum_{n=\tau}^{\infty} (0.9)^{(n-\tau)}(0.9)^n = (0.9)^{-\tau} \sum_{n=\tau}^{\infty} (0.81)^n$$

If the lower limit on the sum was 0, we could use Eq. 5.2.5 to find its value. Since the lower limit is τ, we can write

$$\sum_{n=\tau}^{\infty} a^n = \sum_{n=0}^{\infty} a^n - \sum_{n=0}^{\tau-1} a^n$$

Therefore we need a closed form expression for this last term. To find this, write

$$S = \sum_{n=0}^{\tau-1} a^n = 1 + a + a^2 + \cdots + a^{\tau-1}$$

$$aS = a \sum_{n=0}^{\tau-1} a^n = a + a^2 + a^3 + \cdots + a^{\tau}$$

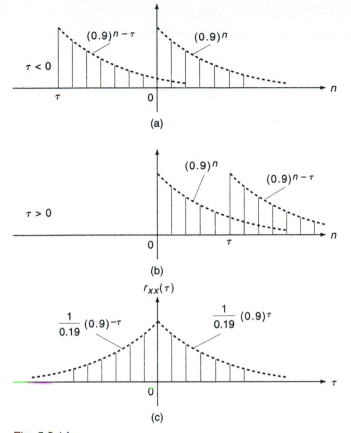

Fig. 5.2.14

Now subtract the bottom expression from the top one to get

$$S - aS = 1 - a^\tau$$

or

$$S = \sum_{n=0}^{\tau-1} a^n = \frac{1 - a^\tau}{1 - a}, \qquad |a| \neq 1 \qquad (5.2.6)$$

For our sum, $a = 0.81$. Note that $(0.81)^\tau = (0.9)^{2\tau}$. From all this we can write

$$r_{xx}(\tau) = \frac{1}{0.19}(0.9)^\tau \qquad \text{for } \tau \geq 0$$

This result is plotted in Fig. 5.2.14c.

Type 4. Discrete-Time Power Signals

The correlation $r_{xy}(\tau)$ between two discrete-time power signals $x(n)$ and $y(n)$ is given by

$$r_{xy}(\tau) = \lim_{N \to \infty} \frac{1}{2N+1} \sum_{n=-N}^{N} x(n)\, y(n-\tau) \qquad (5.2.7a)$$

$$= \lim_{N \to \infty} \frac{1}{2N+1} \sum_{n=-N}^{N} x(n+\tau)\, y(n) \qquad (5.2.7b)$$

If $x(n)$ and $y(n)$ are periodic with the same period N, we may replace Eq. 5.2.7 by the simpler form given by

$$r_{xy}(\tau) = \frac{1}{N} \sum_{n=0}^{N-1} x(n)\, y(n-\tau) \qquad (5.2.8a)$$

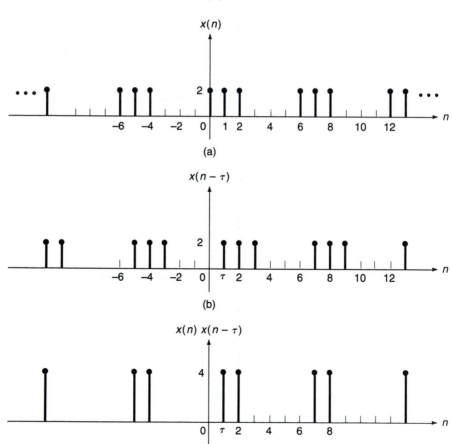

Fig. 5.2.15

$$= \frac{1}{N} \sum_{n=0}^{N-1} x(n + \tau)\, y(n) \qquad (5.2.8b)$$

EXAMPLE 5.2.6.   Find the autocorrelation function for the periodic signal from Example 5.1.5, shown in Fig. 5.2.15a.

SOLUTION: We multiply the shifted function $x(n - \tau)$ in Fig. 5.2.15b by $x(n)$ to produce the product in Fig. 5.2.15c. In this picture, $\tau = 1$, but the general idea is to shift $x(n)$ by any amount τ and multiply to produce the product. For each value of τ the average in Eq. 5.2.8 gives the correlation function plotted in Fig. 5.2.16.

As with continuous-time signals, periodic signals produce periodic auto-correlation functions. Also notice that $r_{xx}(0)$ is the mean square value of the signal. This is a general property of autocorrelation functions. The value at $\tau = 0$ is either the power or the energy in the signal, depending on whether we have a power or an energy signal.

Random Signals

The time correlation function that we have defined for each of the four types of signals applies to both random and deterministic time functions. Therefore we could use these formulas for the coin-toss function in Section 1.2. The trouble is that these formulas are accurate only for $T \to \infty$. The correlation functions in Section 1.2 for the coin-tossing experiment were calculated using the statistical relation between samples, rather than the formulas for time correlation functions that we have presented in this section. Let us now calculate the time correlation functions for the signal $x_0(n)$ for comparison. See Section 1.2 for a description of this signal. Figure 1.2.2 shows a plot of the first 100 samples of $x_0(n)$.

The idea expressed in Eq. 5.2.7 for calculating the autocorrelation function is to multiply $x(n + \tau)$ by $x(n)$, sum over all terms, and divide by the number

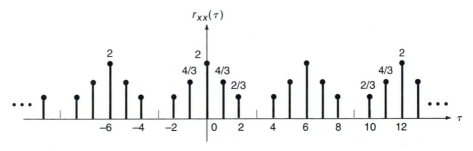

Fig. 5.2.16

of terms. Therefore the appropriate formulas are

$$r(0) = \frac{x(1) \cdot x(1) + x(2) \cdot x(2) + \cdots + x(N) \cdot x(N)}{N}$$

$$r(1) = \frac{x(1) \cdot x(2) + x(2) \cdot x(3) + \cdots + x(N-1) \cdot x(N)}{N-1} \qquad (5.2.9)$$

$$\vdots$$

$$r(\tau) = \frac{x(1) \cdot x(1+\tau) + x(2) \cdot x(2+\tau) + \cdots + x(N-\tau) \cdot x(N)}{N-\tau}$$

Applying these formulas to $x_0(n)$ with $N = 100$ gives the correlation function shown in Fig. 5.2.17. The plot has a maximum value of 39.88 and a minimum value of about 36 over the range $-20 < \tau < 20$.

   The point of all this is that you cannot find the true autocorrelation function for a random signal from one sample of that signal. A better estimate for the autocorrelation function is to use several segments of the signal and average over the several correlation functions obtained from each. An even better way is to use statistical methods if possible.

   Despite the fact that one signal segment generally will not produce an accurate estimate of the autocorrelation function, such an estimate is better than nothing. Instruments are commercially available for calculating the autocorrelation function of a signal (and the cross correlation between two signals). They operate on the idea expressed in Eq. 5.2.9. The diagram in Fig. 5.2.18 shows a system that first forms the product $x(n) \cdot x(n - \tau)$, then clocks this product into the moving average filter of length N. After clocking N such products into the filter, the output $y(n)$ is $N \cdot r(\tau)$ in Eq. 5.2.9.

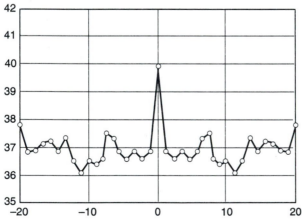

Fig. 5.2.17.   Time autocorrelation function for $x_0(n)$.

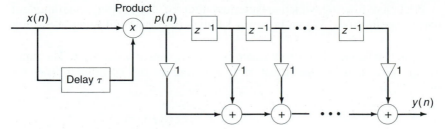

Fig. 5.2.18.   Electronic calculation of $r(\tau)$.

We then repeat this procedure for each τ, thus producing values of the autocorrelation function.

An obvious problem with Fig. 5.2.18 is the large number of delay blocks necessary for any reasonable estimate of $r(\tau)$. In Eq. 5.2.9 we used 100 such delay blocks. But we can solve this problem by converting the moving-average filter into a recursive realization. Consider the filter itself, with input $p(n)$ and output $y(n)$ in Fig. 5.2.18. If

$$y(n) = \sum_{k=n-N}^{n} p(k)$$

then

$$
\begin{aligned}
y(n + 1) &= \sum_{k=n-N+1}^{n+1} p(k) \\
&= \sum_{k=n-N}^{n} p(k) + p(n + 1) - p(n - N + 1) \\
&= y(n) + p(n + 1) - p(n - N + 1)
\end{aligned}
\tag{5.2.10}
$$

which is realized by the first-order system in Fig. 5.2.19. This simple system now replaces the N-delay moving-average filter in Fig. 5.2.18. The chief expense of these measuring systems is in the memory, used to store a long segment of the signal, and in the variable delay and product blocks in the system.

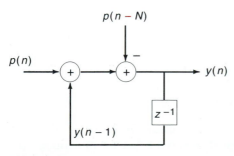

Fig. 5.2.19

Recursive filtering, which is patterned after Eq. 5.2.10, plays an important part in applications. We will discuss Kalman (recursive) filtering in Section 8.6.

Here is a listing of procedure corr, which implements Eq. 5.2.9. The input to this procedure is the signal x, and it returns the correlation $r(\tau)$ for $0 \leq \tau < M$.

```
procedure corr(x)   /* find autocorrelation of x */
define x(N), r(M)
        τ = 0
        while τ < M
        begin
                n = 1
                sum = 0
                while n ≤ (N−τ)
                begin
                        sum = sum + x(n)*x(n+τ)
                        n = n + 1
                end for n
                r(τ) = sum/(N−τ)
                τ = τ + 1
        end for τ
        return(r)
end corr
```

EXAMPLE 5.2.7.   In Section 1.3 we promised to find the correlation between successive numbers from the random number generator there. The autocorrelation function for $\tau = 1$ gives this value. The plot in Fig. 5.2.20 shows the normalized time autocorrelation function averaged over 500 values of this number generator, which was found by the procedure above. At $\tau = 1$ the value of the function is 0.046, a relatively small number compared with 1. If we could trust the results from only one sample of length 500 from the random number generator, we could say that this is a reasonable generator.

Noisy Periodic Sequences

Suppose a signal is periodic with added noise,

$$x(n) = s(n) + w(n) \tag{5.2.11}$$

where $s(n)$ is the signal, $w(n)$ is the noise, and $x(n)$ is the received signal. Because the autocorrelation function of a periodic signal is periodic, the autocorrelation function of $x(n)$ reveals periodic tendencies in the signal. To show why, consider the autocorrelation function of $x(n)$.

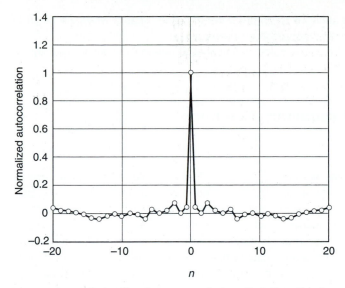

Fig. 5.2.20. Normalized autocorrelation function for the random number generator.

$$r_{xx}(\tau) = \lim_{N \to \infty} \frac{1}{2N+1} \sum_{n=-N}^{N} x(n)\, x(n-\tau) \qquad (5.2.12)$$

Substitute Eq. 5.2.11 into this expression to get

$$r_{xx}(\tau) = \lim_{n \to \infty} \frac{1}{2N+1} \sum_{n=-N}^{N} [s(n) + w(n)][s(n-\tau) + w(n-\tau)]$$

$$= r_{ss}(\tau) + r_{sw}(\tau) + r_{ws}(\tau) + r_{ww}(\tau) \qquad (5.2.13)$$

The first factor is the autocorrelation function of $s(n)$. If $s(n)$ is periodic, this function is periodic. The second and third terms represent cross correlations between signal and noise. If there is no relation between signal and noise, these are 0 for all τ. The last term represents the noise correlation, and this is often 0 for all $\tau \neq 0$. Therefore, under these conditions we can detect periodic components in a received signal by finding its autocorrelation function.

Review

Basically the same operation calculates correlation functions for each class of signal. Integration applies to continuous-time signals, and summation applies to discrete-time signals. For energy signals we integrate or sum over all time, from $-\infty$ to $+\infty$. For power signals we integrate or sum over one period, and then divide by the period length. This works well for periodic signals, but has little meaning for aperiodic signals. This business of letting the period approach ∞

makes good sense until we try to put it into practice. That is when we begin to appreciate the statistical approach for power signals.

You should now be able to calculate the time correlation function for two signals from the same class.

5.3   Time–Frequency Relations _____

Preview

Some people become set in their ways as they grow older, and some become eccentric. Jean Baptiste Joseph Fourier (1768–1830) became eccentric. He pursued a military career in his youth, spending time in Egypt with Napoleon Bonaparte, and also pursued his mathematical interests. He was too good a mathematician to stay in military service, but his stay in Egypt exposed him to the benefits of desert heat. Believing heat essential to health, Fourier spent his later years with his residence overheated and himself swathed in layers of clothes. Perhaps he was right: Isaac Asimov said that Fourier died of a fall down the stairs. (The mathematical historian E.T. Bell said he died of heart disease.)

Fourier's theorem, announced in 1807, states that a periodic function can be expressed as the sum of sinusoidal components. This we now call the Fourier series for continuous-time periodic power signals, and is one of four formulas we call the Fourier transform. There is a Fourier transform for each of the four types of signals, but before defining these four transforms, let us discuss transforms in general.

A transform is a special type of function. In the beginning, the concept of a function applied to numbers. The domain and codomain consisted of numbers. Now the term function applies to the relationship between any two sets if the two properties for a function are satisfied. These are: (1) for each element in the first set (the domain), there corresponds *at least* one element in the second set (the codomain); and (2) for each element in the domain, there corresponds *at most* one element in the codomain. We combine these into one statement by saying that a function is a relationship between two sets such that for each element in the first set there corresponds *exactly* one element in the second set. Here is an example of a function where neither the domain nor the codomain have anything to do with numbers.

Suppose I have a basket full of cards, and on each card is written an instruction. You reach into the basket, select a card, and carry out the instruction written on it. This is a function. The domain is the basket of cards with their instructions. The codomain is the set of all possible actions you could perform. The set of actions that might actually be performed (those on the cards) is called the range, and this is a subset of the codomain.

This distinction between the codomain and range occurred recently in mathematics. We previously called both sets the range, which led to some confusion. In modern parlance the term *range* refers to those elements in the codomain that

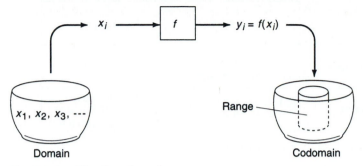

Fig. 5.3.1. The function f.

have an ancestor in the domain. Figure 5.3.1 shows the domain and codomain as baskets. All values of y fall into the subset of the codomain called the range after being ejected from the box labeled f. The function $y = x^2$ provides a specific example. If the domain and codomain are the set of all real numbers, then the range is the set of all nonnegative real numbers. So in this example the range is a proper subset of the codomain.

Now we turn to the idea of a transform. The functions of interest to us here also have no numbers in either the domain or codomain. These special functions go by either of two names, transform or operator. They are functions whose domain and range are themselves sets of functions. If Fig. 5.3.1 represents a transform, then all baskets contain functions instead of numbers. For the forward transform, the domain contains time functions and the codomain contains frequency functions. For the inverse transform, the domain contains frequency functions and the codomain contains time functions.

Recall that we can classify both continuous-time and discrete-time signals as either power or energy signals. This gives us four different categories:

1. Continuous-time energy signals
2. Continuous-time power signals
3. Discrete-time energy signals
4. Discrete-time power signals

These categories were introduced in Section 5.1 so we could discuss the four types of correlation functions in Section 5.2. Now we have the analogous four types of Fourier transforms. There is one Fourier transform for each category. Figure 5.3.2 shows these categories in a format so you can identify which transform goes with which conditions. We now discuss each category in turn, starting with the Fourier transform of continuous-time signals.

Continuous-Time Energy Signals

The forward transform for a continuous-time energy signal $v(t)$ is given by

$$V(\Omega) = \int_{-\infty}^{\infty} v(t)e^{-j\Omega t}\, dt \tag{5.3.1}$$

	Energy Signals	Power Signals
Continuous-time	Fourier transform of continuous-time signals	Fourier series
Discrete-time	Fourier transform of discrete-time signals	Discrete Fourier transform (DFT) and fast Fourier transform (FFT)

Fig. 5.3.2.   The four types of Fourier transforms.

The inverse transform is

$$v(t) = \frac{1}{2\pi} \int_{-\infty}^{\infty} V(\Omega) e^{j\Omega t} \, d\Omega \tag{5.3.2a}$$

$$= \int_{-\infty}^{\infty} V(F) e^{j2\pi F t} \, dF \tag{5.3.2b}$$

We obtain Eq. 5.3.2b from 5.3.2a by a change of variable, $\Omega = 2\pi F$. If for a particular $v(t)$ we have $|V(\Omega)|^2 < \infty$ for all Ω, then we say that the Fourier transform exists. What this really means is that we can evaluate the formula in Eq. 5.3.1 and get a meaningful function $V(\Omega)$. Not all $v(t)$ have a Fourier transform, but all energy signals do.

Parseval's relation states that the energy may be measured in the time or in the frequency domain, and either measurement will give the same result. This relation is given by

$$E = \int_{-\infty}^{\infty} |v(t)|^2 \, dt = \frac{1}{2\pi} \int_{-\infty}^{\infty} |V(\Omega)|^2 \, d\Omega \tag{5.3.3a}$$

$$= \int_{-\infty}^{\infty} |V(F)|^2 \, dF \tag{5.3.3b}$$

The left side measures the energy in the time function, and the right side measures the same energy in the frequency domain. As a rule of thumb, we divide by 2π when the variable of integration is Ω, but not when the variable of integration is F. The energy spectrum, which describes how the energy is distributed in the frequency domain, is given by

$$G(\Omega) = |V(\Omega)|^2 = |V(F)|^2 \tag{5.3.4}$$

It is also true that $|v(t)|^2$ describes how the energy is distributed in the time domain, so we could call this quantity the energy density for the time domain. But there is little use for the energy density in the time domain, so no one ever bothers to make this definition.

Equations 5.3.1 and 5.3.4 give us two ways to represent a continuous-time energy signal as a function of frequency.

1. Fourier transform: A complex-valued function from which we can reconstruct the time function $v(t)$ for all time.

2. Energy spectrum: A real-valued function that represents the energy density at each frequency. This representation has less information than the Fourier transform, so we cannot reconstruct the unique time function $v(t)$ from the energy spectrum.

EXAMPLE 5.3.1.   Find and plot the Fourier transform and the energy spectrum for the decaying exponential signal

$$v(t) = \begin{cases} e^{-at}, & t > 0 \\ 0 & \text{otherwise} \end{cases}$$

SOLUTION: Equation 5.3.1 gives

$$V(\Omega) = \int_0^\infty e^{-at} e^{-j\Omega t}\, dt = \int_0^\infty e^{-(a+j\Omega)t}\, dt = \frac{1}{a + j\Omega}$$

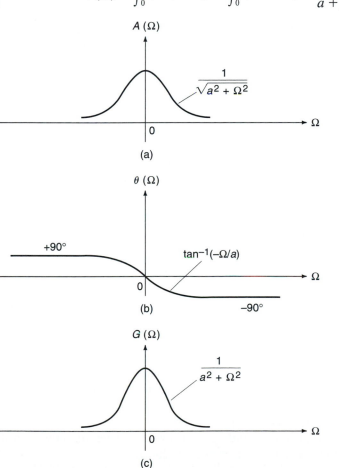

(a)

(b)

(c)

Fig. 5.3.3.   The Fourier transform and energy spectrum of $e^{-at}u(t)$.

Since $V(\Omega)$ is a complex-valued function of Ω, we need three dimensions for the plot, one dimension for the Ω axis, one for the real axis, and one for the imaginary axis. Since this sheet of paper has only two useful dimensions, we will have to improvise and then use our imagination for the missing parts of the plot. Figures 5.3.3a and 5.3.3b show the amplitude and phase plots, which represent $V(\Omega)$ in polar coordinates. That is, for each Ω we represent $V(\Omega)$ as

$$V(\Omega) = A(\Omega)e^{j\theta(\Omega)}$$

where $A(\Omega)$ is the amplitude (both positive and negative magnitude) and $\theta(\Omega)$ is the phase angle. This representation is incomplete because we must use our imagination to put the two parts together to see what it looks like in three dimensions.

The energy spectrum is given by

$$G(\Omega) = |V(\Omega)|^2 = V(\Omega) \cdot V(-\Omega) = \frac{1}{a^2 + \Omega^2}$$

Figure 5.3.3c plots this function.

Continuous-Time Power Signals

We can no longer use the Fourier transform of Eq. 5.3.1 if the signal $v(t)$ is a power signal instead of an energy signal. A problem occurs if a power signal is plugged into the transform, because that makes the integral infinite. That is, for a particular value of Ω, the value of $V(\Omega)$ will be infinite. To avoid this difficulty, we will integrate only over a finite time interval of length T.

The Fourier series for continuous-time signals has the forward transform given by

$$V(k) = \frac{1}{T}\int_0^T v(t)e^{-jk\Omega_1 t}\,dt \tag{5.3.5}$$

with the inverse transform given by

$$v(t) = \sum_{k=-\infty}^{\infty} V(k)e^{jk\Omega_1 t} \tag{5.3.6}$$

where $\Omega_1 = 2\pi/T$. The interval T is often identified with the period of a periodic signal, but these formulas work even if $v(t)$ is aperiodic (not periodic). [Of course, if the signal is not periodic then the resulting $V(k)$ represents the signal only in the interval T.]

Parseval's relation for periodic power signals relates the power in the time domain to the power in the frequency domain,

$$P = \frac{1}{T}\int_0^T |v(t)|^2\,dt = \sum_{k=-\infty}^{\infty} |V(k)|^2 \tag{5.3.7}$$

and the power spectrum is given by $G(k) = |V(k)|^2$. This gives us two ways to represent a power signal as a function of frequency.

1. Fourier series: Complex numbers from which we can reconstruct the time function $v(t)$ over the interval of length T.
2. Power spectrum: Real numbers that represent the power at each frequency. This representation has less information than the Fourier series, and we cannot reconstruct the unique time function $v(t)$ from the power spectrum.

EXAMPLE 5.3.2.   Given the periodic square wave in Fig. 5.3.4, find and plot the Fourier series and the power spectrum.

SOLUTION: Plugging values into Eq. 5.3.5 gives

$$V(k) = \frac{1}{T} \int_0^{T/2} E e^{-jk\Omega_1 t}\, dt$$

where $T = 0.01$ s and $\Omega_1 = 2\pi/T = 200\pi$. Integrating gives

$$V(k) = \frac{E}{-jk\Omega_1 T} e^{-jk\Omega_1 t} \bigg|_0^{T/2} = \frac{E}{j2\pi k}(1 - e^{-jk\pi})$$

This is a perfectly good answer for the Fourier series, but multiplying by $e^{jk\pi/2} \cdot e^{-jk\pi/2} = 1$ followed by some algebra puts it in a better form for plotting:

$$V(k) = \frac{E}{j2\pi k}(1 - e^{-jk\pi}) e^{jk\pi/2} \cdot e^{-jk\pi/2}$$

$$= \frac{E}{k\pi}\left(\frac{e^{jk\pi/2} - e^{-jk\pi/2}}{2j}\right) e^{-jk\pi/2}$$

$$= \frac{E}{2}\frac{\sin(k\pi/2)}{k\pi/2} e^{-jk\pi/2}$$

This allows us to easily plot the amplitude and phase in Figs. 5.3.5a and 5.3.5b. Figure 5.3.5c shows the power spectrum, which is simply the amplitude spectrum squared.

Since the period T is 10 ms, the fundamental frequency is $\Omega_1 = 2\pi/T = 200\pi$. Therefore $k = 1$ in Fig. 5.3.5 is the same as $\Omega = 200\pi$

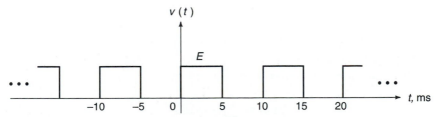

Fig. 5.3.4.   A periodic square wave.

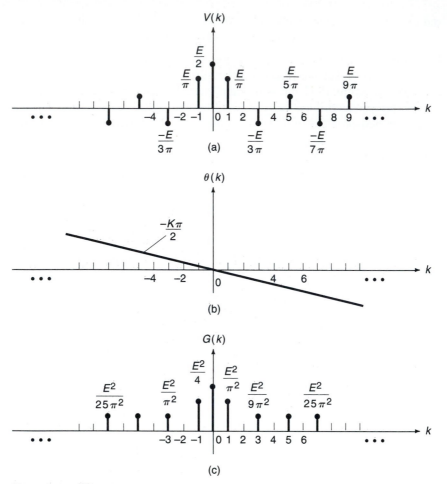

Fig. 5.3.5.   The square wave spectra.

rad/s. We could plot the spectrum versus frequency, and these plots would be identical to Fig. 5.3.5 with discrete frequencies $200k\pi$. To illustrate, the power spectrum is plotted in Fig. 5.3.6 versus frequency Ω.

Discrete-Time Energy Signals

In switching from continuous-time to discrete-time, the time variable changes from t to n, which means we can no longer integrate the time function. We must use sums instead of integrals in dealing with the time function. The forward transform is therefore a sum, given by

$$V(\omega) = \sum_{n=-\infty}^{\infty} v(n)e^{-j\omega n} \tag{5.3.8}$$

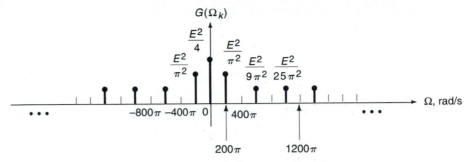

Fig. 5.3.6.   The square wave spectrum plotted versus frequency.

The variable ω is a continuous-valued parameter, so the inverse transform is given by an integral,

$$v(n) = \frac{1}{2\pi} \int_0^{2\pi} V(\omega)e^{j\omega n}\, d\omega \qquad (5.3.9)$$

The infinite sum in Eq. 5.3.8 converges if the time function has finite energy. This means that every discrete-time energy signal has a Fourier transform.

Parseval's relation for discrete-time energy signals says that the energy in the time domain is equal to the energy in the frequency domain,

$$E = \sum_{n=-\infty}^{\infty} |v(n)|^2 = \frac{1}{2\pi} \int_{-\pi}^{\pi} |V(\omega)|^2\, d\omega \qquad (5.3.10)$$

and the energy density spectrum $G(w)$ is given by

$$G(\omega) = |V(\omega)|^2 \qquad (5.3.11)$$

Equations 5.3.8 and 5.3.11 give us two ways to represent a discrete-time energy signal as a function of frequency.

1. Fourier transform: A complex-valued function from which we can reconstruct the time function $v(n)$ for all time.
2. Energy spectrum: A real-valued function that represents the energy density at each frequency. This representation has less information than the Fourier transform, so we cannot reconstruct the unique time function $v(n)$ from the energy spectrum.

EXAMPLE 5.3.3.   Find and plot the Fourier transform and the energy density spectrum for the exponential signal $a^n u(n)$, $0 < a < 1$, shown in Fig. 5.3.7.

SOLUTION: Equation 5.3.8 gives

$$V(\omega) = \sum_{n=0}^{\infty} a^n e^{-j\omega n} = \frac{1}{1 - ae^{-j\omega}}$$

and Eq. 5.3.11 gives the energy spectrum,

$$G(\omega) = |V(\omega)|^2 = \frac{1}{1 - ae^{-j\omega}} \cdot \frac{1}{1 - ae^{j\omega}}$$

$$= \frac{1}{1 - 2a\cos(\omega) + a^2}$$

Figure 5.3.8 shows $V(\omega)$ and $G(\omega)$. Notice that the spectra are periodic with period 2π.

There is a fundamental difference between the frequency variables Ω and ω. The continuous-time frequency Ω has units of radians/second. The discrete-time frequency ω has units of radians/sample. These two need not be related to each other, though they often are. When a continuous-time signal is sampled to produce a discrete-time signal, the frequencies are related by the sampling rate F_s as follows:

$$\Omega(\text{rad/s}) = \omega(\text{rad/sample}) \cdot F_s(\text{samples/s}) \tag{5.3.12}$$

EXAMPLE 5.3.4.   An audio signal is filtered by an analog low-pass filter with 3-dB cutoff frequency of 3 kHz. Suppose we wish to replace this filter by a digital filter $H(z)$ in an A/D-$H(z)$-D/A configuration. If the sampling rate is 10 kHz, find the digital filter cutoff frequency.

SOLUTION: F_s is 10,000 samples/s. Therefore the digital frequency ω_1 we seek is related to the analog cutoff frequency $\Omega_1 = 3(10)^3(2\pi)$ rad/s by

$$\omega = \frac{\Omega}{F_s} = \frac{6\pi(10)^3}{10^4} = 0.6\pi \text{ rad/sample}$$

Since analog frequency has range $-\infty < \Omega < \infty$, and digital frequency ω has range $-\pi < \omega < \pi$, we can obtain correspondence only at this one frequency, 0.6π rad/sample.

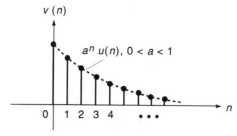

Fig. 5.3.7

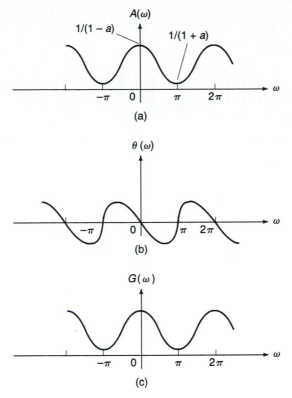

Fig. 5.3.8

Discrete-Time Power Signals

The last of the four forms that we charted in Fig. 5.3.2 applies to discrete-time power signals. The forward transform is given by

$$V(k) = \frac{1}{N} \sum_{n=0}^{N-1} v(n) e^{-j2\pi kn/N} \tag{5.3.13}$$

and the inverse transform is given by

$$v(n) = \sum_{k=0}^{N-1} V(k) e^{j2\pi nk/N} \tag{5.3.14}$$

Parseval's relation for discrete-time power signals says that the power in the time domain is also the power in the frequency domain,

$$P = \frac{1}{N} \sum_{n=0}^{N-1} |v(n)|^2 = \sum_{k=0}^{N-1} |V(k)|^2 \tag{5.3.15}$$

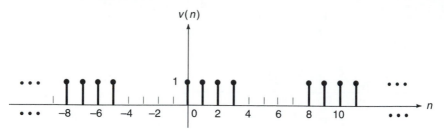

Fig. 5.3.9.   The periodic function $v(n)$.

The power density spectrum is

$$G(k) = |V(k)|^2 \qquad (5.3.16)$$

This gives us two ways to represent a power signal as a function of frequency.

1. Fourier series: Complex numbers from which we can reconstruct the time function $v(n)$ over the interval of length N.
2. Power spectrum: Real numbers that represent the power at each frequency. This representation has less information than the Fourier series, and we cannot reconstruct the unique time function $v(n)$ from the power spectrum.

EXAMPLE 5.3.5.   Find and plot the Fourier coefficients and the power density spectrum for the periodic square wave of period 8 in Fig. 5.3.9.

SOLUTION: From Eq. 5.3.13,

$$
\begin{aligned}
V(k) &= \frac{1}{8} \sum_{n=0}^{7} v(n) e^{-j2\pi kn/8} \\
&= \frac{1}{8} (1 + e^{-jk\pi/4} + e^{-jk\pi/2} + e^{-jk3\pi/4})
\end{aligned}
\qquad (5.3.17)
$$

For each k this formula gives a complex number $V(k)$, shown in Fig. 5.3.10a. Figure 5.3.10b shows the power density spectrum, which is found by squaring each amplitude term in $V(k)$. Although we did not show it, both $V(k)$ and $G(k)$ are periodic with period 8.

Equations 5.3.13 and 5.3.14 are known in the engineering community as the discrete Fourier transform. In other disciplines our third transform for discrete-time energy signals is known as the discrete Fourier transform. This can be confusing unless you restrict your references to the engineering literature, so beware that other disciplines, especially physics, do not use the same notation.

We related frequency Ω to frequency number k for continuous-time frequency in Example 5.3.2. Then Eq. 5.3.12 related discrete-time frequency ω to continuous-time frequency Ω by the sampling rate. Now we have

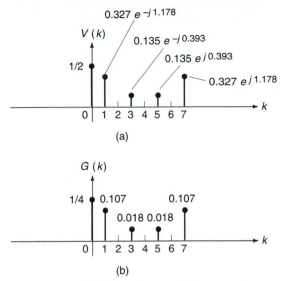

Fig. 5.3.10. The Fourier coefficients and power spectrum for $v(n)$.

another frequency variable k that is related to ω in the following way. For discrete-time frequency, k is the number of cycles per sample of a rotating phasor. The exponential form $e^{-j\omega n}$ describes a vector in the complex plane with unit length at the angle $-\omega n$. As either ω or n changes, the angle changes.

Compare the exponential terms in the two formulas for the discrete-time transforms, Eqs. 5.3.8 and 5.3.13:

$$e^{-j\omega n} \leftrightarrow e^{-j2\pi nk/N}$$

This gives the correspondence $\omega = 2\pi k/N$, leading to the interpretation that k is the number of rotations of the phasor in N samples. Thus k has units of cycles per N samples. When $k = 1$ there are 2π radians per N samples. When $k = 2$ there are 4π radians per N samples, etc. We further illustrate these ideas by reexamining the previous example.

Equation 5.3.17 expresses the coefficients $V(k)$ for each k. If we rewrite this expression for $k = 1$, we get

$$V(1) = \frac{1}{8}(1 + e^{-j\pi/4} + e^{-j\pi/2} + e^{-j3\pi/4})$$

Each term in this equation is a complex number in polar form. These numbers are plotted in the Argand diagram of Fig. 5.3.11a. The sum of these four numbers is $2.613e^{-j3\pi/8}$. If we divide by 8 (as required by the formula for the forward transform) the result is $V(1)$, as shown in Fig. 5.3.11b.

The complex plane is called the Argand diagram in honor of the second person to publish a paper outlining the relation between geometry and com-

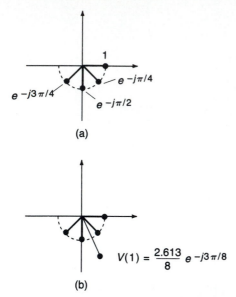

(a)

(b)

Fig. 5.3.11.   The Fourier coefficient $V(1)$.

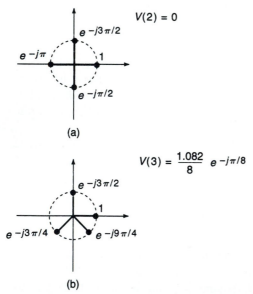

(a)

(b)

Fig. 5.3.12.   The coefficients $V(2)$ and $V(3)$.

plex numbers. The first was Casper Wessell, a Norwegian, who published in 1799. His work received no attention, and the geometric interpretation was rediscovered by Jean Robert Argand, a Parisian bookkeeper, who published his paper in 1806.

Notice that for $k = 1$ the phasors complete one revolution in N samples. In Fig. 5.3.11, if the other four terms were not 0, they would complete the circle with the last ($n = 7$) term at an angle of $\pi/4$. Thus for $k = 1$ we complete one circle in N samples.

When $k = 2$ the phasor completes two cycles in 8 samples, or one cycle in 4 samples. The position of the first four samples is shown in Fig. 5.3.12a; the last four samples are 0. The vector sum of these samples is 0, or $V(2) = 0$.

Figure 5.3.12b shows $V(3)$. Since the phasor completes three revolutions in 8 samples, adjacent samples are $3\pi/4$ radians apart. The vector sum of these samples is $1.082e^{-j\pi/8}$, so $V(3)$ is this number divided by 8.

Review

We need all four forms of the Fourier transform, because no one form applies to every signal. Integration applies to continuous-time signals, but not to discrete-time signals, so two forms use integrals and the other two use sums. If we try to integrate (or sum) a power signal for all time, we get infinity for the answer. If we try to apply the power signal form of the transform to energy signals, we get an answer that depends on the period chosen, not the true transform. So we need all four forms.

Parseval's relations equate power (or energy) in the time domain to the power (or energy) in the frequency domain. One of our chief concerns is power (or energy) in random signals, so Parseval's relations are important to us.

A signal can be continuous in time and frequency, discrete in one but not the other, or discrete in both. It can be periodic or aperiodic, and it can be a power or an energy signal. The following tables summarize the salient features of the four forms of the transform in both the time and frequency domains.

1. Continuous-time energy signals:

Time Domain	Frequency Domain
$V(\Omega) = \displaystyle\int_{-\infty}^{\infty} v(t)e^{-j\Omega t}\, dt$	$v(t) = \dfrac{1}{2\pi}\displaystyle\int_{-\infty}^{\infty} V(\Omega)e^{j\Omega t}\, d\Omega$
$E = \displaystyle\int_{-\infty}^{\infty} \|v(t)\|^2\, dt$	$E = \dfrac{1}{2\pi}\displaystyle\int_{-\infty}^{\infty} \|V(\Omega)\|^2\, d\Omega$
$v(t)$ is aperiodic	$V(\Omega)$ is aperiodic

2. Continuous-time periodic power signals:

Time Domain	Frequency Domain
$V(k) = \frac{1}{T}\int_0^T v(t)e^{-j\Omega_1 t}\, dt$	$v(t) = \sum_{k=-\infty}^{\infty} V(k)e^{jk\Omega_1 t}$
$P = \frac{1}{T}\int_0^T |v(t)|^2\, dt$	$P = \sum_{k=-\infty}^{\infty} |V(k)|^2$
$v(t)$ is periodic	$V(k)$ is aperiodic

3. Discrete-time energy signals:

Time Domain	Frequency Domain
$V(\omega) = \sum_{n=-\infty}^{\infty} v(n)e^{-j\omega n}$	$v(n) = \frac{1}{2\pi}\int_{-\pi}^{\pi} V(\omega)e^{j\omega n}\, d\omega$
$E = \sum_{n=-\infty}^{\infty} |v(n)|^2$	$E = \frac{1}{2\pi}\int_{-\pi}^{\pi} |V(\omega)|^2\, d\omega$
$v(n)$ is aperiodic	$V(\omega)$ is periodic

4. Discrete-time periodic power signals:

Time Domain	Frequency Domain
$V(k) = \frac{1}{N}\sum_{n=0}^{N-1} v(n)e^{-j2\pi nk/N}$	$v(n) = \sum_{k=0}^{N-1} V(k)e^{j2\pi nk/N}$
$P = \frac{1}{N}\sum_{n=0}^{N-1} |v(n)|^2$	$P = \sum_{k=0}^{N-1} |V(k)|^2$
$v(n)$ is periodic	$V(k)$ is periodic

5.4   The Spectrum

Preview

The term *spectrum* refers to the distribution of the signal as a function of frequency. We use it in two ways: as a voltage spectrum and as an energy (or power) spectrum.

The voltage spectrum is a complex-valued function of frequency derived by taking the Fourier transform of the time function. The energy or power spectrum is real-valued, and can be derived by one of the two ways we describe in this section. For an energy signal, the energy spectral density function describes the distribution of energy versus frequency. Likewise, for a power signal, the power spectral density function describes the distribution of power versus frequency. Our interest in this section will be in the real-valued energy and power spectral density functions. These are the quantities we can realistically use when dealing with random signals.

For deterministic signals there are two ways to calculate the power or energy spectrum: (1) Find the Fourier transform, find the magnitude squared, and this gives the spectrum; or (2) find the autocorrelation function, take the transform, and this gives the spectrum. You should get the same answer either way. For random signals we can use the second method if we can either calculate or measure the autocorrelation function. The first method does not apply to random signals, since it requires us to find the Fourier transform of a random signal. (There is nothing wrong with finding the transform of a random signal, it just has little value.) What we really need for random signals is the autocorrelation function so we can apply method 2.

These methods apply to each of the four types of signals:

Continuous-time energy signals
Continuous-time power signals
Discrete-time energy signals
Discrete-time power signals

We will discuss each in turn below for deterministic signals. The following material depends heavily on Sections 5.2 and 5.3.

Continuous-Time Energy Signals

METHOD 1. Find the Fourier transform and square the magnitude to get the energy spectral density function.

$$V(\Omega) = \int_{-\infty}^{\infty} v(t)e^{-j\Omega t}\, dt \qquad (5.4.1)$$

$$G_{vv}(\Omega) = |V(\Omega)|^2 \qquad (5.4.2)$$

METHOD 2. Find the autocorrelation function, and transform this to get the energy spectral density function.

$$r_{vv}(\tau) = \int_{-\infty}^{\infty} v(t)v(t-\tau)\, dt \qquad (5.4.3)$$

$$G_{vv}(\Omega) = \int_{-\infty}^{\infty} r_{vv}(\tau)e^{-j\Omega \tau}\, d\tau \qquad (5.4.4)$$

We should get the same result from either Eq. 5.4.2 or 5.4.4.

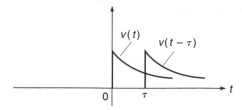

Fig. 5.4.1

EXAMPLE 5.4.1.   Use method 2 to find the energy spectral density function for the decaying exponential signal $v(t)$ given below. Compare results to Example 5.3.1, where we used method 1.

$$v(t) = \begin{cases} e^{-at}, & t > 0 \\ 0 & \text{otherwise} \end{cases}$$

SOLUTION: We first calculate the time autocorrelation function. Figure 5.4.1 pictures $v(t)$ and $v(t - \tau)$ for $\tau > 0$. For this situation, Eq. 5.4.3 gives

$$r_{vv}(\tau) = \int_{\tau}^{\infty} e^{-at} e^{-a(t-\tau)}\, dt = e^{a\tau} \int_{\tau}^{\infty} e^{-2at}\, dt$$

$$= \frac{1}{2a} e^{-a\tau} \qquad \tau > 0$$

For $\tau < 0$, as shown in Fig. 5.4.2, we have

$$r_{vv}(\tau) = \int_{0}^{\infty} e^{-at} e^{-a(t-\tau)}\, dt = e^{a\tau} \int_{0}^{\infty} e^{-2at}\, dt$$

$$= \frac{1}{2a} e^{a\tau} \qquad \tau < 0$$

Take the transform of $r_{vv}(\tau)$ to get $G_{vv}(\Omega)$.

$$G_{vv}(\Omega) = \int_{-\infty}^{\infty} r_{vv}(\tau) e^{-j\Omega\tau}\, d\tau$$

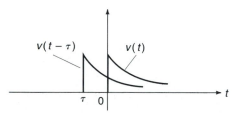

Fig. 5.4.2

$$= \int_{-\infty}^{0} \frac{1}{2a} e^{(a-j\Omega)\tau} d\tau + \int_{0}^{\infty} \frac{1}{2a} e^{-(a+j\Omega)\tau} d\tau$$

$$= \frac{1/2a}{a - j\Omega} + \frac{1/2a}{a + j\Omega}$$

$$= \frac{1}{a^2 + \Omega^2}$$

which checks with the results using method 1 in Example 5.3.1.

Continuous-Time Power Signals

METHOD 1. Find the Fourier transform and square the magnitude to get the power spectral density function.

$$V(k) = \frac{1}{T} \int_{0}^{T} v(t) e^{-jk\Omega_1 t} dt \tag{5.4.5}$$

$$G_{vv}(k) = |V(k)|^2 \tag{5.4.6}$$

METHOD 2. Find the autocorrelation function, and transform this to get the power spectral density function.

$$r_{vv}(\tau) = \frac{1}{T} \int_{0}^{T} v(t) v(t - \tau) dt \tag{5.4.7}$$

$$G_{vv}(k) = \frac{1}{T} \int_{0}^{T} r_{vv}(\tau) e^{-jk\Omega_1 \tau} d\tau \qquad \Omega_1 = \frac{2\pi}{T} \tag{5.4.8}$$

We should get the same result from either Eq. 5.4.6 or 5.4.8.

EXAMPLE 5.4.2.   Use method 2 to find the power spectral density function for the periodic signal in Example 5.3.2. This signal is shown in Fig. 5.3.4. Compare results.

SOLUTION: We calculated the time autocorrelation function for this signal in Example 5.2.3. If you will refer to Fig. 5.2.10, it will save me

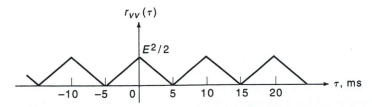

Fig. 5.4.3

a lot of work. Figure 5.4.3 shows this function with $T = 10$ ms. The autocorrelation function is given by

$$r_{vv}(\tau) = \frac{E^2}{T}\left(\tau + \frac{T}{2}\right), \qquad -\frac{T}{2} < t < 0$$

$$= \frac{E^2}{T}\left(\frac{T}{2} - \tau\right), \qquad 0 < t < \frac{T}{2}$$

Substituting this into Eq. 5.4.8 gives

$$G_{vv}(k) = \frac{1}{T}\int_0^T r_{vv}(\tau)e^{-jk\Omega_1\tau}\,d\tau$$

$$= \frac{E^2}{T^2}\left[\int_{-T/2}^0\left(\tau + \frac{T}{2}\right)e^{-jk\Omega_1\tau}\,d\tau + \int_0^{T/2}\left(\frac{T}{2} - \tau\right)e^{-jk\Omega_1\tau}\,d\tau\right]$$

After considerable algebra, this gives

$$G_{vv}(k) = \frac{E^2}{4}\left[\frac{\sin(k\pi/2)}{k\pi/2}\right]^2$$

As before (Example 5.3.2), the fundamental frequency is $\Omega_1 = 200\pi$ rad/s, giving the spectrum shown in Fig. 5.3.6.

Discrete-Time Energy Signals

METHOD 1. Find the Fourier transform and square the magnitude to get the energy spectral density function.

$$V(\omega) = \sum_{n=-\infty}^{\infty} v(n)e^{-j\omega n} \tag{5.4.9}$$

$$G_{vv}(\omega) = |V(\omega)|^2 \tag{5.4.10}$$

METHOD 2. Find the autocorrelation function, and transform this to get the energy spectral density function.

$$r_{vv}(\tau) = \sum_{n=-\infty}^{\infty} v(n)v(n - \tau) \tag{5.4.11}$$

$$G_{vv}(\omega) = \sum_{\tau=-\infty}^{\infty} r_{vv}(\tau)e^{-j\omega\tau} \tag{5.4.12}$$

Equations 5.4.10 and 5.4.12 should give the same result.

EXAMPLE 5.4.3.   Use method 2 to find the energy spectral density function for the decaying exponential signal $v(n) = a^n u(n)$ given in Fig. 5.3.7. Compare results to Example 5.3.3.

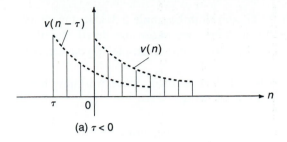

(a) $\tau < 0$

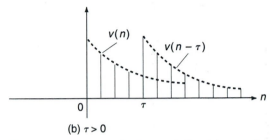

(b) $\tau > 0$

Fig. 5.4.4. $v(n)$ and $v(n - \tau)$ for (a) $\tau < 0$ and (b) $\tau > 0$.

SOLUTION: Figure 5.4.4a shows $v(n)$ and $v(n - \tau)$ for $\tau < 0$. Applying Eq. 5.4.11 gives

$$r_{vv}(T) = \sum_{n=0}^{\infty} a^n a^{(n-\tau)} = a^{-\tau} \sum_{n=0}^{\infty} a^{2n}$$

$$= \frac{a^{-\tau}}{1 - a^2}, \qquad \tau < 0$$

Figure 5.4.4b shows $v(n)$ and $v(n - \tau)$ for $\tau > 0$. Applying Eq. 5.4.11 gives

$$r_{vv}(\tau) = \sum_{n=\tau}^{\infty} a^n a^{(n-\tau)} = a^{-\tau} \sum_{n=\tau}^{\infty} a^{2n}$$

$$= \frac{a^{\tau}}{1 - a^2}, \qquad \tau \geq 0$$

We did this once before. See Example 5.2.5 for the details. Now when we apply Eq. 5.4.12 we get

$$G_{vv}(\omega) = \sum_{\tau=-\infty}^{\infty} r_{vv}(\tau) e^{-j\omega\tau}$$

$$= \frac{1}{1 - a^2} \left(\sum_{\tau=-\infty}^{-1} a^{-\tau} e^{-j\omega\tau} + \sum_{\tau=0}^{\infty} a^{\tau} e^{-j\omega\tau} \right)$$

(5.4.13)

This gives the same answer for $G_{vv}(\omega)$ as in Example 5.3.3. Just to show you that these formulas really do work, we show the details for this example below. You can skip the rest of this example if you are not interested.

The first sum is from $-\infty$ to -1, instead of 0. Since we know how to evaluate the sum from $-\infty$ to 0, we write

$$\sum_{\tau=-\infty}^{-1} a^{-(1+j\omega)\tau} = \sum_{\tau=-\infty}^{0} a^{-(1+j\omega)\tau} - 1 \qquad (\text{because } a^0 = 1)$$

and

$$\sum_{\tau=-\infty}^{0} a^{-(1+j\omega)\tau} = \sum_{\tau=0}^{\infty} a^{(1+j\omega)\tau} = \frac{1}{1 - a^{(1+j\omega)}}$$

Therefore

$$\sum_{\tau=-\infty}^{-1} a^{-(1+j\omega)\tau} = \frac{1}{1 - a^{(1+j\omega)}} - 1 = \frac{a^{(1+j\omega)}}{1 - a^{(1+j\omega)}}$$

The second sum is comparatively easy, given by

$$\sum_{\tau=0}^{\infty} a^{(1-j\omega)\tau} = \frac{1}{1 - a^{(1-j\omega)}}$$

Adding, we get

$$\frac{a \cdot a^{j\omega}}{1 - a \cdot a^{j\omega}} + \frac{1}{1 - a \cdot a^{-j\omega}} = \frac{1 - a^2}{1 - 2a \cos(\omega) + a^2}$$

Substituting this into Eq. 5.4.13 gives the final answer:

$$G_{vv}(\omega) = \frac{1}{1 - 2a \cos(\omega) + a^2}$$

Discrete-Time Power Signals

METHOD 1. Find the Fourier transform, and square the magnitude to get the power spectral density function.

$$V(k) = \frac{1}{N} \sum_{n=0}^{N-1} v(n) e^{-j2\pi nk/N} \qquad (5.4.14)$$

$$G_{vv}(k) = |V(k)|^2 \qquad (5.4.15)$$

METHOD 2. Find the autocorrelation function, and transform this to get the power spectral density function.

$$r_{vv}(\tau) = \frac{1}{N} \sum_{n=0}^{N-1} v(n) v(n - \tau) \qquad (5.4.16)$$

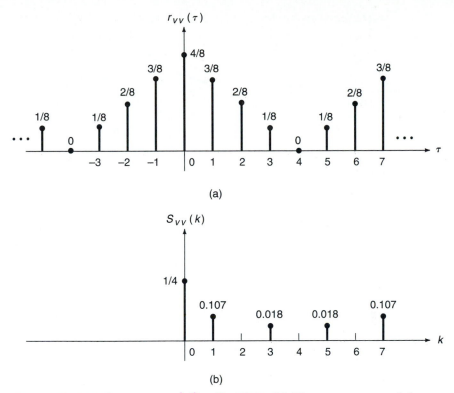

Fig. 5.4.5. (a) The autocorrelation function. (b) The power spectral density for the signal in Fig. 5.3.9.

$$G_{vv}(k) = \frac{1}{N} \sum_{n=0}^{N-1} r_{vv}(\tau)e^{-j2\pi\tau k/N} \qquad (5.4.17)$$

EXAMPLE 5.4.4. Use method 2 to find the power spectral density function for the signal in Fig. 5.3.9. Check the answer with Example 5.3.5.

SOLUTION: Figure 5.4.5a shows the autocorrelation function. We had a similar example in Section 5.2, so see Example 5.2.6 for the details. Figure 5.4.5b shows the result of applying Eq. 5.4.17, and this agrees with Example 5.3.5.

Review

The purpose of this section has been to demonstrate by example that in every case the autocorrelation function and the energy or power spectral density function form a Fourier transform pair. For deterministic signals we can calculate the spectrum by either of two ways: (1) Find the Fourier transform, find the magnitude

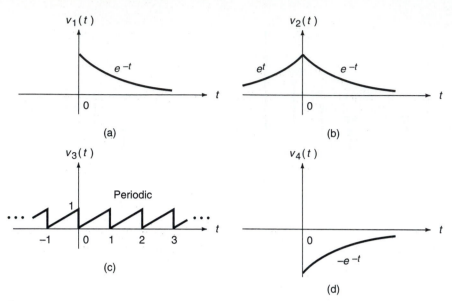

Fig. 5.5.1

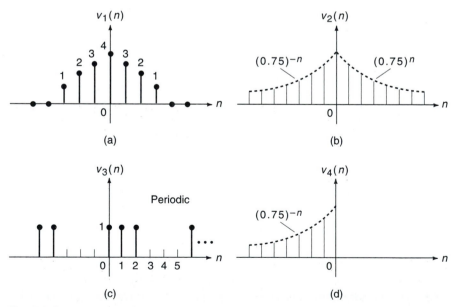

Fig. 5.5.2

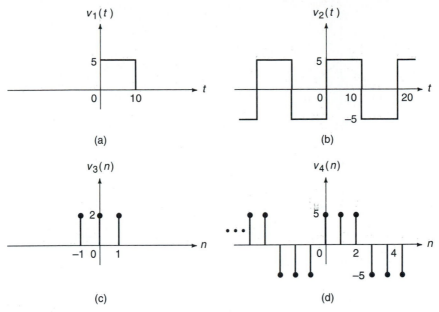

Fig. 5.5.3

squared, and this gives the spectrum; or (2) find the autocorrelation function, take the transform, and this gives the spectrum.

Either of these methods applies to deterministic signals, but only the second applies to random signals. What is more, we will see in the next chapter that this second method applies only to random signals that meet certain restrictions. (They must be wide-sense stationary.)

5.5 Problems

5.1. Find the power and energy in each signal in Fig. 5.5.1. Classify each signal as either a power or an energy signal.

5.2. Find the power and energy in each signal in Fig. 5.5.2. Classify each signal as either a power or an energy signal.

5.3. Find the power and energy in each signal in Fig. 5.5.3. Classify each signal as either a power or an energy signal.

5.4. Find the power and energy in each signal in Fig. 5.5.4. Classify each signal as either a power or an energy signal.

5.5. Find and plot the autocorrelation function for each waveform in Fig. 5.5.1.

5.6. Find and plot the autocorrelation function for each waveform in Fig. 5.5.2.

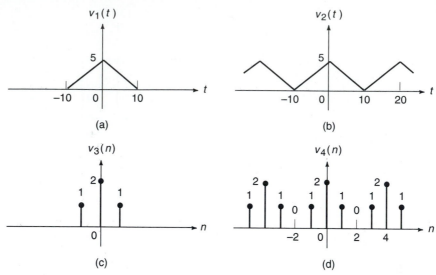

Fig. 5.5.4

5.7. Find and plot the autocorrelation function for each waveform in Fig. 5.5.3.

5.8. Find and plot the autocorrelation function for each waveform in Fig. 5.5.4.

5.9. The functions in Fig. 5.5.5 are reported to be time correlation functions for aperiodic (energy) signals.
   (a) Use your knowledge of correlation to determine which cannot be autocorrelation functions of energy signals.
   (b) Sketch at least two waveforms from which each legitimate autocorrelation could arise.

5.10. (a) Find the Fourier transform for each signal in Fig. 5.5.1.
   (b) Find and plot the energy or power spectral density function.
   (c) Find and plot the autocorrelation function for each signal.
   (d) Find the Fourier transform of each autocorrelation function and show that it equals the spectrum found in part (b).

5.11. Repeat Problem 5.10 for the waveforms in Fig. 5.5.2.

5.12. Repeat Problem 5.10 for the waveforms in Fig. 5.5.3.

5.13. Repeat Problem 5.10 for the waveforms in Fig. 5.5.4.

5.14. In a biological experiment, the electrical activity in a tissue sample is recorded over several minutes. This electrical activity is then converted to a digital signal by sampling at 10 samples/s so the data can be processed by digital computer. If procedure corr in Section 5.2 is used to calculate the autocorrelation function, then the time between correlation values will

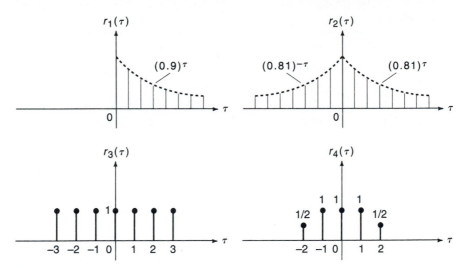

Fig. 5.5.5

be 0.1 s. Suppose we calculate the correlation for 1024 values and take the fast Fourier transform (FFT) of this function. What frequencies are represented by samples of the transform? What analog frequency does the first value, the second value, etc., of the transform represent?

5.15. Find and plot the cross-correlation function $r_{12}(\tau)$ in Fig. 5.5.1. That is, find the cross correlation between $v_1(t)$ and $v_2(t)$. This function is defined by

$$r_{12}(\tau) = \int_{-\infty}^{\infty} v_1(t)v_2(t - \tau) \, dt$$

5.16. Find and plot the cross-correlation function $r_{24}(\tau)$ in Fig. 5.5.2.

5.17. Find and plot the energy spectral density function for each autocorrelation function given below. These represent discrete-time energy signal correlation functions.
(a) $r_{xx}(l) = 4\delta(l)$
(b) $r_{xx}(l) = 0.5\delta(l + 1) + \delta(l) + 0.5\delta(l - 1)$
(c) $r_{xx}(l) = 0.9^{|l|}, \; -\infty < l < \infty$

5.18. Sketch at least one energy signal waveform with the autocorrelation function given in each part of Problem 5.17.

5.19. Computer problem: Use a random number generator to produce a Gaussian discrete-time signal with zero mean and unit variance. Use the "overlap and add" method (see Sections 1.2 and 1.3) to produce a correlation of 0.75 between adjacent samples, 0.5 between samples separated by two time units, and 0.25 between samples spaced three time units apart.

Table 5.5.1

Year	Number	Year	Number	Year	Number	Year	Number
1761	86	1789	118	1816	46	1843	11
1762	61	1790	90	1817	41	1844	15
1763	45	1791	67	1818	30	1845	40
1764	36	1792	60	1819	24	1846	62
1765	21	1793	47	1820	16	1847	98
1766	11	1794	41	1821	7	1848	124
1767	38	1795	21	1822	4	1849	96
1768	70	1796	16	1823	2	1850	66
1769	106	1797	6	1824	8	1851	64
1770	101	1798	4	1825	17	1852	54
1771	82	1799	7	1826	36	1853	39
1772	66	1800	14	1827	50	1854	21
1773	35	1801	34	1828	62	1855	7
1774	31	1802	45	1829	67	1856	4
1775	7	1803	43	1830	71	1857	23
1776	20	1804	48	1831	48	1858	55
1777	92	1805	42	1832	28	1859	94
1778	154	1806	28	1833	8	1860	96
1779	126	1807	10	1834	13	1861	77
1780	85	1808	8	1835	57	1862	59
1781	68	1809	2	1836	122	1863	44
1782	38	1810	0	1837	138	1864	47
1783	23	1811	1	1838	103	1865	30
1784	10	1812	5	1839	86	1866	16
1785	24	1813	12	1840	63	1867	7
1786	83	1814	14	1841	37	1868	37
1787	132	1815	35	1842	24	1869	74
1788	131						

(a) Calculate and plot the autocorrelation function for this signal.

(b) Use Eq. 5.2.9 to estimate the autocorrelation function for this signal. Do your calculated and estimated correlation functions agree?

(c) Calculate the power spectrum by taking the transform of the autocorrelation function from part (a) of this problem.

(d) Estimate the power spectrum by taking the transform of the autocorrelation function from part (b) of this problem. How closely do these last two calculations agree?

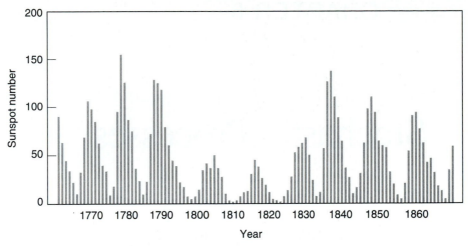

Fig. 5.5.6.   Wolfer's sunspot numbers.

(e) Estimate the cdf of this signal from the samples. Do you think it is Gaussian?

5.20. Computer problem: Use a random number generator to produce a zero-mean, discrete-time signal $x(n)$ whose values are uncorrelated and range uniformly from -0.5 to 0.5. The random number generator described in Section 1.3 will generate this signal. Then process this signal by the formula

$$y(n) = 0.9y(n - 1) + x(n)$$

with $y(0) = 0$. This produces a Gaussian signal with exponential autocorrelation function. Repeat parts (a) through (e) of Problem 5.19 for this signal.

5.21. Computer problem: Wolfer's sunspot numbers in Table 5.5.1 record the sunspot activity for the period 1761–1869. Figure 5.5.6, which shows this same information, displays the periodic nature of sunspot activity. Find and plot the autocorrelation function for this data to determine the period of sunspot activity.

Further Reading

1. DWIGHT F. MIX and NEIL M. SCHMITT, *Circuit Analysis for Engineers,* John Wiley, New York, 1985.
2. WILLIAM M. SIEBERT, *Circuits, Signals, and Systems,* McGraw-Hill, New York, 1986.
3. JOHN G. PROAKIS and DIMITRIS G. MANOLAKIS, *Introduction to Digital Signal Processing,* Macmillan, New York, 1988.

CHAPTER 6 _____

Stochastic Processes

6.1 Definition and Examples _____

Preview

Everything seemed to be fine in Section 3.3 when we applied the concept of random variable to waveforms. The logic used there went something like this: For a given waveform we performed an experiment whose outcome determined a time value, say, t. Then we could define a random variable whose range was the possible values of the waveform at time t. Knowing these possible values we could do several useful things, such as determine the cdf and the pdf of the random variable and, from this, determine the power or energy of the waveform. But closer scrutiny reveals difficulties with this model because it is not general enough. It will not allow us to logically determine correlation and spectral content, at least not as easily as a more general model. This generality is provided by the concept of stochastic processes, a generalization of random variables. We will elaborate more on the advantage of stochastic processes over random variables in our discussion of joint moments in the next section. This section is short, for all we do is define the term stochastic process and give a few examples.

DEFINITION 6.1.1. A stochastic process $X(t, \zeta)$ is a function of two variables, t and ζ, with ζ an element of the sample space. In addition, for fixed t the function $X(t, \zeta)$ must satisfy the definition of a random variable (Definition 3.1.1).

Now, instead of a random variable $X(\zeta)$, we have a function of two variables $X(t, \zeta)$. Although the parameter t is arbitrary, we will identify it with time. And we will use the symbol t for either continuous time or discrete time. The difference between this new concept and a random variable is that now an experimental outcome determines a waveform, as opposed to a number. The word *stochastic* is derived from the Greek *stochastikos,* meaning

242

random. But there is nothing random about a stochastic process, just as there is nothing random about a random variable. The randomness arises from the experiment.

There are four possible interpretations of $X(t, \zeta)$, depending on the nature of t and ζ. (1) If t and ζ are fixed, then $X(t, \zeta)$ is a number. (2) If t is fixed and ζ is variable, then $X(t, \zeta)$ is a random variable. (3) If t is variable and ζ is fixed, then $X(t, \zeta)$ is a time function (also called a sample function). (4) If t and ζ are both variable, then $X(t, \zeta)$ is a stochastic process. (Recall that t can be continuous or discrete, so everything we say here applies equally to continuous-time and discrete-time signals.)

Figure 6.1.1 shows four time functions labeled $X(t, \zeta)$. We can imagine that an experiment with four outcomes, ζ_1, ζ_2, ζ_3, and ζ_4, is used to select one of these time functions. The experiment is performed with outcome ζ_i. We then associate a time function $X(t, \zeta_i)$ with the particular experimental outcome ζ_i. Thus we have a procedure for selecting a time function, and the particular function chosen depends on the experimental outcome ζ_i. This is a stochastic process. The difference between this and a random variable is that we use the experimental outcome to select a function $X(t, \zeta)$, while a random variable uses the experimental outcome to select a number $X(\zeta)$.

A set or collection of time functions such as that shown in Fig. 6.1.1 is called an *ensemble*. In this particular case the ensemble is finite with only four members, but in general we can deal with an infinite number of time functions. We mention again that the interpretation of t as time is a special application. The parameter t can be identified with any appropriate physical attribute, or it can be constant (in which case we are dealing with a random variable). In fact, the term *ensemble* applies to random variables, meaning an ensemble of numbers.

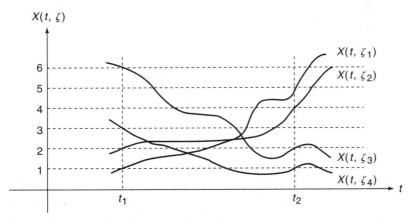

Fig. 6.1.1.   A stochastic process.

Computer-generated random numbers form a random process where ζ_i = ISEED. See Section 1.3 if you have forgotten what ISEED is. Here the idea is that a sequence of random numbers is a digital signal, and different values of ISEED give a different starting point to generate the periodic signal. (Recall that random number sequences generated by our procedure are periodic, although the period may be large.)

For another example of a stochastic process, suppose we toss a coin once every second, starting at time $t = 0$. Let the signal consist of a pulse of value $+1$ if heads occurs on the ith toss, and 0 if tails occurs. Therefore each person who tosses a coin in this manner generates one waveform. To have a stochastic process we would need more than one person willing to toss a coin in this fashion so we could have an ensemble. Of course, what we really do is imagine there are an infinite number of people, each tossing a coin, and use this imaginary infinite ensemble in the discussions to follow. Figure 6.1.2 shows a typical member of this ensemble.

The concept of a stochastic process allows us to logically analyze and design systems with random input signals. In the rest of this chapter we explore the important aspects of stochastic processes, and then begin to formulate the various methods and procedures for using this background in applications to systems. By defining moments of stochastic processes properly, we can find not only the power function, but also the autocorrelation function for random waveforms. We use second-order statistics (the mean, the variance, and correlation functions) in random signal processing, so the remainder of this chapter describes these concepts.

Review

We have defined a stochastic process as a generalization of a random variable. Now the experimental outcome determines a function instead of just a number. The codomain of this function is called an ensemble, the ensemble of all possible waveforms we could possibly get by performing the experiment. These waveforms can be discrete time or continuous time. We will begin to see the difference between this new concept and random variables in the next section.

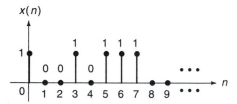

Fig. 6.1.2

6.2   Moments of Stochastic Processes _____

Preview

Section 3.3 introduced the mean, the variance, the mean square value, and higher-order moments of random variables. Knowing the probability density function (pdf) allowed us to calculate these moments. Now we are dealing with the more general concept of stochastic processes, and again the only information needed to calculate moments remains the pdf (or the equivalent cumulative distribution function, the cdf). The difference is that now we can calculate the joint moments at two values of time, and this is all-important.

In this section we define the cdf and the pdf for stochastic processes. They are identical to the cdf and the pdf for random variables, except that we insert the parameter t to account for the time. Moments have the same definition and meaning as before, except that they are now functions of time. The most important of these are the second moments, because they contain frequency information about the signal.

With reference to Fig. 6.1.1, if the probability law is known (i.e., if $P(\zeta_1)$, $P(\zeta_2)$, $P(\zeta_3)$, and $P(\zeta_4)$ are known), then the probability of selecting any one member of the ensemble is known. For a specific value of t, $X(t, \zeta)$ is a random variable. We can therefore speak of the cumulative distribution function (cdf) and the probability density function (pdf) as before:

$$F_X(\alpha; t) = P\{X(t, \zeta) \leq \alpha\} \tag{6.2.1}$$

$$f_X(\alpha; t) = \frac{d}{d\alpha} F_X(\alpha; t) \tag{6.2.2}$$

The cdf and the pdf of a stochastic process can change with time. Instead of writing $F_X(\alpha)$ for the cdf, as we do for a random variable, we write $F_X(\alpha; t)$ to denote the continuous-time or discrete-time dependence. Recall that we are using the symbol t to stand for either the continuous-time or the discrete-time variable, but we will also use the symbol n when we are specifically discussing discrete-time functions. Similarly, we write $f_X(\alpha; t)$ for the pdf.

EXAMPLE 6.2.1.   Refer to Fig. 6.1.1 and suppose that $P(\zeta_i) = 0.25$ for $i = 1, 2, 3, 4$. Find and plot the cdf's $F_X(\alpha; t_1)$ and $F_X(\alpha; t_2)$ along with the pdf's $f_X(\alpha; t_1)$ and $f_X(\alpha; t_2)$.

SOLUTION: The stochastic process X can take on any one of four different values at t_1, and also at t_2. Table 6.2.1 outlines these possible values. Since each experimental outcome ζ_i occurs with equal probability, this gives the cdf's and pdf's in Fig. 6.2.1.

Table 6.2.1

ζ	$X(t_i, \zeta_1)$	$X(t_2, \zeta_i)$
ζ_1	1	5
ζ_2	2	4
ζ_3	6	2
ζ_4	3	1

EXAMPLE 6.2.2.   Toss a coin once each second starting at $n = 0$, and let a time function equal $+1$ at time n when the coin is heads, and 0 otherwise. Find the pdf for any $n > 0$.

SOLUTION: If we think of this as an infinite ensemble of time functions with the experimental outcome determining which of these functions we choose, then we are just as likely to choose a time function that has value $+1$ at some particular time as not. Therefore the pdf is the binary pdf with $P\{X(n, \zeta) = 1\} = 0.5$, and $P\{X(n, \zeta) = 0\} = 0.5$.

We really have two experiments in this example, and that can be confusing. Tossing a coin is one experiment. We imagine a room full of people, each tossing a coin once every second. Selecting one of these people, or the waveform associated with the selected person, is the second experiment. This is the experiment whose outcome is labeled ζ.

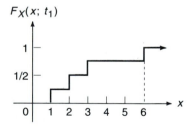

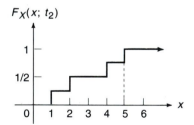

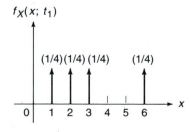

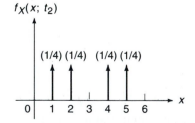

Fig. 6.2.1

Moments

The definition of moments given in Section 3.4 applies here. Of course, these moments are now time-dependent, because the pdf is time-dependent. The kth moment of $X(t, \zeta)$ [or $X(n, \zeta)$] is given by

$$m_k(t) = E[X^k(t, \zeta)] = \int_{-\infty}^{\infty} \alpha^k f_X(\alpha; t)\, d\alpha \qquad (6.2.3)$$

Likewise the kth central moment is given by

$$\mu_k(t) = E\{[X(t, \zeta) - m_X(t)]^k\} = \int_{-\infty}^{\infty} [\alpha - m_X(t)]^k f_X(\alpha; t)\, d\alpha \quad (6.2.4)$$

These definitions follow directly from the corresponding formulas in Section 3.4. The next step is to define joint moments as we did in Section 4.2, and this is where we realize the advantage of stochastic processes in place of random variables. As you can see, to this point we have done nothing more with our new concept of stochastic processes than we did with the concept of random variables. We already applied the concepts above to electrical waveforms in Chapter 4, and all we have done is add the parameter t to the argument of $X(t, \zeta)$. But now we can exploit the advantages of this new concept.

The important difference between stochastic processes and random variables is that we now have the entire waveform. The experimental outcome ζ determines an entire waveform $X(t, \zeta)$, not just a number, as in the case of random variables. Joint moments use this additional information.

To define joint moments for stochastic processes, we first need the joint cdf and pdf, defined by

$$F_{XX}(\alpha_1, \alpha_2; t_1, t_2) = P\{X(t_1) \leq \alpha_1, X(t_2) \leq \alpha_2\} \qquad (6.2.5)$$

$$f_{XX}(\alpha_1, \alpha_2; t_1, t_2) = \frac{\partial^2}{\partial \alpha_1, \partial \alpha_2} F_{XX}(\alpha_1, \alpha_2; t_1, t_2) \qquad (6.2.6)$$

These formulas connect the values of the stochastic process at two different times. For discrete-time processes we use n_1 and n_2 in place of t_1 and t_2. The idea behind Eq. 6.2.5 is to select numbers α_1 and α_2 in advance, then see how many of the functions in the ensemble satisfy both conditions $\{X(t_1) \leq \alpha_1\}$ and $\{X(t_2) \leq \alpha_2\}$. The following examples should make this statement clear.

EXAMPLE 6.2.3.   Find $F_{XX}(\alpha_1, \alpha_2; t_1, t_2)$ in Example 6.2.1 if $\alpha_1 = 2.2$ and $\alpha_2 = 3.5$.

SOLUTION: The answer is 0. To understand this, consider the way we have defined a stochastic process. The experiment is performed with experimental outcome ζ_i. This ζ_i determines which of the four functions we choose. But there is not one function that has value less than or equal to 2.2 at t_1 and also has value less than or equal to 3.5 at t_2. Therefore the probability of choosing a function that satisfies both conditions is 0.

EXAMPLE 6.2.4.   Repeat Example 6.2.3 with $\alpha_1 = 2.2$ and $\alpha_2 = 4.5$.

SOLUTION: There is one function that satisfies these conditions, namely, the function associated with ζ_2. Therefore $F_{XX}(2.2, 4.5; t_1, t_2) = \frac{1}{4}$.

EXAMPLE 6.2.5.   Find $F_{XX}(\alpha_1, \alpha_2; n_1, n_2)$ in Example 6.2.2 (the coin-toss experiment) if $\alpha_1 = 0.7$ and $\alpha_2 = 0.5$. Let $n_1 = 1$ and $n_2 = 10$.

SOLUTION: As always, we perform the experiment and this determines which time function we select. The question is: What is the probability that we will select a time function that satisfies both conditions $\{X(n_1) \leq 0.7\}$ and $\{X(n_2) \leq 0.5\}$? The answer is $\frac{1}{4}$, because we must have tails at time $n = n_1$ and another tails when $n = n_2$.

Having defined the joint cdf and pdf for a stochastic process, we are now in a position to define joint moments. The most useful of these are the second-order moments, the autocorrelation function, and the autocovariance function. The autocorrelation function $R_{XX}(t_1, t_2)$ of a stochastic process $X(t, \zeta)$ is the joint moment of the random variables $X(t_1, \zeta)$ and $X(t_2, \zeta)$. For a stochastic process with continuous values this is given by

$$R_{XX}(t_1, t_2) = E[X(t_1, \zeta) X(t_2, \zeta)]$$
$$= \int_{-\infty}^{\infty} \int_{-\infty}^{\infty} \alpha_1 \alpha_2 f_{XX}(\alpha_1, \alpha_2; t_1, t_2) \, d\alpha_1 \, d\alpha_2 \tag{6.2.7a}$$

For discrete random variables this becomes

$$R_{XX}(t_1, t_2) = \sum_{\alpha_1} \sum_{\alpha_2} \alpha_1 \alpha_2 P[X(t_1, \zeta) = \alpha_1, X(t_2, \zeta) = \alpha_2] \tag{6.2.7b}$$

The autocovariance function $\mathcal{L}_{XX}(t_1, t_2)$ is the joint central moment of the random variables $X(t_1, \zeta)$ and $X(t_2, \zeta)$:

$$\mathcal{L}_{XX}(t_1, t_2) = E\{[X(t_1, \zeta) - m_X(t_1)][X(t_2, \zeta) - m_X(t_2)]\}$$
$$= R_{XX}(t_1, t_2) - m_X(t_1) m_X(t_2) \tag{6.2.8}$$

We use double subscripts on all these terms F_{XX}, f_{XX}, R_{XX}, and $\mathcal{L}_{XX}$ because we will also be interested in relations between two different stochastic processes X and Y. In that case we use subscripts such as f_{XY} and R_{XY}. This notation makes it easier to distinguish between these different functions.

    The domain of the two functions $R_{XX}(t_1, t_2)$ and $\mathcal{L}_{XX}(t_1, t_2)$ is the two values of time, t_1 and t_2. More specifically, the domain is $(T \times T)$ (the set cross product), where T is the set of all values of time. The codomain is a set of real numbers, usually $-\infty < R < \infty$. If $t_1 = t_2$, then $R_{XX}(t_1, t_2) =$

$m_2(t_1)$, the second moment (mean square value) of the process evaluated at $t = t_1$.

Of course all this applies equally to discrete-time stochastic processes. Simply substitute n for t if the process is discrete-time.

EXAMPLE 6.2.6.   Find $R_{XX}(t_1, t_2)$ and $\mathscr{L}_{XX}(t_1, t_2)$ for the stochastic process in Fig. 6.1.1.

SOLUTION: The two-dimensional pdf is given by

$$f_{XX}(\alpha_1, \alpha_2; t_1, t_2) = \tfrac{1}{4}[\delta(\alpha_1 - 1)\,\delta(\alpha_2 - 5) + \delta(\alpha_1 - 2)\,\delta(\alpha_2 - 4)$$
$$+ \delta(\alpha_1 - 6)\,\delta(\alpha_2 - 2) + \delta(\alpha_1 - 3)\,\delta(\alpha_2 - 1)]$$

Therefore, from Eq. 6.2.7b we have

$$R_{XX}(t_1, t_2) = E[X(t_1, \zeta) X(t_2, \zeta)]$$
$$= \tfrac{1}{4}(1 \cdot 5 + 2 \cdot 4 + 6 \cdot 2 + 3 \cdot 1) = 7$$

In order to calculate $\mathscr{L}_{XX}(t_1, t_2)$ we need the mean values at t_1 and t_2.

$$m_X(t_1) = \tfrac{1}{4}(1 + 2 + 6 + 3) = 3$$
$$m_X(t_2) = \tfrac{1}{4}(5 + 4 + 2 + 1) = 3$$

so

$$\mathscr{L}_{XX}(t_1, t_2) = R_{XX}(t_1, t_2) - m_X(t_1)\,m_X(t_2) = 7 - 9 = -2$$

EXAMPLE 6.2.7.   Find $R_{XX}(n_1, n_2)$ and $\mathscr{L}_{XX}(n_1, n_2)$ for the "coin-toss" stochastic process of Example 6.2.2 if $n_1 = 1$ and $n_2 = 10$.

SOLUTION: (a) The random variables $X(n_1, \zeta)$ and $X(n_2, \zeta)$ are independent for these values of time, giving

$$R_{XX}(n_1, n_2) = E[X(n_1, \zeta)] E[X(n_2, \zeta)] = \tfrac{1}{4}$$

and since $m_X(n) = \tfrac{1}{2}$ for any n,

$$\mathscr{L}_{XX}(n_1, n_2) = R_{XX}(n_1, n_2) - m_X(n_1)\,m_X(n_2) = 0$$

Review

We have done a lot in this section:

Defined one-dimensional cdf and pdf
Defined joint (two-dimensional) cdf and pdf
Calculated moments

The idea of a stochastic process is important in finding the two-dimensional cdf and pdf, and in finding the autocorrelation and autocovariance functions. The two-dimensional cdf is a probability, the probability that the particular function we select from the ensemble will have value less than or equal to α_1 at time t_1 (or n_1) and have value less than or equal to α_2 at time t_2 (or n_2). Both these conditions must be satisfied—not one or the other, but both. From this cdf we find the pdf, and use the pdf to calculate the autocorrelation or autocovariance function. From this chain we see that the concept of stochastic processes really is different from that of random variables, and it is important in signal processing. The autocorrelation function contains frequency information about the process. It is this frequency information that we use to design filters.

6.3   Characterizing Stochastic Processes _____

Preview

There are several ways to characterize stochastic processes. They can be periodic or aperiodic, stationary or nonstationary, ergodic or nonergodic, discrete or continuous, and member functions can be power or energy signals. There is also the important matter of how a stochastic process is specified. Our analysis of both signals and systems depends on our ability to find the joint cdf or pdf for a stochastic process at arbitrary times. We say that a stochastic process is specified if we can find these joint functions for any set of time instants. All of these topics will be important to us in the future, so let us pause to introduce the new topics in this list—specifying a stochastic process, stationary processes, and ergodic processes. We begin with the concept of specifying a random process.

For a process $X(t, \zeta)$, the experiment is performed with outcome ζ, which in turn determines the particular sample function $X(t)$. Let us suppose that we sample $X(t)$ k times, giving a vector of values

$$X = \begin{bmatrix} X(t_1) \\ X(t_2) \\ \vdots \\ X(t_k) \end{bmatrix}$$

This random vector has a joint cdf, F_X, and pdf, f_X. All our discussion in this section applies equally to discrete-time processes: Simply substitute n for t. With this notation in mind, we define the concept of specifying a stochastic process as follows.

DEFINITION 6.3.1.   A stochastic process $X(t, \zeta)$ is said to be specified if F_X can be determined for every set of time instants $t_1, t_2, \ldots, t_k$.

This means that for any finite set of time instants (say, 1000 of them), we can write down F_X, by no means a trivial undertaking in the general case. Three methods are commonly used to specify stochastic processes.

Method 1.   State the rule for determining F_X directly. For this to be possible, the joint cdf must depend on the time instants in a known elementary way. It is a special case when this is possible, but fortunately the Gaussian process can be specified in this manner, as we now show.

A stochastic process is called Gaussian if F_X is Gaussian for every finite set of time instants $t_1, t_2, \ldots, t_k$. Recall that knowing the means, variances, and covariances allows us to write down the Gaussian pdf. Therefore, if a rule can be stated to determine these moments for any set of samples, the Gaussian process is specified. The mean function describes the mean as a function of time.

$$m_X(t) = E[X(t, \zeta)] = \int_{-\infty}^{\infty} \alpha(t) f_X(\alpha; t)\, d\alpha \qquad (6.3.1)$$

If this mean function and either the correlation or covariance function are given, then the mean vector and covariance matrix can be determined in the Gaussian pdf. Thus the Gaussian process is specified if $m_X(t)$ and $\mathscr{L}_{XX}(t_1, t_2)$ are known. Here is an example.

EXAMPLE 6.3.1.   Determine an expression for f_X for three samples taken from a Gaussian process at times $t_1 = 0$, $t_2 = 1$, and $t_3 = 2$ s. The mean is 0, and the covariance function is

$$\mathscr{L}_{XX}(t_i, t_j) = e^{-|t_i - t_j|} \qquad \text{for all } i, j \qquad (6.3.2)$$

SOLUTION: The pdf $f_X(\alpha)$ is of the form

$$f_X(\alpha) = \frac{1}{(2\pi)^{3/2}|\Lambda|^{1/2}} \exp\left(-\frac{1}{2}\alpha^t \Lambda_X^{-1}\alpha\right) \qquad (6.3.3)$$

The covariance matrix is given by

$$\Lambda_X = \begin{bmatrix} \mathscr{L}_{XX}(t_1, t_1) & \mathscr{L}_{XX}(t_1, t_2) & \mathscr{L}_{XX}(t_1, t_3) \\ \mathscr{L}_{XX}(t_2, t_1) & \mathscr{L}_{XX}(t_2, t_2) & \mathscr{L}_{XX}(t_2, t_3) \\ \mathscr{L}_{XX}(t_3, t_1) & \mathscr{L}_{XX}(t_3, t_2) & \mathscr{L}_{XX}(t_3, t_3) \end{bmatrix} \qquad (6.3.4)$$

This is the general form. Substituting values of t into Eq. 6.3.2 gives

$$\Lambda_X = \begin{bmatrix} 1 & e^{-1} & e^{-2} \\ e^{-1} & 1 & e^{-1} \\ e^{-2} & e^{-1} & 1 \end{bmatrix}$$

Substituting this into Eq. 6.3.3 completes the solution.

Method 2.   Transformation of variables. A time function involving one or more parameters is given, where the parameters are random variables with known distribution. The joint cdf F_X can be determined from knowledge of the distribution of the random variables. Here is an example.

EXAMPLE 6.3.2.   A sample function from a stochastic process is given by

$$x(t) = At, \qquad -\infty < t < \infty$$

where A is a random variable with uniform distribution (0, 1). That is,

$$f_A(\alpha) = \begin{cases} 1, & 0 < \alpha < 1 \\ 0 & \text{otherwise} \end{cases}$$

Find the two-dimensional pdf for $[X(t_1, \zeta), X(t_2, \zeta)]$ at times $t_1 = 1$ and $t_2 = 2$.

SOLUTION: A particular experimental outcome ζ determines a particular value of A (call it α) and also a particular function $x(t) = \alpha t$. The value of this function at $t_1 = 1$ is tied to the value at $t_2 = 2$ by the relationship

$$x(t_1) = \left(\frac{t_1}{t_2}\right) \cdot x(t_2)$$

If $x(t_1) = \alpha_1$ and $x(t_2) = \alpha_2$, then the two-dimensional pdf is given by

$$f_{XX}(\alpha_1, \alpha_2; t_1, t_2) = f_X[\alpha_1 | X(t_2) = \alpha_2] \cdot f_X(\alpha_2; t_2)$$

$$= \delta\left[\alpha_1 - \left(\frac{t_1}{t_2}\right)\alpha_2\right] \cdot f_X(\alpha_2; t_2)$$

Since A is uniform (0, 1), X is uniform (0, t). With $t_1 = 1$ and $t_2 = 2$, the two-dimensional pdf is given by

$$f_{XX}(\alpha_1, \alpha_2; t_1, t_2) = \delta(\alpha_1 - 0.5\alpha_2) f_X(\alpha_2; 2)$$

where $f_X(\alpha_2; 2)$ is uniform (0, 2).

Method 3.   System input–output. Suppose that we are given a specified stochastic process, one for which we can find the k-dimensional cdf. Then if a sample function is applied to the input of a linear time-invariant system, the output is a sample function from another stochastic process. In this case we may be able to specify the output from a knowledge of the input and system transformation. We will postpone examples of this procedure until we use these methods.

These are the three primary methods of specifying stochastic processes. Of course any method that allows us to write down F_X may be used. We next turn our attention to stationary and ergodic processes.

Stationary Processes

The statistics of a stationary stochastic process do not change with time. This implies that the mean, the variance, and the first-order pdf for $X(t_1, \zeta)$

are equal to those for $X(t_2, \zeta)$ for every t_1 and t_2. That is, the statistics determined for $X(t, \zeta)$ are equal to those for $X(t + \varepsilon, \zeta)$ for every ε. But this is only one type (or one degree) of stationarity. To be more precise, we make the following definitions.

DEFINITION 6.3.2.   A stochastic process is stationary of order k if

$$f_X(\alpha_1, \ldots, \alpha_k; t_1, \ldots, t_k) = f_X(\alpha_1, \ldots, \alpha_k; t_1 + \varepsilon, \ldots, t_k + \varepsilon) \text{ for all } \varepsilon$$
$$(6.3.5)$$

Thus, if $f_X(\alpha; t_1) = f_X(\alpha; t_1 + \varepsilon)$, the process is stationary of order 1. If $f_X(\alpha_1, \alpha_2; t_1, t_2) = f_X(\alpha_1, \alpha_2; t_1 + \varepsilon, t_2 + \varepsilon)$, the process is stationary of order 2.

Since there are several types of stationarity, some special terminology has arisen. A process is strictly stationary if it is stationary for any order, $k = 1, 2, \ldots$. A process is called *wide-sense* (or weakly) stationary if its mean value is a constant and its autocorrelation function depends only on $\tau = t_2 - t_1$. Wide-sense stationary processes are important to us because of the Wiener–Khinchine theorem: If a process is wide-sense stationary, the autocorrelation function and the power spectral density function form a Fourier transform pair. Therefore, if we know—or can somehow measure— the autocorrelation function, we can find which frequencies contain the power in the signal. This is the information we need to design frequency-selective filters for processing random signals.

The process in Example 6.3.1 is wide-sense stationary because the mean is constant and the covariance depends on the time difference $t_2 - t_1$. The process in Example 6.3.2 is not stationary of any order, because the first-order pdf changes with time.

Ergodic Processes

The idea of ergodicity arises if we have only one sample function from a stochastic process, instead of the entire ensemble. A single sample function will often provide little information about the statistics of the process. However, if the process is ergodic, then all statistical information can be derived from just one sample function.

DEFINITION 6.3.3.   A stochastic process is ergodic if time averages equal ensemble averages.

When a process is ergodic, any one sample function represents the entire process. A little thought should convince you that the process must necessarily be stationary for this to occur. Thus ergodicity implies stationarity. There are levels of ergodicity, just as there are levels (degrees) of stationarity. We will discuss two levels of ergodicity: ergodicity in the mean and correlation.

Level 1.   A process is *ergodic* in the mean if

$$\langle x(t) \rangle = \lim_{T \to \infty} \frac{1}{2T} \int_{-T}^{T} x(t)\, dt = E[X(t, \zeta)] \qquad (6.3.6)$$

We use the notation $\langle x(t) \rangle$ to stand for the time average of $x(t)$, and $E[X(t, \zeta)]$ to denote the ensemble average of the process. We can compute the left side of Eq. 6.3.6 by first selecting a particular member function $x(t)$ and then averaging in time. To compute the right-hand side, we must know the first-order pdf $f_X(\alpha; t)$. The left-hand side of Eq. 6.3.6 is a number. Hence the mean

$$m_X(t) = E[X(t, \alpha)]$$

must be a constant. Therefore, ergodicity of the mean implies stationarity of the mean. Stationarity of the mean does not imply ergodicity of the mean, as our next example indicates.

Level 2.   A process is *ergodic in autocorrelation* if

$$r(\tau) = \langle x(t)\, x(t - \tau) \rangle = \lim_{T \to \infty} \frac{1}{2T} \int_{-T}^{T} x(t)\, x(t - \tau)\, dt$$
$$= E[X(t)\, X(t - \tau)] = R_{XX}(\tau) \qquad (6.3.7)$$

We use lowercase r to denote the time autocorrelation function and R_{XX} to denote the statistical autocorrelation function. We can compute the left side of Eq. 6.3.7 by using a particular function $x(t)$. To compute the right side, we must know the second-order pdf.

EXAMPLE 6.3.3.   Consider a basket full of batteries. There are some flashlight batteries, some car batteries, and several other kinds of batteries. Suppose that a battery is selected at random and its voltage measured. This battery voltage $v(t)$ is a member function from a class of constant battery voltages, that is, a member function from a stochastic process.

   This process is stationary but not ergodic in the mean or correlation. The time average is equal to the particular battery voltage selected (say, 1.5 V). The statistical average $E[X(t, \zeta)]$ is some other number, depending on what is in the basket. Thus Eq. 6.3.6 does not hold.

EXAMPLE 6.3.4.   Let $x(t) = \cos(\omega t + \theta)$ be a member function from a stochastic process specified by a transformation of variables. Let Θ be a random variable with uniform distribution over the interval $0 < \theta \leq 2\pi$.

$$f_\Theta(\theta) = \frac{1}{2\pi}, \qquad 0 < \theta \leq 2\pi$$

Then each θ determines a time function $x(t)$, which means that the stochastic process is specified by a transformation of variables. This stochastic process is ergodic in both the mean and autocorrelation. You can see that the time average of $x(t)$ is 0. The ensemble average at any one time is over an infinite variety of sinusoids of all phases, and so must also be 0. Since the time average equals the ensemble average, the process is ergodic in the mean. It is also true that Eq. 6.3.7 holds, so the process is ergodic in correlation. For any other distribution of Θ the process is not stationary, and hence not ergodic.

EXAMPLE 6.3.5.   Suppose that a periodic square wave has random starting time t_0 uniformly distributed over one period:

$$f_T(\alpha) = \frac{1}{T_0}, \qquad 0 < t < T$$

Since this waveform can be expressed as the sum of sinusoids, as in the example above, the process is ergodic. Again, for any other distribution of T_0, the process is not stationary.

Review

A stochastic process is specified if we know the statistics of the process. This is another way of saying that we can write down the k-dimensional cdf for any set of k time instants. What this really means for us is that we can find the autocorrelation function, because all we need is the two-dimensional pdf for arbitrary time instants t_1 and t_2.

The idea of a stationary process is important, because the autocorrelation function and power density spectrum form a Fourier transform pair for wide-sense (or more) stationary processes. An important point: Sample functions from stationary processes are power signals. Stationarity (of any order) implies that each sample function in the process has been present since before the world was formed, and will be present long after we are gone. Look at the definition for any order stationary process to see that this is true; an energy signal fails to satisfy the definition for stationarity because it does not last forever.

Ergodic processes are important for two reasons. An ergodic process is stationary, and we can measure statistical information from one sample function.

6.4   System Input–Output _____

Preview

Systems operate on random signals to suppress noise or enhance particular aspects of the signal. By necessity, this means that systems destroy information in the sense defined in Section 2.4. But they destroy unwanted randomness (which is technically called information) and retain the desired randomness in a signal. And of course any information-bearing signal must be random, for if the signal is deterministic we can predict its future values, so there is no need to process it.

In this section we discuss linear time-invariant systems and their effect on the statistics of a stochastic process. This will lay the foundation for our future investigation of the various ways to process random signals for particular purposes. Of particular interest will be stationary signals, because this simplifies time-domain analysis, and we know how to apply frequency analysis to these signals. A stationary signal is a power signal, because energy signals are, by their nature, nonstationary.

Continuous-Time Stationary Processes

The linear time-invariant (LTI) system shown in Fig. 6.4.1 produces the output $y(t)$ from the input $x(t)$. This process is described by the convolution integral given by

$$y(t) = \int_{-\infty}^{\infty} h(t - \lambda)\, x(\lambda)\, d\lambda \qquad (6.4.1)$$

When we say that the input is a random process $X(t, \zeta)$, we mean that the experiment to select a sample function has been performed with outcome ζ. This determines the sample function, and this particular function is applied to the system input. In the following we suppress the ζ parameter and use the notation $X(t)$ for the input process. Lowercase $x(t)$ stands for a particular input function.

We cannot predict which sample function will be chosen, so the best way we know to treat this situation is to find averages. The mean-value output function is given by

$$E[Y(t)] = E\left[\int_{-\infty}^{\infty} h(t - \lambda)\, X(\lambda)\, d\lambda \right] = \int_{-\infty}^{\infty} h(t - \lambda)\, E[X(\lambda)]\, d\lambda$$

or

$$m_Y(t) = \int_{-\infty}^{\infty} h(t - \lambda)\, m_X(\lambda)\, d\lambda \qquad (6.4.2)$$

We restrict our discussion in this section to wide-sense stationary processes for two reasons: They are easier to deal with, and we will have little need

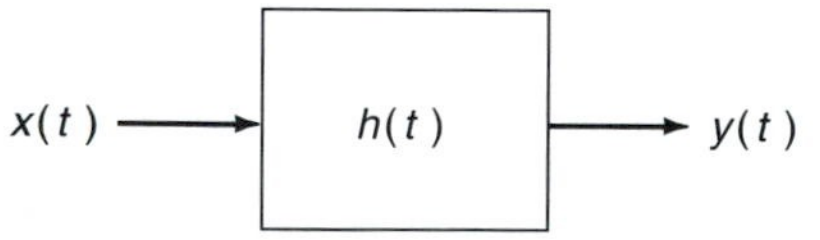

Fig. 6.4.1. A linear time-invariant system.

for nonstationary processes in all that follows. Thus the input mean is constant with $m_X(t) = m_X$, which simplifies Eq. 6.4.2.

$$m_Y = m_X \int_{-\infty}^{\infty} h(t)\, dt \tag{6.4.3}$$

The output mean is also a constant equal to the input mean multiplied by the area under the impulse response.

The output autocorrelation function can be obtained by multiplying Eq. 6.4.1 by $y(t)$ and taking the expected values.

$$E[Y(t_1)\, Y(t_2)] = E\left\{\left[\int_{-\infty}^{\infty} h(t_1 - \lambda)\, X(\lambda)\, d\lambda\right] Y(t_2)\right\}$$

$$= \int_{-\infty}^{\infty} h(t_1 - \lambda)\, E[X(\lambda)\, Y(t_2)]\, d\lambda \tag{6.4.4}$$

$$= \int_{-\infty}^{\infty} h(t_1 - \lambda)\, R_{XY}(\lambda, t_2)\, d\lambda$$

This says that the output correlation function depends on the impulse response and the cross-correlation function between input and output. But what we wanted was to be able to calculate the output statistics from a knowledge of the input statistics, so we are not quite there yet. We can calculate the cross-correlation function in Eq. 6.4.4 from the input statistics as follows. First we calculate R_{YX}, and then use the relation

$$R_{XY}(t_1, t_2) = R_{YX}(t_2, t_1) \tag{6.4.5}$$

To calculate R_{YX} multiply Eq. 6.4.1 by x (instead of y) and take the expected value.

$$E[Y(t_1)\, X(t_2)] = \int_{-\infty}^{\infty} h(t_1 - \lambda)\, E[X(\lambda)\, X(t_2)]\, d\lambda$$

or

$$R_{YX}(t_1, t_2) = \int_{-\infty}^{\infty} h(t_1 - \lambda)\, R_{XX}(\lambda, t_2)\, d\lambda \tag{6.4.6}$$

Now we can get there from here. Knowing the input autocorrelation function, apply Eq. 6.4.6 to get $R_{YX}(t_1, t_2)$. Then apply Eq. 6.4.5 and substitute the resulting R_{XY} function into Eq. 6.4.4 to get the output autocorrelation function.

If $X(t)$ is wide-sense stationary, then we can write

$$R_{XX}(\lambda, t_2) = R_{XX}(\lambda - t_2)$$

and

$$R_{XY}(\lambda, t_2) = R_{XY}(\lambda - t_2)$$

Substituting these and similar relations into the formulas above gives the following three steps to calculate the output autocorrelation function from the input autocorrelation function.

$$R_{YY}(t_1 - t_2) = \int_{-\infty}^{\infty} h(t_1 - \lambda)\, R_{XY}(\lambda - t_2)\, d\lambda \qquad (6.4.7)$$

$$R_{XY}(t_1 - t_2) = R_{YX}(t_2 - t_1) \qquad (6.4.8)$$

and

$$R_{YX}(t_1 - t_2) = \int_{-\infty}^{\infty} h(t_1 - \lambda)\, R_{XX}(\lambda - t_2)\, d\lambda \qquad (6.4.9)$$

For stationary processes, we calculate the output autocorrelation function by applying Eqs. 6.4.9, 6.4.8, and 6.4.7 in that order.

EXAMPLE 6.4.1.   White Gaussian noise (WGN) with mean 0 and variance 1 is applied to an RC low-pass filter, as shown in Fig. 6.4.2. Find the output mean and autocorrelation function.

SOLUTION: The output mean is 0 since the input mean is 0. White noise has autocorrelation function given by

$$R_{XX}(t_1 - t_2) = \sigma^2\, \delta(t_1 - t_2)$$

Here the variance $\sigma^2 = 1$. The fact that the noise is Gaussian has no bearing on this problem. The RC low-pass filter has impulse response given by

$$h(t) = e^{-t} u(t)$$

Substituting into Eqs. 6.4.9, 6.4.8, and 6.4.7 gives

$$R_{YX}(t_1 - t_2) = \int_{-\infty}^{\infty} e^{-(t_1 - \lambda)}\, \delta(\lambda - t_2)\, d\lambda = e^{-(t_1 - t_2)} u(t_1 - t_2)$$

$$R_{XY}(t_1 - t_2) = e^{-(t_2 - t_1)} u(t_2 - t_1)$$

Figures 6.4.3a and 6.4.3b show these functions along with $h(t_1 - \lambda)$ (Fig. 6.4.3c) in preparation for the last step, substituting into Eq. 6.4.7. For $t_1 > t_2$ (as shown in Fig. 6.4.3), we multiply the functions in Figs. 6.4.3b and 6.4.3c and integrate to find $R_{YY}(t_1 - t_2)$:

$$R_{YY}(t_1 - t_2) = \int_{-\infty}^{t_2} e^{-(t_2 - \lambda)} e^{-(t_1 - \lambda)}\, d\lambda = e^{-(t_2 + t_1)} \int_{-\infty}^{t_2} e^{2\lambda}\, d\lambda$$

$$= \frac{1}{2} e^{-(t_1 - t_2)}, \qquad t_1 > t_2$$

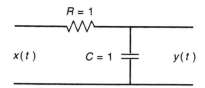

Fig. 6.4.2.   An RC low-pass filter.

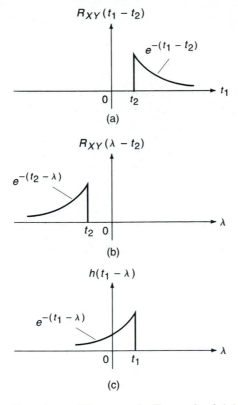

Fig. 6.4.3.   The steps in Example 6.4.1.

For $t_1 < t_2$ the upper limit on the integral is t_1 instead of t_2. Using identical steps as above, this gives

$$R_{YY}(t_1 - t_2) = \frac{1}{2} e^{(t_1 - t_2)}, \qquad t_1 < t_2$$

We can combine these last two formulas into one by using the absolute value in the exponent.

$$R_{YY}(t_1 - t_2) = \frac{1}{2} e^{-|t_1 - t_2|}$$

This output autocorrelation function is plotted in Fig. 6.4.4. It is customary to define $\tau = t_1 - t_2$ and label the abscissa in Fig. 6.4.4 as τ.

Equation 6.4.9 suggests a useful application to system identification. Among the several ways to measure the characteristics of an unknown system (a black box) are the following.

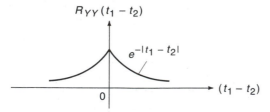

Fig. 6.4.4.   The output autocorrelation function in Example 6.4.1.

1. Apply a steady-state sinusoidal input signal to the unknown system and measure the amplitude and phase of the response for each input frequency, $0 < \Omega < \infty$. In practice we would make the measurements at selected frequencies and interpolate to obtain the frequency response. The problem here is time: We must wait for the transients to die out after applying each new frequency.

2. Apply an impulse and measure the response. Practical limitations impose restrictions on the accuracy of this method. An approximation to an impulse is difficult to generate because we need a tall, narrow pulse, and this may damage the system.

3. Often the best solution is to use Eq. 6.4.9. Notice that if the input is white noise (so that R_{XX} is a delta function), then the cross-correlation function is the impulse response. We can easily measure the cross correlation between system input and output, and white noise will not stress the system. This has a surprising number of applications, from on-line monitoring of nuclear power plants to measuring the structural integrity of aircraft wings.

Let us derive a practical form of Eq. 6.4.9 for this purpose. First let $\tau = t_1 - t_2$ and write Eq. 6.4.9 as

$$R_{YX}(\tau) = \int_{-\infty}^{\infty} h(\tau + t_2 - \lambda) R_{XX}(\lambda - t_1 + \tau) \, d\lambda$$

Now change the variable of integration by letting $\lambda = t_1 - \gamma$. This gives

$$R_{YX}(\tau) = \int_{-\infty}^{\infty} h(\gamma) R_{XX}(\tau - \gamma) \, d\gamma = h(\tau) * R_{XX}(\tau) \qquad (6.4.10)$$

which is just the convolution of the impulse response with the input autocorrelation function. Any time we convolve a function $h(\tau)$ with an impulse $R_{XX}(\tau) = \sigma^2 \, \delta(t)$, the result is just $\sigma^2 h(\tau)$. This means that $h(\tau) = (1/\sigma^2) R_{YX}(\tau)$ when the input is white noise. Therefore we can measure the impulse response of a linear system by applying white noise of known variance and measuring the cross correlation between the system input and output.

While we are at it, you can see that Eqs. 6.4.8 and 6.4.7 can be written in terms of τ as

$$R_{XY}(\tau) = R_{YX}(-\tau) \tag{6.4.11}$$

$$R_{YY}(\tau) = \int_{-\infty}^{\infty} h(\gamma) R_{XY}(\tau - \gamma) \, d\gamma \tag{6.4.12}$$

We can combine Eqs. 6.4.10, 6.4.11, and 6.4.12 into one convenient form and, at the same time, define the important "second-order impulse response function." From Eq. 6.4.10,

$$R_{YX}(\tau) = h(\tau) * R_{XX}(\tau)$$

where $*$ stands for convolution. Since $R_{YX}(\tau) = R_{XY}(-\tau)$, this can be written

$$R_{XY}(\tau) = h(-\tau) * R_{XX}(-\tau)$$

But R_{XX} is even, giving

$$R_{XY}(\tau) = h(-\tau) * R_{XX}(\tau)$$

Substituting this into $R_{YY}(\tau) = h(\tau) * R_{XY}(\tau)$ gives

$$R_{YY}(\tau) = h(\tau) * h(-\tau) * R_{XX}(\tau) = r_{hh}(\tau) * R_{XX}(\tau) \tag{6.4.13}$$

Note that $h(\tau)$ convolved with $h(-\tau)$ is the same as $h(\tau)$ correlated with itself, so let us define the second-order impulse response as

$$r_{hh}(\tau) = h(\tau) * h(-\tau) = h(\tau) \otimes h(\tau) \tag{6.4.14}$$

where $\otimes$ stands for correlation. It is the impulse response correlated with itself. Notice that we use lowercase r for the time correlation function and uppercase R for the ensemble correlation. This is consistent with the notation used in Chapter 5.

EXAMPLE 6.4.2.   The analog system in Fig. 6.4.5a has impulse response $h(t) = 1$ for $0 < t < T$, and 0 elsewhere, as shown in Fig. 6.4.5b. If zero-mean white noise with variance σ^2 is applied to this system, find the output autocorrelation function.

SOLUTION: We first calculate $r_{hh}(\tau) = \int h(t)h(t - \tau) \, dt$. The details are shown in Fig. 6.4.6a, first for $\tau < -T$, giving

$$r_{hh}(\tau) = 0, \qquad \tau < -T$$

Next,

$$r_{hh}(\tau) = \int_{0}^{\tau+T} (1) \, dt = \tau + T, \qquad -T < \tau < 0$$

Then,

$$r_{hh}(\tau) = \int_{\tau}^{T} (1) \, dt = T - \tau, \qquad 0 < \tau < T$$

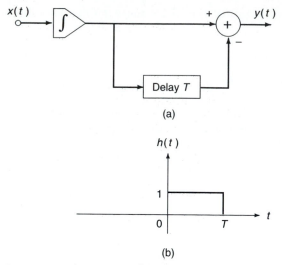

(a)

(b)

Fig. 6.4.5.   A system with impulse response $h(t)$.

Finally,

$$r_{hh}(\tau) = 0, \qquad \tau > T$$

This gives the function shown in Fig. 6.4.6b.

Since the input autocorrelation function is $\sigma^2 \delta(\tau)$, the output autocorrelation function is simply the function in Fig. 6.4.6b multiplied by σ^2.

Discrete-Time Stationary Processes

The convolution summation expresses the relation between system input $x(n)$ and output $y(n)$.

$$y(n) = \sum_{k=-\infty}^{\infty} h(n-k)\, x(k) \tag{6.4.15}$$

Taking the expected value of both sides gives

$$E[Y(n)] = E\left[\sum_{k=-\infty}^{\infty} h(n-k)\, X(k)\right] = \sum_{k=-\infty}^{\infty} h(n-k)\, E[X(k)]$$

or

$$m_Y(n) = \sum_{k=-\infty}^{\infty} h(n-k)\, m_X(k) \tag{6.4.16}$$

When the input process is stationary the input mean is a constant, giving

$$m_Y = m_X \sum_{k=-\infty}^{\infty} h(k) \tag{6.4.17}$$

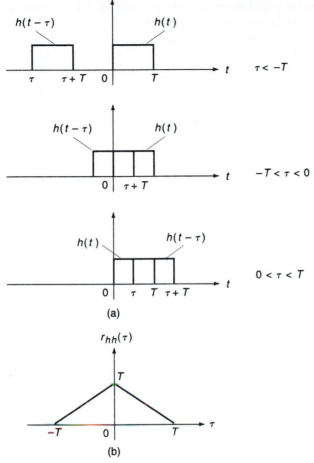

Fig. 6.4.6.   (a) Steps in the correlation. (b) The second-order impulse response $r_{hh}(\tau)$.

The output mean is also a constant equal to the input mean multiplied by the area under the impulse response.

The output autocorrelation function can be obtained by multiplying Eq. 6.4.15 by $y(n)$ and taking the expected values.

$$E[Y(n_1)Y(n_2)] = E\left\{\left[\sum_{k=-\infty}^{\infty} h(n_1 - k)X(k)\right] Y(n_2)\right\}$$

$$= \sum_{k=-\infty}^{\infty} h(n_1 - k)\, E[X(k)Y(n_2)] \qquad (6.4.18)$$

$$= \sum_{k=-\infty}^{\infty} h(n_1 - k)R_{XY}(k, n_2)$$

This expresses the output correlation function in terms of the impulse response and the cross correlation function between input and output. As in the continuous-time case, we want to calculate the output statistics as a function of the input statistics, so we are not quite there yet. Repeating the steps in the continuous-time case, we write

$$R_{XY}(n_1, n_2) = R_{YX}(n_2, n_1) \tag{6.4.19}$$

Now multiply Eq. 6.4.15 by x and take the expected value to get

$$E[Y(n_1)X(n_2)] = \sum_{k=-\infty}^{\infty} h(n_1 - k)E[X(k)X(n_2)]$$

or

$$R_{YX}(n_1, n_2) = \sum_{k=-\infty}^{\infty} h(n_1 - k)R_{XX}(k, n_2) \tag{6.4.20}$$

If $X(t)$ is wide-sense stationary, we can write

$$R_{XX}(n_1, n_2) = R_{XX}(n_1 - n_2)$$

Substituting this and similar expressions into the formulas above leads to

$$R_{YY}(n_1 - n_2) = \sum_{k=-\infty}^{\infty} h(n_1 - k)R_{XY}(k - n_2) \tag{6.4.21a}$$

$$R_{XY}(n_1 - n_2) = R_{YX}(n_2 - n_1) \tag{6.4.22a}$$

$$R_{YX}(n_1 - n_2) = \sum_{k=-\infty}^{\infty} h(n_1 - k)R_{XX}(k - n_2) \tag{6.4.23a}$$

For stationary processes, we calculate the output autocorrelation function by applying Eqs. 6.4.23, 6.4.22, and 6.4.21 in that order.

By the substitutions $l = n_1 - n_2$, and $j = k - n_2$ (similar to the derivation of Eqs. 6.4.10, 6.4.11, and 6.4.12), we can cast these three equations in a simpler form, given by

$$R_{YY}(l) = \sum_{k=-\infty}^{\infty} h(l - k)R_{XY}(k) \tag{6.4.21b}$$

$$R_{XY}(l) = R_{YX}(-l) \tag{6.4.22b}$$

$$R_{YX}(l) = \sum_{k=-\infty}^{\infty} h(l - k)R_{XX}(k) \tag{6.4.23b}$$

These three equations can be combined (see the derivation of Eq. 6.4.13) to obtain

$$R_{YY}(l) = r_{hh}(l) * R_{XX}(l) \tag{6.4.24}$$

(Note that we use the symbol $l = n_1 - n_2$ for the discrete-time shift parameter, in place of τ. We will continue to use τ for continuous-time signals, and for theoretical discussions that apply to both types of signals. We do this not to add complexity, but to agree with most of the literature, and also to make it clear when we are concerned only with discrete-time signals.)

One more point: Since the covariance function is the correlation function for a random process with the mean removed, the covariance function must satisfy a similar set of equations. These are given by

$$\mathcal{L}_{YY}(n_1 - n_2) = \sum_{k=-\infty}^{\infty} h(n_1 - k)\mathcal{L}_{XY}(k - n_2) \qquad (6.4.25)$$

$$\mathcal{L}_{XY}(n_1 - n_2) = \mathcal{L}_{YX}(n_2 - n_1) \qquad (6.4.26)$$

$$\mathcal{L}_{YX}(n_1 - n_2) = \sum_{k=-\infty}^{\infty} h(n_1 - k)\mathcal{L}_{XX}(k - n_2) \qquad (6.4.27)$$

With $l = n_1 - n_2$ we can combine these to obtain

$$\mathcal{L}_{YY}(l) = r_{hh}(l) * \mathcal{L}_{XX}(l) \qquad (6.4.28)$$

We did not mention these relations for continuous-time systems, but of course they exist.

EXAMPLE 6.4.3.   White Gaussian noise with mean m and covariance function $\mathcal{L}_{XX}(n_1 - n_2) = \sigma^2 \, \delta(n_1 - n_2)$ is applied to the FIR moving-average filter whose impulse response is shown in Fig. 6.4.7. Find the output mean and correlation function.

SOLUTION: The output mean value is the input mean times the sum of terms in the impulse response. This sum is 1, so $m_Y = m_X$. To find the output correlation function we first apply Eq. 6.4.27. Figure 6.4.8a shows

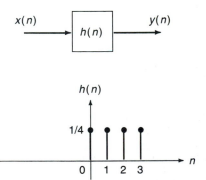

Fig. 6.4.7.   A moving-average FIR filter.

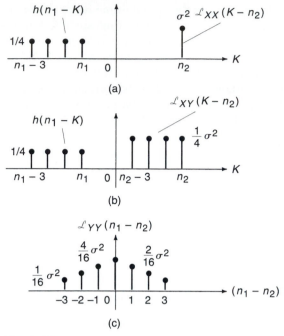

Fig. 6.4.8

$h(n_1 - k)$ and $\mathcal{L}_{XX}(k - n_2)$ for $n_2 > n_1$. You can see that as $n_1 - n_2$ increases, Eq. 6.4.26 gives

$$\mathcal{L}_{YX}(n_1 - n_2) = \begin{cases} \frac{1}{4}\sigma^2 & \text{for } n_1 - n_2 = 0, 1, 2, 3 \\ 0 & \text{otherwise} \end{cases}$$

Now when we apply Eq. 6.4.26, we get the function $\mathcal{L}_{XY}$ shown in Fig. 6.4.8b. Equation 6.4.24 then multiplies the two functions in Fig. 6.4.8b and sums to get the output covariance function $\mathcal{L}_{YY}(n_1 - n_2)$ shown in Fig. 6.4.8c.

EXAMPLE 6.4.4.   Suppose that white Gaussian noise with mean 0 and variance σ^2 is applied to the first-order filter in Fig. 6.4.9. Find the output autocorrelation function.

SOLUTION: The second-order system impulse response is found by correlating the first-order impulse response with itself:

$$r_{hh}(l) = \sum_{n=-\infty}^{\infty} h(n)\, h(n - l)$$

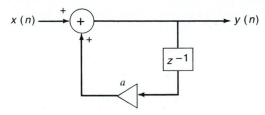

Fig. 6.4.9   A first-order filter.

where $h(n) = a^n u(n)$. Figure 6.4.10a shows the situation for $l < 0$. The product of the two impulse response functions is nonzero for $n > 0$, giving

$$r_{hh}(l) = \sum_{n=0}^{\infty} a^n a^{(n-l)} = a^{-l} \sum_{n=0}^{\infty} a^{2n} = \frac{a^{-l}}{1 - a^2}, \qquad l < 0$$

Figure 6.4.10b depicts $l \geq 0$. For this situation we have

$$r_{hh}(l) = \sum_{n=l}^{\infty} a^n a^{(n-l)} = a^{-l} \sum_{n=l}^{\infty} a^{2n}, \qquad l \geq 0 \qquad (6.4.29)$$

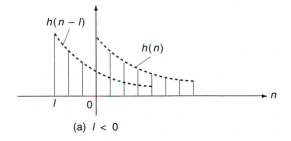

(a) $l < 0$

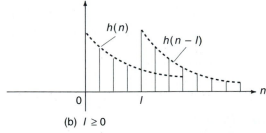

(b) $l \geq 0$

Fig. 6.4.10.   Correlating $h(n)$ with itself.

Now we have the problem of evaluating the sum from $n = l$ to ∞, instead of the familiar limits from 0 to ∞. We can write

$$\sum_{n=0}^{\infty} (a^2)^n = \sum_{n=0}^{l-1} (a^2)^n + \sum_{n=l}^{\infty} (a^2)^n$$

or

$$\frac{1}{1 - a^2} = \frac{1 - a^{2l}}{1 - a^2} + \sum_{n=l}^{\infty} (a^2)^n$$

Substituting this into Eq. 6.4.29 gives

$$r_{hh}(l) = a^{-l} \sum_{n=l}^{\infty} a^{2n} = \frac{a^l}{1 - a^2}, \qquad l \geq 0$$

Since $R_{XX}(l) = \sigma^2\, \delta(l)$, the output correlation function is given by

$$R_{YY}(l) = \frac{\sigma^2}{1 - a^2}\, a^{|l|}$$

Review

The second-order impulse response $r_{hh}(\tau)$ is to random signal analysis as the impulse response $h(t)$ is to deterministic signal analysis. For wide-sense stationary processes, the relationship between input and output autocorrelation functions is given by Eq. 6.4.13 (continuous time) and by Eq. 6.4.24 (discrete time). Equation 6.4.28 provides these same relations for the covariance function.

6.5 Spectral Relations for LTI Systems _____

Preview

We use one of three methods to analyze linear systems with deterministic inputs: convolution, differential (or difference) equations, and frequency analysis (transforms). These same three methods apply to linear systems with random inputs, but we use second-order descriptions. In the previous section we introduced what could be called second-order convolution. In this section we will repeat Section 6.4, except that we will do so in the frequency domain. We skip any discussion of differential or difference equations, but they apply to linear systems with random inputs in much the same way that convolution and transform analysis apply.

Continuous-Time Stationary Processes

Equation 6.4.10 says that the cross-correlation function R_{YX} is the convolution of R_{XX} with h. If we take transforms of each term, we get

$$G_{YX}(\Omega) = H(\Omega)G_{XX}(\Omega) \tag{6.5.1}$$

This is just an application of the convolution property of Fourier transforms. The power spectral density function for the input X is related to the cross spectral density function between input and output by the system transfer function.

From Eq. 6.4.11 it follows that G_{XY} is the complex conjugate of G_{YX}. Using this in Eq. 6.5.1 gives

$$G_{XY}(\Omega) = H(-\Omega)G_{XX}(\Omega) \tag{6.5.2}$$

From Eq. 6.4.12 we get

$$G_{YY}(\Omega) = H(\Omega)G_{XY}(\Omega) \tag{6.5.3}$$

Substituting Eq. 6.5.2 into this gives

$$G_{YY}(\Omega) = H(\Omega)H(-\Omega)G_{XX}(\Omega) = |H(\Omega)|^2 G_{XX}(\Omega) \tag{6.5.4}$$

The term $|H(\Omega)|^2$ is often called the *second-order system function* or the *power transfer function*. It is to random signal analysis what the transfer function $H(\Omega)$ is to deterministic signal analysis. It is the transform of $r_{hh}(\tau)$ from Eq. 6.4.14. Another way to derive Eq. 6.5.4 is to transform Eq. 6.4.13.

We can obtain more general relations corresponding to Eqs. 6.5.1–6.5.4 by substituting $s = j\Omega$:

$$G_{YX}(s) = H(s)G_{XX}(s) \tag{6.5.5}$$

$$G_{XY}(s) = H(-s)G_{XX}(s) \tag{6.5.6}$$

$$G_{YY}(s) = H(s)G_{XY}(s) \tag{6.5.7}$$

$$G_{YY}(s) = H(s)H(-s)G_{XX}(s) = |H(s)|^2 G_{XX}(s) \tag{6.5.8}$$

EXAMPLE 6.5.1.   White noise with mean 0 and variance 1 is applied to an RC low-pass filter with time constant 1 (Fig. 6.4.2). Find the output power spectral density function as a function of both Ω and s.

SOLUTION: Notice that "power" spectral density function is specified. That is because the concepts in this section apply only to power signals, i.e., to stationary processes. To find the power transfer function we find the first-order transfer function by taking the Fourier transform of $h(t)$ in Example 6.4.1. This gives

$$H(\Omega) = \frac{1}{1 + j\Omega}$$

Then

$$|H(\Omega)|^2 = \frac{1}{1 + \Omega^2}$$

The input white noise has constant spectrum σ^2. Multiplying the constant spectrum by the power transfer function produces the output power spectral density function as shown in Fig. 6.5.1.

Now, to obtain the power transfer function as a function of s, let $s = j\Omega$, or $s^2 = -\Omega^2$, to get

$$|H(s)|^2 = \frac{1}{1 - s^2}$$

Discrete-Time Stationary Processes

When we take the transform of Eq. 6.4.23b we get

$$G_{YX}(\omega) = H(\omega)G_{XX}(\omega) \tag{6.5.9}$$

Some authors write these functions in terms of $e^{j\omega}$ rather than in terms of ω, so they would write

$$G_{YX}(e^{j\omega}) = H(e^{j\omega})G_{XX}(e^{j\omega})$$

which is equivalent to Eq. 6.5.9. This different notation arises because in terms of the z transform, Eq. 6.5.9 would be written as

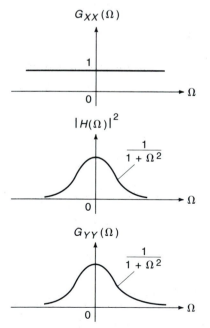

Fig. 6.5.1. The power transfer function and input–output spectra for Example 6.5.1.

$$G_{YX}(z) = H(z)G_{XX}(z)$$

and for sinusoidal analysis, which is the basis for frequency selective filter design, we let $z = e^{j\omega}$. The exponential notation is probably clearer, because it reminds us that we are restricted to the unit circle in the z domain. We will stick with the notation in Eq. 6.5.9, but you should be aware that we are on the unit circle in the z plane.

Now, if we take the conjugate of Eq. 6.5.9 and apply Eq. 6.4.22b, we get

$$G_{XY}(\omega) = H(-\omega)G_{XX}(\omega) \qquad (6.5.10)$$

From Eq. 6.4.21b,

$$G_{YY}(\omega) = H(\omega)G_{XY}(\omega) \qquad (6.5.11)$$

Substituting Eq. 6.5.10 into 6.5.11 gives

$$G_{YY}(\omega) = H(\omega)H(-\omega)G_{XX}(\omega) = |H(\omega)|^2 G_{XX}(\omega) \qquad (6.5.12)$$

These equations are identical in form to those derived above for the continuous-time case. The only difference is between ω and Ω. But this is a big difference. The units of these parameters are:

Ω, radians per second
ω, radians per sample

and Ω ranges over the infinite interval, while ω is restricted to a 2π interval.

In terms of the z transform, Eqs. 6.5.9, 6.5.10, 6.5.11, and 6.5.12 are given by

$$G_{YX}(z) = H(z)G_{XX}(z) \qquad (6.5.13)$$
$$G_{XY}(z) = H(z^{-1})G_{XX}(z) \qquad (6.5.14)$$
$$G_{YY}(z) = H(z)G_{XY}(z) \qquad (6.5.15)$$
$$G_{YY}(z) = H(z)H(z^{-1})G_{XX}(z) = |H(z)|^2 G_{XX}(z) \qquad (6.5.16)$$

EXAMPLE 6.5.2.   The problem in Example 6.4.3 was stated in the time domain. Here we consider the same problem, but in the frequency domain. White Gaussian noise with mean m and covariance function $\mathcal{L}_{XX}(l) = \sigma^2 \delta(l)$ is applied to the FIR moving-average filter whose impulse response is shown in Fig. 6.4.7. Find the input and output power spectral density function.

SOLUTION: The mean m is the dc component in the input signal. Therefore the power spectral density has a discrete component of magnitude m^2 at zero frequency. Also, the transform of the covariance function is a constant with frequency of magnitude σ^2. Combining these facts gives the input spectrum shown in Fig. 6.5.2.

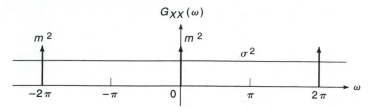

Fig. 6.5.2.   The input power spectrum in Example 6.5.2.

In order to find the output spectrum by Eq. 6.5.12, we need the system power transfer function. So we first calculate $H(\omega)$, given by

$$H(\omega) = \sum_{n=0}^{\infty} h(n)e^{-j\omega n} = 0.25[1 + e^{-j\omega} + e^{-j2\omega} + e^{-j3\omega}]$$
$$= 0.25[1 + e^{-j\omega} + e^{-j2\omega} + e^{-j3\omega}]e^{j1.5\omega}e^{-j1.5\omega}$$

where the last two exponential terms are attached to make the next step easier. Notice that their product equals 1, so we are simply multiplying by 1. Multiplying the first term (with the positive exponent) inside the brackets gives

$$H(\omega) = 0.25[e^{j1.5\omega} + e^{j0.5\omega} + e^{-j0.5\omega} + e^{-j1.5\omega}]e^{-j1.5\omega}$$
$$= 0.5[\cos(1.5\omega) + \cos(0.5\omega)]e^{-j1.5\omega}$$

This gives the power transfer function

$$|H(\omega)|^2 = 0.25[\cos(1.5\omega) + \cos(0.5\omega)]^2$$

Multiplying this by the input spectrum in Fig. 6.5.2 gives the output spectrum in Fig. 6.5.3.

Review

The output spectrum of a system is the input spectrum shaped by the system. This basic idea shapes the design of systems for any application with random input. Equations 6.5.8, 6.5.12, and 6.5.16 all say this.

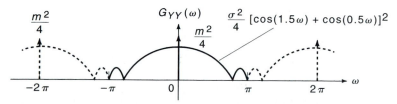

Fig. 6.5.3.   The output power spectrum in Example 6.5.2.

6.6   The Crux of the Matter _____

Preview

In every discipline there are a few concepts that are central to understanding, that transcend mere formulas, and that provide true insight. Such a central concept for random signal processing is provided by the topics in this section. Here we take material from the previous two sections and put it all together. The crucial point here is that the output correlation/spectrum is shaped by the system. As we saw in Section 6.4, when we correlate the (first-order) impulse response $h(t)$ with itself, the result is the second-order impulse response function $r_{hh}(\tau)$ for continuous-time systems, or $r_{hh}(l)$ for discrete-time systems. In either event, this function characterizes the system. Likewise, we saw in Section 6.5 that the power transfer function $|H(\Omega)|^2$ or $|H(\omega)|^2$ characterizes the system in the frequency domain. The output spectrum is shaped by the power transfer function.

Now consider the following question: Suppose we wish to design a system to remove noise from a signal with a known autocorrelation function. What type of system would serve our purpose best? Although we have yet to say what is "best" or anything else about optimum systems, we can gain considerable insight by considering this problem in a general way. In this problem the system input is signal $s(t)$ plus noise $n(t)$, where the signal and noise have known autocorrelation functions, and therefore known power spectra. This means that in the frequency domain there should be some frequencies where the signal has large power compared with the noise, and other frequencies where just the opposite occurs. Therefore, in the design of frequency selective filters, we should put the filter where the signal is present and where the noise is absent.

Thus system designers use knowledge of the input spectrum, combined with their experience in knowing what the system does to the input signal, in order to design useful systems. In the time domain the system shapes the input autocorrelation function R_{XX} into the output correlation R_{YY}. In the frequency domain the system reshapes the input spectrum G_{XX} to produce the output spectrum G_{YY}. This section consists of little more than a series of examples designed to illustrate these ideas, but do not underestimate the importance of these concepts.

We will first study the simple RC low-pass filter shown in Fig. 6.6.1. The relationship between the input $v_1(t)$ and the output $v_2(t)$ can be established by any one of numerous circuit analysis techniques familiar to all. To pick one, let us write

$$V_1(s) = RI(s) + \frac{I(s)}{sC}$$

Substituting $I(s) = sC \cdot V_2(s)$ gives

$$V_1(s) = (sRC + 1)V_2(s)$$

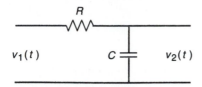

Fig. 6.6.1.   A first-order low-pass filter.

With $a = 1/RC$ and $H(s) = V_2(s)/V_1(s)$, we get

$$H(s) = \frac{a}{s + a} \leftrightarrow h(t) = ae^{-at}u(t) \qquad (6.6.1)$$

The impulse response $h(t)$ is the (Laplace) transform of the transfer function $H(s)$, and that is what the double-headed arrow stands for. These first-order functions, $H(\Omega)$ and $h(t)$, are plotted in Fig. 6.6.2, where $H(\Omega) = H(s)$ with $s = j\Omega$.

We can derive the second-order system description from $h(t)$ and $H(s)$ as follows. First, in the time domain, correlate $h(t)$ with itself to get

$$r_{hh}(\tau) = \int_{-\infty}^{\infty} h(t)\, h(t - \tau)\, dt = \frac{a}{2}\, e^{-a|\tau|} \qquad (6.6.2)$$

The second-order transfer function is given by

$$|H(s)|^2 = H(s) \cdot H(-s) = \frac{a}{s + a}\ \frac{a}{-s + a} = \frac{a^2}{-s^2 + a^2}$$

which we can find from the first-order transfer function $H(s)$, or we can find it by taking the transform of $r_{hh}(\tau)$. With $s = j\Omega$ this gives

$$|H(\Omega)|^2 = \frac{a^2}{\Omega^2 + a^2} \qquad (6.6.3)$$

Now, if we apply white noise with variance σ^2 to this system, the output will be colored noise with autocorrelation and power spectrum shaped by the system as illustrated in Fig. 6.6.3. These shapes are described by Eqs. 6.6.2 and 6.6.3 when scaled by the input power σ^2.

The important concept to be gained from this and the following examples is that the system output correlation and spectrum functions are shaped by the system. In Fig. 6.6.3, the input autocorrelation function is $\delta(\tau)$, and the output autocorrelation function is the double-sided exponential function $R_{YY}(\tau)$. The system function $r_{hh}(\tau)$ determines the shape of $R_{YY}(\tau)$. In the frequency domain, the flat input spectrum $G_{XX}(\Omega)$ is altered by the system function $|H(\Omega)|^2$ to produce $G_{YY}(\Omega)$. Thus the system shapes the output.

For a second continuous-time example, consider the system shown in Fig. 6.6.4. A delayed version of the input signal $x(t)$ is subtracted from $x(t)$ and

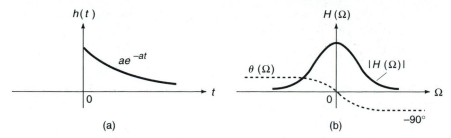

Fig. 6.6.2. The impulse response (a) and frequency response (b) of the RC low-pass filter.

integrated to obtain the output $y(t)$. The impulse response is a square pulse as shown in Fig. 6.6.5a. You can derive this response by applying an impulse $\delta(t)$ to the system, so the input to the integrator is $\delta(t) - \delta(t - T)$. Since the integral of an impulse is a step, the impulse response is the difference $u(t) - u(t - T)$, or the square pulse in Fig. 6.6.5a. This gives

$$h(t) = u(t) - u(t - T) \leftrightarrow H(s) = \frac{1}{s}(1 - e^{-sT}) \qquad (6.6.4)$$

Then $H(\Omega) = H(s)$ with $s = j\Omega$, or

$$\begin{aligned} H(\Omega) &= \frac{1}{j\Omega}(1 - e^{-j\Omega T}) \cdot e^{j\Omega T/2} \cdot e^{-j\Omega T/2} \\ &= \frac{2}{\Omega}\left(\frac{e^{j\Omega T/2} - e^{-j\Omega T/2}}{2j}\right) e^{-j\Omega T/2} \qquad (6.6.5) \\ &= T\left(\frac{\sin(\Omega T/2)}{\Omega T/2}\right) e^{-j\Omega T/2} \end{aligned}$$

This is the function plotted in Fig. 6.6.5b.

When we correlate $h(t)$ with itself to obtain $r_{hh}(\tau)$, the result is a triangular pulse of maximum height T and width $2T$ centered at the origin. When white noise with variance σ^2 (Fig. 6.6.6a) is applied to the system, the output autocorrelation function is $r_{hh}(\tau)$ multiplied by the variance, as shown in Fig. 6.6.6b.

The output spectrum can be found by either of two ways: The transform of $R_{YY}(\tau)$ gives $G_{YY}(\Omega)$, and $H(\Omega) H(-\Omega)$ multiplied by the variance gives $G_{YY}(\Omega)$. Either method gives the output spectrum shown in Fig. 6.6.6d, which is found by multiplying the input spectrum in Fig. 6.6.6c by $|H(\Omega)|^2$.

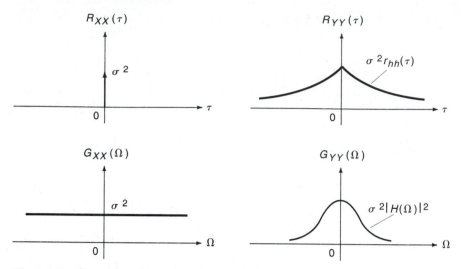

Fig. 6.6.3.   Second-order response to white noise.

Figure 6.6.7 shows the digital counterpart to the RC low-pass filter. For $0 < a < 1$, this system is a low-pass filter with impulse response given by

$$h(n) = a^n u(n) \leftrightarrow H(z) = \frac{1}{1 - az^{-1}} \tag{6.6.6}$$

Substituting $z = e^{j\omega}$ into the transfer function gives

$$H(\omega) = \frac{1}{1 - ae^{-j\omega}} \tag{6.6.7}$$

These functions are plotted in Fig. 6.6.8.

We correlate the impulse response with itself to obtain the second-order impulse response $r_{hh}(l)$. When white noise with mean 0 and variance σ^2 is applied to the system, the result is shown in Fig. 6.6.9. The output power spectrum shown there can be obtained in either of two ways: Transform $R_{YY}(l)$ or multiply the power transfer function by the input power spectral density function. The diagram depicts the second method.

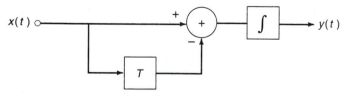

Fig. 6.6.4

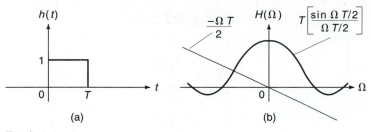

Fig. 6.6.5

For our final example, consider the FIR moving-average filter in Fig. 6.6.10. There are N taps and therefore $N - 1$ delays. The impulse response is shown in Fig. 6.6.11a. The frequency response is given by

$$H(z) = \sum_{n=0}^{N-1} \frac{1}{N} z^{-N} = \frac{1}{N}[1 + z^{-1} + z^{-2} + \cdots + z^{-(N-1)}]$$

The plot in Fig. 6.6.11b is for $N = 2$, where $H(\omega)$ is found by substituting $z = e^{j\omega}$ into the above expression for $H(z)$, or

$$H(\omega) = \frac{1}{2}(1 + e^{-j\omega}) = \cos\left(\frac{\omega}{2}\right) e^{-j\omega/2}$$

Figure 6.6.12 shows the output correlation and power spectrum for the averaging filter with $N = 2$ when white noise is applied. If we add more

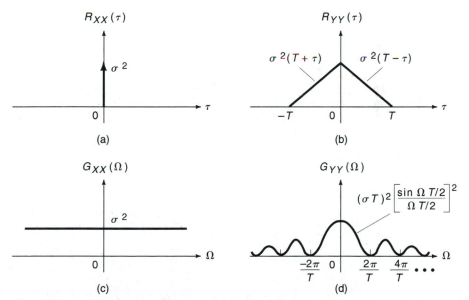

Fig. 6.6.6.   Second-order response to white noise.

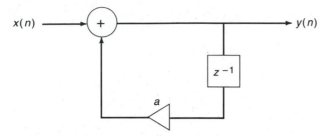

Fig. 6.6.7. A first-order IIR system.

filter taps (increase N), the output correlation and power spectrum will change, and this change will be dictated by the system.

Review

The point of all these examples is that the output is shaped by the system. If white noise is applied at the system input, the output correlation and spectrum assume the characteristic shape of the system. But if the input is not white, the output is still shaped by the system. So in general we can say that the output shape is determined by both the system and the input signal. This idea is central to the design of random signal processing systems. We will expand these concepts in Section 8.4, where we introduce spectral factorization and explore further the relation between the power spectrum and the system function.

6.7 Problems _____

6.1. Define the term stochastic process.

6.2. The stochastic process in Fig. 6.7.1 has three sample functions. Their probabilities are $P(\zeta_1) = \frac{1}{2}$, $P(\zeta_2) = \frac{1}{3}$, $P(\zeta_3) = \frac{1}{6}$.
(a) Find and plot the cdf $F_X(\alpha, t_1)$
(b) What is the probability $P\{X(t_1) \le 4, X(t_2) \le 3\}$?
(c) Calculate $R_{XX}(t_1, t_2)$.

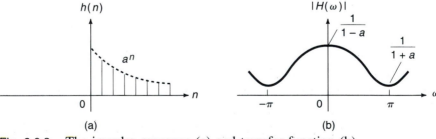

Fig. 6.6.8.   The impulse response (a) and transfer function (b).

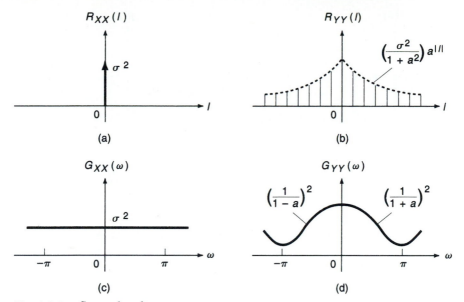

Fig. 6.6.9.   Second-order system response.

6.3.   A stochastic process is generated as follows: Starting at $t = 0$, we toss a coin every second. If heads comes up, the random variable X is set equal to 1. If tails comes up, $X = 0$. Calculate:
(a) The mean, $m_X(t)$
(b) The correlation, $R_{XX}(0.5, 0.6)$
(c) The correlation, $R_{XX}(0.5, 1.5)$

6.4.   A stochastic process is generated as follows: Starting at $t = 0$, we toss a die every second. If the one-spot or two-spot comes up, the random variable X is set equal to 1. Otherwise, $X = 0$. Calculate:
(a) The mean, $m_X(t)$
(b) The correlation, $R_{XX}(0.5, 0.6)$
(c) The correlation, $R_{XX}(0.5, 1.5)$

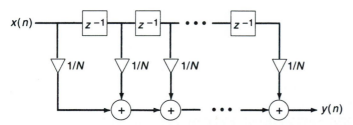

Fig. 6.6.10.   An averaging filter with N taps and $N - 1$ delays.

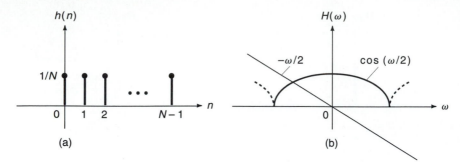

Fig. 6.6.11.   Impulse response and transfer function (for $N = 2$) for the averaging filter.

6.5.   A stochastic process consists of three time functions as shown in Fig. 6.7.2. The probabilities are $P(\zeta_1) = \frac{1}{2}, P(\zeta_2) = \frac{1}{4}, P(\zeta_3) = \frac{1}{4}$.
(a) Find $F_{XX}(3, 4; t_1, t_2)$
(b) Find $F_{XX}(4, 3; t_1, t_2)$
(c) Find $R_{XX}(t_1, t_2)$

6.6.   A random process consists of three sample functions,

$$X(t, \zeta_1) = 1 \qquad X(t, \zeta_2) = \sin t \qquad X(t, \zeta_3) = \cos t$$

each occurring with equal probability.

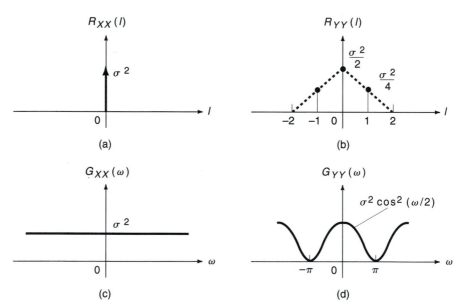

Fig. 6.6.12.   Output autocorrelation and power spectral density functions for the averaging filter for $N = 2$.

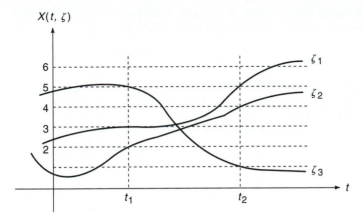

Fig. 6.7.1

(a) Is the process stationary? Why?

(b) Calculate the mean $m_X(t)$.

(c) Calculate the autocorrelation $R_{XX}(t_1, t_2)$.

6.7. The sample function $x(n)$ in Table 6.7.1 is a segment from a stochastic process. Estimate the values of $R_{XX}(0)$ and $R_{XX}(1)$.

6.8. Determine an expression for the pdf of a Gaussian random process for three samples taken at times $t_1 = 0$, $t_2 = 1$, and $t_3 = 3$ s. The mean is 0 and the correlation function is given by

$$R_{XX}(t_1, t_2) = \begin{cases} 3, & |t_2 - t_1| = 0 \\ 2, & |t_2 - t_1| = 1 \\ 1, & |t_2 - t_1| = 2 \\ 0 & \text{otherwise} \end{cases}$$

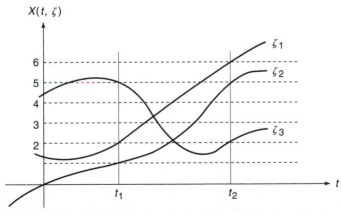

Fig. 6.7.2

6.9 Two statistically independent, stationary random processes $X(t)$ and $Y(t)$ with mean 0 have the following autocorrelation functions:

$$R_{XX}(\tau) = e^{-|\tau|} \qquad R_{YY}(\tau) = \cos(2\pi\tau)$$

Find the autocorrelation function of
(a) $Z(t) = X(t) + Y(t)$
(b) $W(t) = X(t) - Y(t)$
(c) Find the cross-correlation function $R_{ZW}(\tau)$.

6.10. A stochastic process is specified by a transformation of variables as

$$X(t) = A \cos(\omega t + \theta)$$

where A and ω are constants but θ is a random variable uniformly distributed between 0 and π. Determine whether the process is stationary.

6.11. A Gaussian process has mean $m_X(t) = 2$ and covariance function $\mathcal{L}_{XX}(\tau) = 8 \cos(\pi\tau)$. Samples are taken $\frac{1}{2}$s apart. Write the second-order density function.

6.12. A stationary Gaussian process with mean 0 has autocorrelation function

$$R_{XX}(\tau) = \frac{\sin \pi\tau}{\pi\tau}$$

Write the third-order density function for samples taken at $t_1 = 0$, $t_2 = \frac{1}{2}$, $t_3 = 1$.

Computer Assignments

6.13. Produce discrete-time white Gaussian noise with unit variance using a pseudo-random number generator. (See Section 1.3.) Let X_i, $i = 1, 2, \ldots$ be the sequence of uniformly distributed random numbers with mean 0 and range from -0.5 to 0.5. The Gaussian numbers are given by

$$y_1 = \sum_{i=1}^{12} x_i, \qquad y_2 = \sum_{i=13}^{24} x_i, \qquad \ldots, \qquad y_n = \sum_{i=12n-11}^{12n} x_i, \qquad \ldots$$

Table 6.7.1

$x(n)$	1	2	2.5	1	2.5	-2	0.6	-2.2	-1.5
n	1	2	3	4	5	6	7	8	9

(a) Generate 1000 uncorrelated Gaussian numbers this way.
(b) Calculate the autocorrelation function for the Gaussian numbers using the methods described in Eq. 5.2.9 to check that the signal is indeed white.
(c) Find and plot an empirical cdf for these numbers and check that they are Gaussian.
(d) Check to see that the power spectrum is flat. We have not discussed any formal methods for estimating spectra, but you should refer to Eqs. 5.4.15 and 5.4.17 for guidance. (Your results here should make you appreciate better methods of spectral estimation.)

6.14. Simulate a moving-average filter of length 4 (see Fig. 6.4.7) and pass the white Gaussian noise for Problem 6.13 through this filter.
(a) Find and plot the filter output autocorrelation function for $\tau = 0$, 1, ..., 10. Do these results agree with Fig. 6.4.8?
(b) Find an empirical cdf for the filter output signal. Is the signal still Gaussian?
(c) Estimate the output spectrum.

6.15. Produce discrete-time correlated Gaussian noise with unit variance using a pseudo-random number generator. (See Section 1.3.) Let X_i, $i = 1, 2, \ldots$ be the sequence of uniformly distributed random numbers with mean 0 and range from -0.5 to 0.5. The Gaussian numbers are given by

$$y_1 = \sum_{i=1}^{12} x_i, \qquad y_2 = \sum_{i=2}^{13} x_i, \qquad \ldots, \qquad y_n = \sum_{i=n}^{n+11} x_i, \qquad \ldots$$

(a) Generate 1000 correlated Gaussian numbers this way.
(b) Calculate the autocorrelation function for the Gaussian numbers using the methods described in Eq. 5.2.9 to check that the signal has correlation like $x_{11}(n)$ in Fig. 1.2.6.
(c) Find and plot an empirical cdf for these numbers and check that they are Gaussian.
(d) Find the power spectrum. Estimate it from the samples, and calculate it using Eq. 5.4.17. Do they agree?

6.16. Simulate a moving-average filter of length 4 (see Fig. 6.4.7) and pass the signal from Problem 6.15 through this filter.
(a) Find and plot the filter output autocorrelation function for $\tau = 0$, 1, ..., 10. Do these results agree with Fig. 6.4.8?
(b) Find an empirical cdf for the filter output signal. Is the signal still Gaussian?
(c) Estimate the output spectrum.

Further Reading ─────────────────────────────

1. ATHANASIOS PAPOULIS, *Probability, Random Variables, and Stochastic Processes,* 3rd ed., McGraw-Hill, New York, 1991.
2. CHARLES W. THERRIEN, Discrete Random Signals and Statistical Signal Processing, Prentice Hall, Englewood Cliffs, NJ, 1992.
3. MICHAEL O'FLYNN, *Probabilities, Random Variables, and Random Processes,* Harper & Row, New York, 1982.

CHAPTER 7 _____

Least-Squares Techniques

7.1 Least-Squares Estimation _____

Preview

Least-squares techniques are used to make estimates of every sort. We call this technique "data-driven" or "deterministic," meaning that the mean, the variance, the pdf, etc., of the data are not used. Instead we base any estimates derived by this technique on samples.

Least squares applies not only to line fitting, which we describe in this section, but also to many other data-driven problems. As a result, numerous techniques have been developed that enhance and simplify its application. We will describe some of these in Sections 7.2 and 7.3. We begin with an introduction to the idea behind least-squares estimation.

Least squares applies to problems where data are given and we can define an error that involves the quantity to be estimated. The simplest problem is estimating a constant from noisy measurements. Let y be the unknown constant, let $\hat{y}$ be the estimate, and let $x_1, x_2, \ldots, x_n$ be the noisy measurements.

$$x_1 = y + e_1$$
$$x_2 = y + e_2$$
$$\vdots$$
$$x_n = y + e_n$$

The e_i terms could be additive noise, measurement error, or some other source of uncertainty. In any event, if we solve for each error term and square the result, we have a list of squared errors.

$$e_1^2 = (x_1 - y)^2$$
$$e_2^2 = (x_2 - y)^2$$
$$\vdots$$
$$e_n^2 = (x_n - y)^2$$

The least-squares criterion specifies that we should choose as our estimate the value $\hat{y}$ that minimizes the sum of squared errors. Let ε denote this sum.

$$\varepsilon = \sum_{i=1}^{n} e_i^2 = \sum_{i=1}^{n} (x_i - y)^2 \qquad (7.1.1)$$

We solve for the value of $y = \hat{y}$ that minimizes this sum, which means we should set the derivative of ε with respect to y equal to 0.

$$\frac{d}{dy} \varepsilon = \frac{d}{dy} \sum_{i=1}^{n} (x_i - y)^2 = -2 \sum_{i=1}^{n} (x_i - y) = 0$$

Solving for y gives the least-squares estimate $\hat{y}$:

$$\hat{y} = \frac{1}{n} \sum_{i=1}^{n} x_i \qquad (7.1.2)$$

Notice that this is the average of the observations, as we would expect.

Least squares often applies to situations where there is a relation between two or more variables. For example, weight is related to height in adult males, stress is related to strain in metal, and force is related to displacement in a spring. None of these relationships is exact, but some sort of curve should provide a reasonable model for many applications.

Table 7.1.1 shows stress (y_i in psi $\times 10^3$) versus strain (x_i in strain $\times 10^{-3}$) for a material. Figure 7.1.1 shows these same data plotted on a scatter diagram.

We often model data like these by a straight-line relationship:

$$\hat{y} = c_0 + c_1 x \qquad (7.1.3)$$

where y is the stress, $\hat{y}$ is our guess of the stress, x is the strain, and c_0 and c_1 are the intercept and the slope of the line. The problem in least-squares estimation is to choose the line that best fits the data. The measure for deciding which line is best is the sum of square errors.

Let the error be given by

$$e = y - \hat{y} = y - c_0 - c_1 x$$

For the given data we have an error for each point:

$$e_1 = y_1 - c_0 - c_1 x_1$$
$$e_2 = y_2 - c_0 - c_1 x_2$$
$$\vdots$$
$$e_n = y_n - c_0 - c_1 x_n$$

Table 7.1.1.   Stress (y_i) Versus Strain (x_i)

i	x_i	y_i
1	0.25	10
2	0.5	14
3	0.75	20
4	1.0	22
5	1.25	29
6	1.5	31
7	1.75	34
8	2.0	35
9	2.25	37
10	2.5	38

Define the sum of the squared errors as ε:

$$\varepsilon = \sum_{i=1}^{n} e_i^2 = \sum_{i=1}^{n} (y_i - c_0 - c_1 x_i)^2 \qquad (7.1.4)$$

With these definitions the procedure for choosing the parameters of the straight line is straightforward. Choose c_0 and c_1 to minimize ε. Setting derivatives to 0, we get

$$\frac{\partial \varepsilon}{\partial c_0} = -2 \sum_{i=1}^{n} (y_i - c_0 - c_1 x_i) = 0$$

giving

$$\sum_{i=1}^{n} y_i = n c_0 + c_1 \sum_{i=1}^{n} x_i \qquad (7.1.5)$$

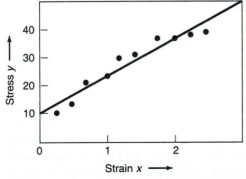

Fig. 7.1.1.   Scatter diagram of data in Table 7.7.1.

Also

$$\frac{\partial \varepsilon}{\partial c_1} = -2 \sum_{i=1}^{n} x_i(y_i - c_0 - c_1 x_i) = 0$$

giving

$$\sum_{i=1}^{n} x_i y_i = c_0 \sum_{i=1}^{n} x_i + c_1 \sum_{i=1}^{n} x_i^2 \qquad (7.1.6)$$

Equations 7.1.5 and 7.1.6 constitute two independent equations in two unknowns, c_0 and c_1. Displayed in matrix form, they become

$$\begin{bmatrix} n & \Sigma x_i \\ \Sigma x_i & \Sigma x_i^2 \end{bmatrix} \begin{bmatrix} c_0 \\ c_1 \end{bmatrix} = \begin{bmatrix} \Sigma y_i \\ \Sigma x_i y_i \end{bmatrix} \qquad (7.1.7)$$

EXAMPLE 7.1.1   Find the line giving the least-squares fit to the data in Table 7.1.1.

SOLUTION: Calculating the sums in Eq. 7.1.7 gives

$$\begin{bmatrix} 10 & 13.75 \\ 13.75 & 24.06 \end{bmatrix} \begin{bmatrix} c_0 \\ c_1 \end{bmatrix} = \begin{bmatrix} 270 \\ 437 \end{bmatrix}$$

Solving for c_0 and c_1 gives $c_0 = 9.46$, $c_1 = 12.76$. Therefore the least-squares fit to the data is

$$\hat{y} = 9.46 + 12.76x$$

which is plotted as the straight line in Fig. 7.1.1.

The principle of least squares applies to many situations in signal processing. Here is an example of one such situation. Sample values from a stochastic process with mean 0, mean square value of 1, and correlation coefficient between successive values of 0.5 are plotted in Fig. 7.1.2. This signal could represent the erratic flight path of a pilot under enemy fire, the values of a radio signal, or the fluctuations of the Dow–Jones stock average. Let us suppose we wish to predict the next value of this signal given the past values. Then we can use the given signal values as data and formulate a least-squares solution as follows.

For a particular time, think of the previous value of the signal $y(n - 1)$ as the data x_i, and the value to be estimated, $y(n)$, as the data y_i in the previous example. Then we can construct a table similar to Table 7.1.1, where the column labeled x_i consists of the first 40 values of the signal (there are 41 values of this signal plotted in Fig. 7.1.2), and the signal values 2 through 41 form the column labeled y_i. This is shown in Table 7.1.2.

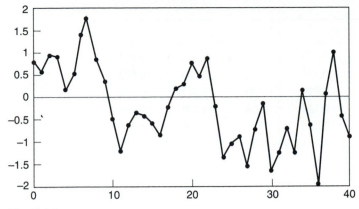

Fig. 7.1.2

We now formulate the estimate as

$$\hat{y}(n) = c_0 + c_1 y(n - 1)$$

and proceed to calculate the sums in Eq. 7.1.7. This gives

$$\begin{bmatrix} 40 & -7.86 \\ -7.86 & 32.86 \end{bmatrix} \begin{bmatrix} c_0 \\ c_1 \end{bmatrix} = \begin{bmatrix} -7.86 \\ 20.66 \end{bmatrix}$$

Solving for c_0 and c_1 gives $c_0 = -0.088$, $c_1 = 0.6098$. Therefore we use the straight-line estimate

$$\hat{y}(n) = -0.088 + 0.6098 y(n - 1)$$

To see how well this works, the signal $y(n)$ and the values predicted for $y(n)$ using the past value in this equation are plotted in Fig. 7.1.3. For a particular time n, the signal $y(n)$ is the curve marked by open dots, and the predicted value for that n is marked by closed dots. As you can see, the estimates are reasonable, although they differ from the actual values to some extent. The sum of square errors for the data in Fig. 7.1.3 is 19.89.

Generalized Least Squares

We have restricted the development above to a straight-line fit. There is no reason for this, except to keep things simple. A curve will obviously fit the data in Fig. 7.1.1 better than the straight line there. For this we can use the second-order model

$$\hat{y} = c_0 + c_1 x + c_2 x^2 \qquad (7.1.8)$$

Repeating the development that led to the least-squares formula in Eq. 7.1.7,

Table 7.1.2

i	x_i	y_i	i	x_i	y_i
1	0.755	0.550	21	0.719	0.400
2	0.550	0.950	22	0.400	0.842
3	0.950	0.929	23	0.842	-0.229
4	0.929	0.154	24	-0.229	-1.398
5	0.154	0.476	25	-1.398	-1.064
6	0.476	1.434	26	-1.064	-0.939
7	1.434	1.754	27	-0.939	-1.545
8	1.754	0.850	28	-1.545	-0.718
9	0.850	0.324	29	-0.718	-0.107
10	0.324	-0.534	30	-0.107	-1.673
11	-0.534	-1.250	31	-1.673	-1.189
12	-1.250	-0.657	32	-1.189	-0.741
13	-0.657	-0.406	33	-0.741	-1.227
14	-0.406	-0.456	34	-1.227	0.140
15	-0.456	-0.581	35	0.140	-0.661
16	-0.581	-0.868	36	-0.661	-1.967
17	-0.868	-0.214	37	-1.967	0.074
18	-0.214	0.169	38	0.074	1.000
19	0.169	0.259	39	1.000	-0.461
20	0.259	0.719	40	-0.461	-0.896

we start with the error, given by

$$e = y - \hat{y} = y - c_0 - c_1 x - c_2 x^2$$

The total squared error is the sum ε, given by

$$\varepsilon = \sum_{i=1}^{n} e_i^2 = \sum_{i=1}^{n} (y_i - c_0 - c_1 x_i - c_2 x_i^2)^2 \qquad (7.1.9)$$

Setting derivatives with respect to the c_i parameters to 0 gives

$$\frac{\partial \varepsilon}{\partial c_0} = -2 \sum_{i=1}^{n} (y_i - c_0 - c_1 x_i - c_2 x_i^2) = 0$$

giving

$$\sum_{i=1}^{n} y_i = n c_0 + c_1 \sum_{i=1}^{n} x_i + c_2 \sum_{i=1}^{n} x_i^2 \qquad (7.1.10)$$

$$\frac{\partial \varepsilon}{\partial c_1} = -2 \sum_{i=1}^{n} x_i (y_i - c_0 - c_1 x_i - c_2 x_i^2) = 0$$

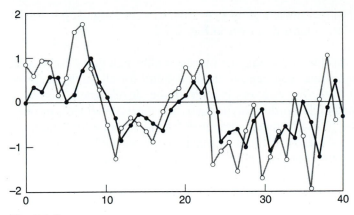

Fig. 7.1.3

giving

$$\sum_{i=1}^{n} x_i y_i = c_0 \sum_{i=1}^{n} x_i + c_1 \sum_{i=1}^{n} x_i^2 + c_2 \sum_{i=1}^{n} x_i^3 \qquad (7.1.11)$$

and

$$\frac{\partial \varepsilon}{\partial c_2} = -2 \sum_{i=1}^{n} x_i^2 (y_i - c_0 - c_1 x_i - c_2 x_i^2) = 0$$

giving

$$\sum_{i=1}^{n} x_i^2 y_i = c_0 \sum_{i=1}^{n} x_i^2 + c_1 \sum_{i=1}^{n} x_i^3 + c_2 \sum_{i=1}^{n} x_i^4 \qquad (7.1.12)$$

This gives three simultaneous linear equations, which can be expressed in matrix form as

$$\begin{bmatrix} n & \Sigma x_i & \Sigma x_i^2 \\ \Sigma x_i & \Sigma x_i^2 & \Sigma x_i^3 \\ \Sigma x_i^2 & \Sigma x_i^3 & \Sigma x_i^4 \end{bmatrix} \begin{bmatrix} c_0 \\ c_1 \\ c_2 \end{bmatrix} = \begin{bmatrix} \Sigma y_i \\ \Sigma x_i y_i \\ \Sigma x_i^2 y_i \end{bmatrix} \qquad (7.1.13)$$

EXAMPLE 7.1.2.   Solve for the quadratic curve to give a least-squares fit to the data in Table 7.1.1.

SOLUTION: We have already calculated most of the sums in Eq. 7.1.13 in the previous example. Those combined with the remaining sums give the matrix equation

$$\begin{bmatrix} 10 & 13.75 & 24.06 \\ 13.75 & 24.06 & 47.27 \\ 24.06 & 47.27 & 98.96 \end{bmatrix} \begin{bmatrix} c_0 \\ c_1 \\ c_2 \end{bmatrix} = \begin{bmatrix} 270 \\ 437 \\ 821 \end{bmatrix}$$

Solving for the c_i parameters gives $c_0 = 2.88$, $c_1 = 25.90$, and $c_2 = -4.78$. This least-squares curve is shown in Fig. 7.1.4.

Having gone this far, and seeing the success with which this second-order polynomial fits the data, you can see the possibility for all kinds of curves to fit different data. We can easily extend the development above to third-, fourth-, and higher-order polynomials, and there are all sorts of other curves we can use. These include Lagrange polynomials and spline functions. We will not need these more complex functions in our applications, so we will not discuss them, but you should be aware of their existence.

Multiple Regression

Regression is another name for least-squares procedures. This odd term originated with a biometrician of the late 1800s named Galton. In an effort to study heredity in an objective manner, he obtained data on the heights of fathers and their sons. Galton found that tall fathers, on the average, had sons who were shorter than they, although not as short as the average man. Short fathers tended to have taller sons, but again they were shorter than the population mean. Galton regarded this result as a "regression" toward the mean of the population, and he called the line fitted to predict the son's height the "regression of the son's height on the father's height." Thus, in our work, when we find the regression of y on x, we fit a curve $y = f(x)$ onto the data x. Multiple regression fits a function onto multivariate data x, $z, \ldots, w$.

The general least-squares procedure applies to any problem where data are given and we wish to discover the relation between one variable (y) and one or more other variables $(x, z, \ldots, w)$. First define a model such as

$$y = c_0 + c_1 x + c_2 z + \cdots + c_k w \qquad (7.1.14)$$

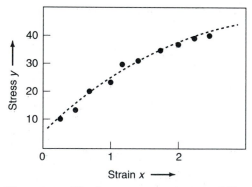

Fig. 7.1.4.   The least-squares curve of Example 7.1.2.

Then the error is $e = y - \hat{y} = y - c_0 - c_1 x + c_2 z - \cdots - c_k w$. Now list the errors.

$$e_1 = y_1 - c_0 - c_1 x_1 - c_2 z_1 - \cdots - c_k w_1$$
$$e_2 = y_2 - c_0 - c_1 x_2 - c_2 z_2 - \cdots - c_k w_2$$
$$\vdots$$
$$e_n = y_n - c_0 - c_1 x_n - c_2 z_n - \cdots - c_k w_n$$

Square each error term and sum to get the sum of squared errors ε.

$$\varepsilon = \sum e_i^2 = \sum (y_i - c_0 - c_1 x_i - c_2 z_i - \cdots - c_k w_i)^2$$

Now solve for the parameters $c_0, c_1, \ldots, c_k$ that minimize the sum. To do this set the partial derivatives equal to 0,

$$\frac{\partial \varepsilon}{\partial c_i} = 0$$

and solve the resulting equations for the unknown parameters. Then Eq. 7.1.14 gives the least-squares fit. We won't do more with this concept here because the pseudoinverse of the next section gives us an easier way to attack this problem, but the outline above describes the procedure we use for multiple regression.

Linearization

Polynomial functions easily apply to least-squares curve fitting, but many nonlinear relationships cannot be modeled adequately by polynomials. Natural phenomena such as radioactive decay (exponential) and sunspot activity (sinusoidal) provide examples of such relationships. However, we can sometimes linearize such models by a mathematical transformation and then apply least-squares techniques. To illustrate, consider the exponential model given by

$$y = ae^{bx} \qquad (7.1.15)$$

Take logarithms of both sides to get

$$\ln y = \ln a + bx \qquad (7.1.16)$$

Let

$$v = \ln y \qquad c_0 = \ln a \qquad c_1 = b \qquad (7.1.17)$$

Then Eq. 7.1.16 becomes

$$v = c_0 + c_1 x \qquad (7.1.18)$$

EXAMPLE 7.1.3.   Table 7.1.3 displays $x_i - y_i$ data that follow an exponentially decaying curve. Fit an exponential curve given by Eq. 7.1.15 to this data.

SOLUTION: The first step is to replace each y_i value by its logarithm. This column is also listed in Table 7.1.3. Now we simply fit a straight line to the $x_i - v_i$ data, where $v_i = \ln y_i$, using Eq. 7.1.18. Applying Eq. 7.1.17 with v_i replacing y_i gives $c_0 = 6.32$ and $c_1 = -0.21$. Solving for a and b in Eq. 7.1.17 gives $y = 553e^{-0.21x}$. This is all displayed in Fig. 7.1.5, where the smooth curve is the least-squares fit.

Review

Least-squares estimation applies to those situations where we must make an estimate from data. We say that these procedures are data-driven. No statistical knowledge is required, so we call them deterministic. In Chapter 8 we will discuss mean-square estimation, which does require some knowledge about the statistics of the data. You should attempt to keep these estimation techniques separated in your thinking, and knowing that the least-squares technique is data-driven will help in this effort.

Least-squares techniques are basically straightforward. They consist of defining an estimate $\hat{y}$ in terms of given n-dimensional data $x_1, x_2, \ldots, x_n$, such as

$$\hat{y} = a_0 + a_1 x_1 + a_2 x_2 + \cdots + a_n x_n$$

For a given sample of k data vectors $(k > n)$, we sum the square errors $(y_i - \hat{y}_i)^2$ and minimize this sum. In this section we have presented the basic way to do this, namely, set the derivatives of the sum with respect to the unknown parameters $a_0, a_1, \ldots, a_n$ equal to 0 and solve for the parameters. In the next section we will present another way to accomplish this same objective with the pseudoinverse.

Table 7.1.3

x_i	y_i	$\ln y_i$	x_i	y_i	$\ln y_i$
1	375	5.93	11	70	4.25
2	305	5.72	12	55	4.01
3	250	5.52	13	50	3.91
4	225	5.42	14	45	3.81
5	200	5.30	15	25	3.22
6	150	5.01	16	20	3.00
7	125	4.83	17	15	2.71
8	120	4.79	18	12	2.48
9	100	4.61	19	10	2.30
10	75	4.32	20	5	1.61

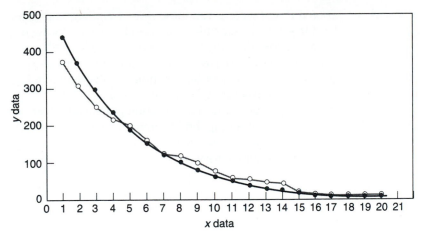

Fig. 7.1.5

7.2 Least Squares and the Pseudoinverse _____

Preview

When dealing with simultaneous equations, there are three possibilities: more equations than unknowns, the same number of equations as unknowns, and fewer equations than unknowns. The overdetermined problem has more equations than unknowns. The exactly determined problem has the same number of linearly independent equations as unknowns. The underdetermined problem has fewer equations than unknowns. We use the pseudoinverse for the overdetermined problem, the matrix inverse for the exactly determined problem, and we often apply linear programming to the underdetermined problem. The pseudoinverse also applies to one form of the underdetermined problem. We are interested in the pseudoinverse because it accomplishes least-squares estimation in the overdetermined problem. This means that the major topic of this section is the overdetermined problem, but we will point out the other form of the pseudoinverse and its application to the underdetermined problem.

The Overdetermined Problem

Suppose that $x_1 = 6$ and $x_2 = 7$. Then the following three equations are all true:

$$x_1 - 2x_2 = -8$$
$$2x_1 - x_2 = 5 \qquad (7.2.1)$$
$$x_1 + x_2 = 13$$

How would you proceed if you were given these three equations and asked to solve for x_1 and x_2? If you were sensible, you would probably throw one of these equations away (it doesn't matter which one) and use the other two to solve algebraically. Or you could plot all three equations in the x_1–x_2 plane and their intersection would denote the solution. This is shown in Fig. 7.2.1, where all three lines intersect at $x_1 = 6$ and $x_2 = 7$.

If we can use all three equations to find the solution graphically, then we should be able to use all three in an algebraic solution. This is effectively what Descartes said when he originated analytic geometry. So let us try: Writing the three equations in matrix form gives

$$\begin{bmatrix} 1 & -2 \\ 2 & -1 \\ 1 & 1 \end{bmatrix} \begin{bmatrix} x_1 \\ x_2 \end{bmatrix} = \begin{bmatrix} -8 \\ 5 \\ 13 \end{bmatrix}$$

which is of the form $Ax = y$. Premultiply both sides by A-transpose to get

$$A'Ax = A'y$$

Solve for x to get

$$x = (A'A)^{-1} A'y$$

The matrix that multiplies y is called the *pseudoinverse* $A^{\#}$,

$$A^{\#} = (A'A)^{-1} A' \tag{7.2.2}$$

It exists whenever $A'A$ has an inverse.

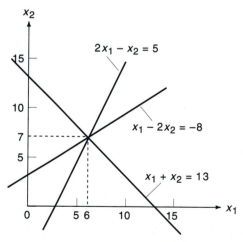

Fig. 7.2.1. The solution of three equations.

EXAMPLE 7.2.1.   Solve the three equations given above using the pseudoinverse; that is, find x_1 and x_2 from the equations

$$\begin{bmatrix} 1 & -2 \\ 2 & -1 \\ 1 & 1 \end{bmatrix} \begin{bmatrix} x_1 \\ x_2 \end{bmatrix} = \begin{bmatrix} -8 \\ 5 \\ 13 \end{bmatrix}$$

SOLUTION:

$$A^tA = \begin{bmatrix} 1 & 2 & 1 \\ -2 & -1 & 1 \end{bmatrix} \begin{bmatrix} 1 & -2 \\ 2 & -1 \\ 1 & 1 \end{bmatrix} = \begin{bmatrix} 6 & -3 \\ -3 & 6 \end{bmatrix}$$

Then

$$(A^tA)^{-1} = \frac{1}{27} \begin{bmatrix} 6 & 3 \\ 3 & 6 \end{bmatrix}$$

Postmultiply this by A^t to get the pseudoinverse, giving

$$x = A^{\#}y = \begin{bmatrix} x_1 \\ x_2 \end{bmatrix} = \frac{1}{3} \begin{bmatrix} 0 & 1 & 1 \\ -1 & 0 & 1 \end{bmatrix} \begin{bmatrix} -8 \\ 5 \\ 13 \end{bmatrix} = \begin{bmatrix} 6 \\ 7 \end{bmatrix}$$

This solution is the algebraic equivalent of Fig. 7.2.1.

In order to show the connection between least squares and the pseudoinverse, suppose we change one of the three equations in Eq. 7.2.1, say the last one, to give

$$\begin{aligned} x_1 - 2x_2 &= -8 \\ 2x_1 - x_2 &= 5 \\ x_1 + x_2 &= 16 \end{aligned} \qquad (7.2.3)$$

No one pair of values for x_1 and x_2 will satisfy all three equations. The first two still have the solution $x_1 = 6$, $x_2 = 7$. We did not change them, so they still have the same solution. The first and third have the solution $x_1 = 8$, $x_2 = 8$, and the second and third have the solution $x_1 = 7$, $x_2 = 9$. This is shown in Fig. 7.2.2, where the three equations intersect at three different locations.

How does least squares enter into this? Suppose that the original three equations (Eq. 7.2.1) represent measurements of linear combinations of the two variables, x_1 and x_2. In our first measurement the sum $x_1 - 2x_2$ was -8. In the second measurement we got 5 for the sum of $2x_1 - x_2$, and the third measurement resulted in $x_1 + x_2 = 13$. These values represent noiseless or exact measurements. But noise or some other type of error could cause these three equations to differ, as they do in Eq. 7.2.3. Now we have a situation where the "best" estimate is needed, and we can use the principle of least squares.

The principle of least squares is equivalent to the assumption that the average is the best estimate. We will not show the truth of this statement here, but you can find a discussion or proof in almost any statistics textbook. This means that the best estimate for x_1 is the average of the three values 6, 7, and 8 obtained from the three pairs or equations, giving 7 as the best estimate. Likewise, the best estimate for x_2 is the average of the three values 7, 8, and 9, or 8.

Having said all this, let us now apply the pseudoinverse to this problem. The three equations in Eq. 7.2.3 are the same as Eq. 7.2.1 on the left. Therefore the pseudoinverse matrix is the same as before. Only the y vector is different, giving

$$x = A^{\#}y = \frac{1}{3}\begin{bmatrix} 0 & 1 & 1 \\ -1 & 0 & 1 \end{bmatrix}\begin{bmatrix} -8 \\ 5 \\ 16 \end{bmatrix} = \begin{bmatrix} 7 \\ 8 \end{bmatrix}$$

Notice that the solutions $x_1 = 7$ and $x_2 = 8$ are those obtained by averaging the solutions from the three pairs of simultaneous equations. Therefore the pseudoinverse accomplishes least-squares estimation in a compact and easily used form.

To demonstrate this point further, let us return to curve fitting and begin with the following equations from Section 7.1.

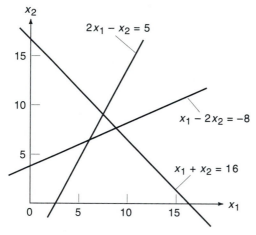

Fig. 7.2.2.   Three incompatible equations.

$$c_0 + c_1 x_1 = y_1$$
$$c_0 + c_1 x_2 = y_2$$
$$\vdots$$
$$c_0 + c_1 x_n = y_n$$

or

$$\begin{bmatrix} 1 & x_1 \\ 1 & x_2 \\ \vdots & \vdots \\ 1 & x_n \end{bmatrix} \begin{bmatrix} c_0 \\ c_1 \end{bmatrix} = \begin{bmatrix} y_1 \\ y_2 \\ \vdots \\ y_n \end{bmatrix} \tag{7.2.4}$$

This is of the form $Xc = Y$. Premultiply both sides by the X-transpose to get

$$X'Xc = X'Y \tag{7.2.5}$$

Using the X matrix in Eq. 7.2.4, we get for $X'X$,

$$X'X = \begin{bmatrix} n & \Sigma x_i \\ \Sigma x_i & \Sigma x_i^2 \end{bmatrix} \tag{7.2.6}$$

The right side of Eq. 7.2.5 is

$$X'Y = \begin{bmatrix} \Sigma y_i \\ \Sigma x_i y_i \end{bmatrix} \tag{7.2.7}$$

Substituting Eqs. 7.2.6 and 7.2.7 into 7.2.5 gives Eq. 7.1.7. Thus we have arrived at the same equation as before using the pseudoinverse.

EXAMPLE 7.2.2.   Repeat Example 7.1.1 using the pseudoinverse; that is, find the line giving the least-squares fit to the data in Table 7.1.1 using the pseudoinverse.

SOLUTION: Substituting the data from Table 7.1.1 into Eq. 7.2.4 gives

$$\begin{bmatrix} 1 & 0.25 \\ 1 & 0.5 \\ \vdots & \vdots \\ 1 & 2.5 \end{bmatrix} \begin{bmatrix} c_0 \\ c_1 \end{bmatrix} = \begin{bmatrix} 10 \\ 14 \\ \vdots \\ 38 \end{bmatrix}$$

Premultiplying both sides by X' gives

$$\begin{bmatrix} 10 & 13.75 \\ 13.75 & 24.06 \end{bmatrix} \begin{bmatrix} c_0 \\ c_1 \end{bmatrix} = \begin{bmatrix} 270 \\ 437 \end{bmatrix}$$

which is the same result obtained in Example 7.1.1. The least-squares fit is the same as before,

$$\hat{y} = 9.46 + 12.76x$$

Generalized Least Squares

The pseudoinverse applies to all least-squares problems. To show the connection to the quadratic fit, we will arrive at Eq.7.1.13 by the pseudoinverse method. To do this we begin with Eq. 7.1.8 applied to each point:

$$c_0 + c_1x_1 + c_2x_1^2 = y_1$$
$$c_0 + c_1x_2 + c_2x_2^2 = y_2$$
$$\vdots$$
$$c_0 + c_1x_n + c_2x_n = y_n$$

or

$$\begin{bmatrix} 1 & x_1 & x_1^2 \\ 1 & x_2 & x_2^2 \\ \vdots & \vdots & \vdots \\ 1 & x_n & x_n^2 \end{bmatrix} \begin{bmatrix} c_0 \\ c_1 \\ c_2 \end{bmatrix} = \begin{bmatrix} y_1 \\ y_2 \\ \vdots \\ y_n \end{bmatrix} \tag{7.2.8}$$

This is the familiar form $Xc = Y$. Premultiply both sides by the X-transpose to get

$$X'Xc = X'Y \tag{7.2.9}$$

Substituting Eq.7.2.8 into this gives

$$\begin{bmatrix} n & \Sigma x_i & \Sigma x_i^2 \\ \Sigma x_i & \Sigma x_i^2 & \Sigma x_i^3 \\ \Sigma x_i^2 & \Sigma x_i^3 & \Sigma x_i^4 \end{bmatrix} \begin{bmatrix} c_0 \\ c_1 \\ c_2 \end{bmatrix} = \begin{bmatrix} \Sigma y_i \\ \Sigma x_i y_i \\ \Sigma x_i^2 y_i \end{bmatrix} \tag{7.2.10}$$

which is recognized as the formula for the quadratic least-squares fit from Section 7.1. See Eq. 7.1.13.

The Three Possible Cases

In order to give some perspective to our discussion of the pseudoinverse, here is what we are doing. Consider a general system of n equations in m unknowns, $x_1, x_2, \ldots, x_m$. The general form for these equations is given by

$$Ax = y \tag{7.2.11a}$$

or, in expanded form,

$$\begin{bmatrix} a_{11} & a_{12} & \cdots & a_{1m} \\ a_{21} & a_{22} & \cdots & a_{2m} \\ \vdots & & & \\ a_{n1} & a_{n2} & \cdots & a_{nm} \end{bmatrix} \begin{bmatrix} x_1 \\ x_2 \\ \vdots \\ x_m \end{bmatrix} = \begin{bmatrix} y_1 \\ y_2 \\ \vdots \\ y_n \end{bmatrix} \qquad (7.2.11b)$$

The familiar situation is when $m = n$ and det $A \neq 0$. The solution of this set of equations is

$$x = A^{-1}y \qquad (7.2.12)$$

where A^{-1} is the inverse of matrix A. When $m < n$, the set of equations is overdetermined and the solution is given by

$$x = A^{\#}y$$

where

$$A^{\#} = (A^t A)^{-1} A^t \qquad (7.2.13)$$

This form of the pseudoinverse exists whenever $A^t A$ has an inverse. When the equations are not exact, this solution provides the least-squares solution in the sense that it minimizes the norm $\|Ax - y\|$. (The norm $\|x\|$ is the length of the vector x. See Section 8.2 for a definition of the norm.)

In the case when $m > n$, Eq. 7.2.11 is underdetermined and has an infinite number of solutions. However, if we are interested in the solution that has the smallest norm $\|x\|$, then the second form of the pseudoinverse provides the solution. This second form of the pseudoinverse is given by

$$A^{\#} = A^t (AA^t)^{-1} \qquad (7.2.14)$$

This pseudoinverse exists whenever AA^t has an inverse. The following discussion introduces this second form.

The Underdetermined Problem

Suppose that $x_1 = 6$, $x_2 = 7$, and $x_3 = 4$. Then the following two equations are true:

$$\begin{aligned} x_1 - 2x_2 + x_3 &= -4 \\ 2x_1 - x_2 - x_3 &= 1 \end{aligned} \qquad (7.2.15)$$

The problem here is that there are an infinite number of values for the three x_i's that satisfy these two equations in addition to the values 6, 7, and 4. The pseudoinverse comes into play if we wish to find the solution x_1, x_2, and x_3 that has the smallest norm.

$$\|x\| = [x_1^2 + x_2^2 + x_3^2]^{1/2} \qquad (7.2.16)$$

To illustrate, let us apply Eq. 7.2.14 to this problem. In matrix form, Eqs. 7.2.15 are

$$\begin{bmatrix} 1 & -2 & 1 \\ 2 & -1 & -1 \end{bmatrix} \begin{bmatrix} x_1 \\ x_2 \\ x_3 \end{bmatrix} = \begin{bmatrix} -4 \\ 1 \end{bmatrix}$$

Calculating the terms in Eq. 7.2.14, we have

$$AA^t = \begin{bmatrix} 1 & -2 & 1 \\ 2 & -1 & -1 \end{bmatrix} \begin{bmatrix} 1 & 2 \\ -2 & -1 \\ 1 & -1 \end{bmatrix} = \begin{bmatrix} 6 & 3 \\ 3 & 6 \end{bmatrix}$$

$$(AA^t)^{-1} = \frac{1}{27} \begin{bmatrix} 6 & -3 \\ -3 & 6 \end{bmatrix}$$

so

$$A^\# = \frac{1}{27} \begin{bmatrix} 1 & 2 \\ -2 & -1 \\ 1 & -1 \end{bmatrix} \begin{bmatrix} 6 & -3 \\ -3 & 6 \end{bmatrix} = \frac{1}{27} \begin{bmatrix} 0 & 9 \\ -8 & 0 \\ 9 & -9 \end{bmatrix} \qquad (7.2.17)$$

Solving for x gives

$$\begin{bmatrix} x_1 \\ x_2 \\ x_3 \end{bmatrix} = \frac{1}{27} \begin{bmatrix} 0 & 9 \\ -8 & 0 \\ 9 & -9 \end{bmatrix} \begin{bmatrix} -4 \\ 1 \end{bmatrix} = \frac{1}{27} \begin{bmatrix} 9 \\ 36 \\ -45 \end{bmatrix} \qquad (7.2.18)$$

Geometrically, we can interpret these results as follows: The two equations 7.2.15 each define a plane in three-dimensional space. The intersection of these two planes defines a line. Any point on this line satisfies the two equations. The pseudoinverse gives that point on the line closest to the origin.

The Rest of the Story

We have described two forms of the pseudoinverse, one that holds when A^tA has an inverse, and one that holds when AA^t has an inverse. It is also possible for neither of these conditions to hold. Moore and Penrose succeeded in defining a pseudoinverse for that situation also. Their procedure generates the same solution as Eq. 7.2.2 for the first situation, and generates the same solution as Eq. 7.2.11 for the second case. For more information on the Moore–Penrose pseudoinverse, see Chapter 21 of reference 3 listed at the end of this chapter.

Review

In the overdetermined problem, the pseudoinverse of an $m \times n$ matrix A is given by the $n \times m$ matrix of Eq. 7.2.2:

$$A^{\#} = (A^tA)^{-1} A^t$$

It exists whenever A^tA has an inverse. We have demonstrated (but not proven) that this pseudoinverse accomplishes least-squares estimation. This means that we can do the same thing two ways, by the conventional problem setup in Section 7.1, or by the pseudoinverse as demonstrated in this section. The advantage of the pseudoinverse is in the notation. It simplifies both the notation and the procedures used in data-driven problems where least squares is the criterion. For this reason you will see it in many theoretical discussions of estimation theory.

The second form for the pseudoinverse applies to the underdetermined problem, where $m > n$ as in Eq. 7.2.11:

$$A^{\#} = A^t(AA^t)^{-1}$$

This form finds the unique solution closest to the origin from the infinite number of possible solutions.

7.3   Iterative Methods

Preview

When asked how to reach Mt. Olympus, Socrates replied, "Make every step you take go in that direction." This is the basic idea behind iterative methods. A goal is established, and the procedure makes every step go in that direction.

Iterative methods apply to all sorts of problems, but the two most common are root finding and function minimization or maximization. In root finding we have an equation of the form $g(x) = 0$, and we wish to find the values of x that make the equation true. Of course, the proper way to solve this problem is to set the derivatives with respect to the various x_i equal to 0 and solve for the values that make it so. But iterative methods accomplish the task when this is not possible.

A closely related problem is function minimization. Given a function $f(x_1, x_2, \ldots, x_n)$, find the vector $x^* = (x_1, x_2, \ldots, x_n)$ if it exists such that $f(x^*)$ is a minimum. Least-squares estimation is a problem in function minimization. You can see that there is a close relationship between root finding and function minimization if we set

$$g(x) = \frac{\partial}{\partial x} f(x) = 0$$

The values of x that make $f(x)$ an extremum make $g(x)$ equal to 0.

The most understandable and popular iterative method is steepest descent. If you find yourself on a hillside at night and wish to reach the valley floor, you should walk downhill. But suppose you are on a pogo stick and are able to leap great distances. How far should you leap each time? If you jump too far you may land on the opposite hillside, and if your steps are small it will take a long time to reach the bottom. In the method of steepest descent (also called the gradient method) we attempt to answer two questions at each step:

Question 1: What is the best direction?
Question 2: What is the best step size?

In order to give you some feel for the problem and solution, we discuss gradient descent before concentrating on least squares.

Our problem is to find those values of $x = (x_1, x_2, \ldots, x_n)$ that minimize $f(x_1, x_2, \ldots, x_n)$. Let x^* be the optimum solution, the value of $x = (x_1, x_2, \ldots, x_n)$ that minimizes $f(x_1, x_2, \ldots, x_n)$. Then x^* has the property that

$$f(x^*) \le f(x) \qquad \text{for all } x \qquad (7.3.1)$$

So x^* represents the solution to our problem. In the gradient method we compute the gradient vector at each step k,

$$\text{gradient vector at } x_k = \frac{\partial}{\partial x} f(x)\big|_{x=x_k} \qquad (7.3.2)$$

For example, if $f(x) = x_1^2 + x_2^2$ as shown in Fig. 7.3.1a, then

$$\frac{\partial f}{\partial x_1} = 2x_1 \qquad \frac{\partial f}{\partial x_2} = 2x_2$$

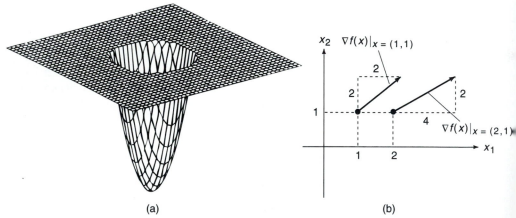

(a)   (b)

Fig. 7.3.1

so the gradient vector is given by

$$\nabla f(x) = \begin{bmatrix} 2x_1 \\ 2x_2 \end{bmatrix}$$

Evaluated at $x = (1, 1)$, the gradient has components $(2, 2)$; and at $x = (2, 1)$ the gradient has components $(4, 2)$, as shown in Fig. 7.3.1b.

Figure 7.3.2a shows another function, $f(x) = x_1^2 + x_2^2 - 1.8x_1x_2$. The gradient is

$$\nabla f(x) = \begin{bmatrix} \dfrac{\partial f}{\partial x_1} \\ \dfrac{\partial f}{\partial x_2} \end{bmatrix} = \begin{bmatrix} 2x_1 - 1.8x_2 \\ 2x_2 - 1.8x_1 \end{bmatrix}$$

which has components at locations given by

Location	Gradient Vector
(1, 1)	(0.2, 0.2)
(2, 1)	(2.2, −1.6)
(1, 0)	(2, −1.8)

These are plotted in Fig. 7.3.2b. Notice that the size of the gradient vector depends on its location.

As you can see from these examples, the gradient vector points in the direction of steepest ascent (most uphill), so the best direction to jump is in the opposite or most negative direction. If we find ourselves at x_k, and if we call the next location x_{k+1}, then these two are related by the step size

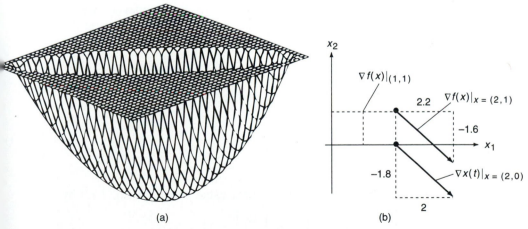

(a)   (b)

Fig. 7.3.2

and direction. The step is $x_{k+1} - x_k$, and this is proportional to $-\partial f(x)/\partial x$. Therefore

$$x_{k+1} - x_k = -\alpha_k \frac{\partial}{\partial x} f(x)\big|_{x=x_k}$$

where α_k is a scaling factor, which together with the magnitude of the gradient determines the step size. Solving for x_{k+1} gives

$$x_{k+1} = x_k - \alpha_k \frac{\partial}{\partial x} f(x)\big|_{x=x_k} \tag{7.3.3}$$

This basic equation describes steepest descent. The present position is x_k and the next position is x_{k+1} as determined by Eq. 7.3.3.

To illustrate some of the features of this method we will compare steepest descent for two functions f_1 and f_2, given by

$$f_1(x_1, x_2) = x_1^2 + x_2^2, \qquad f_2(x_1, x_2) = x_1^2 + x_2^2 - 1.8x_1x_2$$

The first function is bowl-shaped with a minimum at the origin as shown in Fig. 7.3.1. The second function also has a minimum at the origin, but it is shaped like a trough or an elongated valley running at a 45° angle to the x_1 and x_2 axes. Figure 7.3.2 illustrates this troughlike shape.

If we place a marble at any location in f_1 and let go, it will travel directly toward the bottom. If we place a marble out near the end of the trough and up on the side in Fig. 7.3.2, however, it will not fall directly toward the lowest point at the origin. Of course it will eventually get there, but by a circuitous route. The method of steepest descent has no problem with f_1, but f_2 creates problems similar to, and sometimes worse than, those of the marble. If the step size is too big, steepest descent will simply oscillate from one side of the trough to the other. Only when the step size is small can steepest descent sense the necessary change in direction to converge to the minimum. Here we have the conflicting criteria that are the hallmark of any good engineering problem. Small step size gives better accuracy and guaranteed convergence, while large step size converges faster if at all. (A common ploy is to start the initial step size large and let it decrease with each step.)

Now let us apply steepest descent to f_1. The derivative is given by

$$\frac{\partial}{\partial x_1} f_1(x) = 2x_1$$

Likewise,

$$\frac{\partial}{\partial x_2} f_1(x) = 2x_2$$

Therefore, at each x we calculate the next value of x from Eq. 7.3.3 by

$$\begin{bmatrix} x_1 \\ x_2 \end{bmatrix}_{k+1} = \begin{bmatrix} x_1 \\ x_2 \end{bmatrix}_k - \alpha_k \begin{bmatrix} 2x_1 \\ 2x_2 \end{bmatrix}_k = (1 - 2\alpha_k) \begin{bmatrix} x_1 \\ x_2 \end{bmatrix}_k$$

For $\alpha_k = 0.2$ (a constant) and starting at $\begin{bmatrix} 2 \\ 1 \end{bmatrix}$, this algorithm gives the sequence

$$\begin{bmatrix} 2 \\ 1 \end{bmatrix} \quad \begin{bmatrix} 1.2 \\ 0.6 \end{bmatrix} \quad \begin{bmatrix} 0.72 \\ 0.36 \end{bmatrix} \quad \begin{bmatrix} 0.432 \\ 0.216 \end{bmatrix} \cdots$$

which converges exponentially to the correct solution at the origin.

Now for f_2. The derivatives are given by

$$\frac{\partial}{\partial x_1} f_2(x) = 2x_1 - 1.8x_2$$

$$\frac{\partial}{\partial x_2} f_2(x) = 2x_2 - 1.8x_1$$

Therefore, at each x we use Eq. 7.3.3 to calculate the next value of x by

$$\begin{bmatrix} x_1 \\ x_2 \end{bmatrix}_{k+1} = \begin{bmatrix} x_1 \\ x_2 \end{bmatrix}_k - \alpha_k \begin{bmatrix} 2x_1 - 1.8x_2 \\ 2x_2 - 1.8x_1 \end{bmatrix}_k$$

$$= (1 - 2\alpha_k) \begin{bmatrix} x_1 \\ x_2 \end{bmatrix}_k + 1.8\alpha_k \begin{bmatrix} x_2 \\ x_1 \end{bmatrix}_k$$

For $\alpha_k = 0.2$ and starting at $\begin{bmatrix} 2 \\ 1 \end{bmatrix}$, this algorithm gives the sequence

$$\begin{bmatrix} 2 \\ 1 \end{bmatrix} \quad \begin{bmatrix} 1.56 \\ 1.32 \end{bmatrix} \quad \begin{bmatrix} 1.411 \\ 1.354 \end{bmatrix} \quad \begin{bmatrix} 1.334 \\ 1.320 \end{bmatrix} \cdots$$

This sequence does approach the origin, but as you can see it does so slowly. The reason for the snail's pace is the function f_2 itself. Any flat-bottom troughlike function such as f_2 will have slow convergence, so it is customary to test any new algorithm on a function of this type. (Most algorithms are concerned with selecting the step size, because the direction is determined by the gradient.)

All of the discussion above has been directed toward minimizing a function. Now, to apply this to least squares we must have a function to minimize. In least squares we minimize ε, the sum of squared errors. In this estimation problem we are given A and y, and the relationship

$$Ax = y \tag{7.3.4}$$

We are asked to choose x to make Eq. 7.3.4 true, or at least as true as possible, meaning we choose x using the least-squares criterion. (See Examples 7.2.1 and 7.2.2 for examples in this form.) Thus we choose x so that our estimate for y, given by

$$\hat{y} = Ax \tag{7.3.5}$$

minimizes the sum of squared error terms.

In our matrix notation, the sum of squared errors is given by

$$\varepsilon = (y - Ax)^t(y - Ax) \qquad (7.3.6)$$

Taking the gradient with respect to x gives (see Appendix C)

$$\nabla\varepsilon = 2A^t(Ax - y) \qquad (7.3.7)$$

In our introduction of the general method above, we used the gradient in

$$x_{k+1} = x_k - \alpha_k \nabla\varepsilon \qquad (\nabla\varepsilon \text{ evaluated at } x = x_k)$$

so,

$$x_{k+1} = x_k - \alpha_k A^t(Ax_k - y) \qquad (7.3.8)$$

Notice that this is Eq. 7.3.3 with the gradient given in the appropriate form for least-squares estimation.

EXAMPLE 7.3.1.   Apply this algorithm to the three equations in Eq. 7.2.3.

SOLUTION: Let $\alpha_k = 1/k$. Choose $x_0 = (1, 1)^t$. Then the algorithm proceeds as follows: Set $x_k = x_0$ and begin.

Step 1: Multiply A times x_k.

$$Ax_k = \begin{bmatrix} 1 & -2 \\ 2 & -1 \\ 1 & 1 \end{bmatrix} \begin{bmatrix} 1 \\ 1 \end{bmatrix} = \begin{bmatrix} -1 \\ 1 \\ 2 \end{bmatrix}$$

Step 2: Subtract y from Ax_k.

$$Ax_k - y = \begin{bmatrix} -1 \\ 1 \\ 2 \end{bmatrix} - \begin{bmatrix} -8 \\ 5 \\ 16 \end{bmatrix} = \begin{bmatrix} 7 \\ -4 \\ -14 \end{bmatrix}$$

Step 3: Multiply A^t times $Ax_k - y$.

$$A^t(Ax_k - y) = \begin{bmatrix} 1 & 2 & 1 \\ -2 & -1 & 1 \end{bmatrix} \begin{bmatrix} 7 \\ -4 \\ -14 \end{bmatrix} = \begin{bmatrix} -15 \\ -24 \end{bmatrix}$$

Step 4: Use Eq. 7.3.8 to calculate $x_{k+1} = x_1$.

$$x_1 = \begin{bmatrix} 1 \\ 1 \end{bmatrix} - \begin{bmatrix} -15 \\ -24 \end{bmatrix} = \begin{bmatrix} 16 \\ 25 \end{bmatrix}$$

Step 5: Let $k = k + 1$ and go to step 1.

If this algorithm is followed, with $\alpha_k = 1/k$, the sequence of x vectors is

$$\begin{bmatrix} 1 \\ 1 \end{bmatrix} \begin{bmatrix} 16 \\ 25 \end{bmatrix} \begin{bmatrix} 14.5 \\ -12.5 \end{bmatrix} \begin{bmatrix} -21 \\ 36 \end{bmatrix} \begin{bmatrix} 42 \\ -27 \end{bmatrix} \begin{bmatrix} -21 \\ 36 \end{bmatrix}$$

$$\begin{bmatrix} 21 \\ -6 \end{bmatrix} \begin{bmatrix} 3 \\ 12 \end{bmatrix} \begin{bmatrix} 7.5 \\ 7.5 \end{bmatrix} \begin{bmatrix} 7 \\ 8 \end{bmatrix} \begin{bmatrix} 7 \\ 8 \end{bmatrix} \dots$$

So the algorithm reaches the final solution in nine steps.

EXAMPLE 7.3.2.   Find the line in Examples 7.1.1 and 7.2.1 by iteration.

SOLUTION: The algorithm will converge if the step size α_k is given by $\alpha_k = \alpha_1/k$, where α_1 is a positive constant. However, this example provides one case where this method converges slowly. Better convergence is obtained with $\alpha_k = 0.05$ for each step. We begin with the equations in matrix form from Example 7.2.1.

$$\begin{bmatrix} 1 & 0.25 \\ 1 & 0.5 \\ \vdots & \vdots \\ 1 & 2.5 \end{bmatrix} \begin{bmatrix} c_0 \\ c_1 \end{bmatrix} = \begin{bmatrix} 10 \\ 14 \\ \vdots \\ 38 \end{bmatrix}$$

which is of the form $Ax = y$. We wish to choose x so that our estimate for y, given by

$$\hat{y} = Ax$$

minimizes the sum of squared error terms in Eq. 7.3.6. Applying Eq. 7.3.8 with $\alpha_k = 0.05$ and the initial

$$x = \begin{bmatrix} c_0 \\ c_1 \end{bmatrix} = \begin{bmatrix} 1 \\ 1 \end{bmatrix}$$

gives the following steps:

Step 1: Multiply A times x_k.

$$\begin{bmatrix} 1 & 0.25 \\ 1 & 0.5 \\ \vdots & \vdots \\ 1 & 2.5 \end{bmatrix} \begin{bmatrix} c_0 \\ c_1 \end{bmatrix} = \begin{bmatrix} c_0 + 0.25c_1 \\ c_0 + 0.5c_1 \\ \vdots \\ c_0 + 2.5c_1 \end{bmatrix} = \begin{bmatrix} 1.25 \\ 1.5 \\ \vdots \\ 3.5 \end{bmatrix}$$

Step 2: Subtract y from Ax_k.

$$
\begin{bmatrix} 1.25 \\ 1.5 \\ \vdots \\ 3.5 \end{bmatrix} - \begin{bmatrix} 10 \\ 14 \\ \vdots \\ 38 \end{bmatrix} = \begin{bmatrix} -8.75 \\ -12.5 \\ \vdots \\ -34.5 \end{bmatrix}
$$

Step 3: Multiply A' times $(Ax_k - y)$.

$$
\begin{bmatrix} 1 & 1 & \cdots & 1 \\ 0.25 & 0.5 & \cdots & 2.5 \end{bmatrix} \begin{bmatrix} -8.75 \\ -12.5 \\ \vdots \\ -34.5 \end{bmatrix} = \begin{bmatrix} -246.25 \\ -399.19 \end{bmatrix}
$$

Step 4: Use Eq. 7.3.8 to calculate x_{k+1} with $\alpha_k = 0.05$.

$$
x_1 = \begin{bmatrix} 1 \\ 1 \end{bmatrix} - 0.05 \begin{bmatrix} -246.25 \\ -399.19 \end{bmatrix} = \begin{bmatrix} 13.3 \\ 20.96 \end{bmatrix}
$$

Step 5: Increment k and go to step 1.

This procedure gives the following sequence:

$$
\begin{bmatrix} 1 \\ 1 \end{bmatrix} \quad \begin{bmatrix} 13.3 \\ 20.96 \end{bmatrix} \quad \begin{bmatrix} 5.75 \\ 8.44 \end{bmatrix} \quad \begin{bmatrix} 10.57 \\ 16.18 \end{bmatrix} \quad \begin{bmatrix} 7.66 \\ 11.29 \end{bmatrix}
$$

$$
\begin{bmatrix} 9.56 \\ 14.29 \end{bmatrix} \quad \begin{bmatrix} 8.46 \\ 12.37 \end{bmatrix} \quad \cdots \quad \begin{bmatrix} 9.466 \\ 12.75 \end{bmatrix}
$$

It converges to the correct solution in about 100 steps or so, indicating the need for computer solution.

Here is the pseudocode to implement iterative least squares. This is a straightforward implementation of the formulas above, where we assume that the A matrix and the y vector are given and we solve iteratively for the x vector.

```
procedure least(a,y)   /* a(i,j) and y(i) are given */
define a(M,N), y(M), x(N)   /* M = 10, N = 2 in      */
define y1(M), x1(N)    /* Example 7.3.2        */
    j = 1
    while j ≤ N /* initialize x */
    begin
        x(j) = 1
        j = j + 1
    end for j
    ak = 0.05     /* define increment ak */
```

```
    i = 1
    while i ≤ 100
    begin
        call a_times_x(a,x)    /* multiply a*x     */
        call y1my(y1,y)        /* sets y1 = y1 - y */
        call at_times_y(a,y1)/* multiply a-transpose*y */
        call xmx1(x,x1,ak)     /* sets x = x - ak*x1   */
        i = i + 1
    end for i
end least

procedure a_times_x(a,x)      /* multiply a*x   */
define a(M,N), x(N), y1(M)
    i = 1
    while i ≤ M
    begin
        sum = 0
        j = 1
        while j ≤ N
            sum = sum + a(i,j)*x(j)
            j = j + 1
        end for j
        y1(i) = sum
        i = i + 1
    end for i
    return(y1)
end a_times_x

procedure y1my(y1,y) /* sets y1 = y1 - y  */
define y1(M), y(M)
    i = 1
    while i ≤ M
    begin
        y1(i) = y1(i) - y(i)
        i = i + 1
    end for i
    return(y1)
end y1my

procedure at_times_y(a,y1)    /* multiplies a-transpose * y  */
define a(M,N), y1(M), x1(N)
    j = 1
    while j ≤ N
    begin
```

```
            sum = 0
            i = 1
            while i ≤ M
            begin
                sum = sum + a(i,j)*y1(i)
                i = i + 1
            end for i
            x1(j) = sum
            j = j + 1
        end for j
        return(x1)
end at_times_y

procedure xmx1(x,x1,ak)       /* sets x = x - ak * x1   */
define x(N), x1(N)
    j = 1
    while j ≤ N
    begin
        x(j) = x(j) - ak*x1(j)
        j = j + 1
    end for j
    return(x)
end xmx1
```

Review

We have done little more than scratch the surface in our introduction of iterative solutions to least-squares problems, but we can do little more without embarking on some fundamental theory that would lead us far from our goal. I hope, however, that you can see the basic idea of gradient descent applied to least-squares problems. The gradient of the function we wish to minimize is used to determine the direction for the next step, always going downhill or in the negative direction determined by the gradient. The step size is another matter. We have said little about it, leaving this subject to texts on iterative methods. My suggestion is to program the problem on a computer and then experiment with various step sizes. You can always use the ploy of decreasing the step size as k increases by the formula $\alpha_k = \alpha_1/k$, where α_1 is any positive constant.

7.4   The Orthogonality Principle _____

Preview

The orthogonality principle applies to a wide variety of linear estimation problems, and we will use it extensively in the next chapter. We introduce it here because the vectors we will use are the familiar $n \times 1$ column vectors, and the procedures are deterministic. Then, in Chapter 8, we will take a closer look at orthogonality and show that it applies to random variables. As we have stated several times, the least squares technique uses measured data rather than known statistical quantities to make estimates. This means we need not appeal to statistics in this section, but only algebra.

The problem addressed here is to estimate some desired quantity $d(n)$ from a related sequence $x(n)$. Both given sequences are used to design a finite-impulse response filter, which means we find the optimum coefficients as before. The new concepts we introduce in this section show a connection between vectors and estimation. We will expand on this theme in the next chapter.

Figure 7.4.1 shows a finite-impulse response (FIR) filter with $p + 1$ taps $h_0, h_1, \ldots, h_p$. The input is data $x(n)$ and the output is an estimate for some desired quantity $d(n)$. For example, we were trying to estimate the future value $y(n + 1)$ from the present and past values in the example associated with Fig. 7.1.3, so $d(n) = y(n + 1)$ there. As another example, we are often given signal plus noise,

$$x(n) = s(n) + w(n) \tag{7.4.1}$$

and asked to estimate the signal, so $d(n) = s(n)$.

The general problem we are concerned with here is to estimate $d(n)$ from some related sequence $x(n)$. For a least-squares solution we must assume that both $d(n)$ and $x(n)$ have been measured and recorded, so that these sequences can be used to design the filter. This is the usual situation where we use known samples to design the system, and then hope our design works well later when $d(n)$ is not given. Let the given data be

$$x(0), x(1), \ldots, x(N)$$
$$d(0), d(1), \ldots, d(N)$$

The filter in Fig. 7.4.1 has input $x(n)$ and output

$$\hat{d}(n) = \sum_{k=0}^{p} h(k)\, x(n - k) \tag{7.4.2}$$

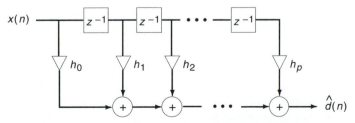

Fig. 7.4.1.   A finite-impulse response (FIR) filter with $p + 1$ taps.

The error is

$$e(n) = d(n) - \hat{d}(n) = d(n) - \sum_{k=0}^{p} h(k)\, x(n-k) \qquad (7.4.3)$$

The least-squares approach is to design the filter to minimize the sum of squared errors:

$$\varepsilon = \sum_{n=0}^{N} e^2(n) \qquad (7.4.4)$$

Now consider the action of the filter. When the data sequence $x(n)$ is first supplied to the filter, the output sequence is in the transient mode until the registers are filled, i.e., until time p. The first nontransient output therefore occurs at time p and is given by

$$\hat{d}(p) = x(0)h_p + x(1)h_{p-1} + \cdots + x(p)h_0$$

$$= [x(0) \quad x(1) \quad \ldots \quad x(p)] \begin{bmatrix} h_p \\ h_{p-1} \\ \vdots \\ h_0 \end{bmatrix}$$

At the next time instant the output is given by

$$\hat{d}(p+1) = x(1)h_p + x(2)h_{p-1} + \cdots + x(p+1)h_0$$

$$= [x(1) \quad x(2) \quad \ldots \quad x(p+1)] \begin{bmatrix} h_p \\ h_{p-1} \\ \vdots \\ h_0 \end{bmatrix}$$

The output terms continue to be generated in this manner, so we can express the vector of output terms as

$$\begin{bmatrix} \hat{d}(p) \\ \hat{d}(p+1) \\ \vdots \\ \hat{d}(p+N) \end{bmatrix} = \begin{bmatrix} x(0) & x(1) & \ldots & x(p) \\ x(1) & x(2) & \ldots & x(p+1) \\ \vdots & & & \vdots \\ x(N) & x(N+1) & \ldots & x(p+N) \end{bmatrix} \begin{bmatrix} h_p \\ h_{p-1} \\ \vdots \\ h_0 \end{bmatrix}$$

or

$$\hat{d} = X\overline{h} \qquad (7.4.5)$$

where we use the bar over h for the reversal operation, that is, $\overline{h}$ is just h upside down. The matrix X is called the data matrix and has dimension $(N+1) \times (p+1)$. We assume that $N > p$.

The error vector has dimension $N + 1$ and is given by

$$e = d - \hat{d} = d - X\bar{h} \tag{7.4.6}$$

The problem is to minimize

$$\varepsilon = e^t e \tag{7.4.7}$$

We will solve this problem in two ways: (1) by substituting Eq. 7.4.6 into 7.4.7 and differentiating to find the optimum h vector; and (2) by applying the orthogonality principle.

1. Substituting Eq. 7.4.6 into 7.4.7, we get

$$\varepsilon = (d - X\bar{h})^t(d - X\bar{h}) \tag{7.4.8}$$

Setting the derivative with respect to the vector $\bar{h}$ equal to 0 gives

$$X^t X \bar{h} = X^t d \tag{7.4.9}$$

This is called the *least-squares Wiener–Hopf equation*. Solving for $\bar{h}$ gives the familiar pseudoinverse

$$\bar{h} = (X^t X)^{-1} X^t y = X^\# d \tag{7.4.10}$$

This solution exists when $(X^t X)$ has an inverse.

This completes the solution of the problem the first way. While we are at it, we can find a different expression for the sum of square errors ε by writing

$$\begin{aligned}
\varepsilon &= (d - X\bar{h})^t(d - X\bar{h}) \\
&= d^t(d - X\bar{h}) - (X\bar{h})^t(d - X\bar{h}) \\
&= d^t d - d^t X\bar{h} - \bar{h}^t(X^t d - X^t X\bar{h})
\end{aligned}$$

The last term is 0 because the optimum filter satisfies Eq. 7.4.9, giving

$$\varepsilon = d^t d - d^t X\bar{h} = d^t e \tag{7.4.11}$$

The sum of squared errors is the inner product of the error with the desired quantity. Now here is an example before we look at the second way to solve this problem.

EXAMPLE 7.4.1.   Design a two-tap filter for the data and desired sequences $x(n)$ and $d(n)$ given by

$$x(n) = \{x(0), x(1), \ldots, x(p)\} = \{1, -1, 2, -2, 3, -3\}$$
$$d(n) = \{d(0), d(1), \ldots, d(p)\} = \{1, -1, 1, -1, 1, -1\}$$

SOLUTION: For a two-tap filter, $p = 1$. The X matrix and d vector in Eq. 7.4.10 are given by

$$X = \begin{bmatrix} x(0) & x(1) \\ x(1) & x(2) \\ \vdots & \vdots \\ x(p-1) & x(p) \end{bmatrix} = \begin{bmatrix} 1 & -1 \\ -1 & 2 \\ 2 & -2 \\ -2 & 3 \\ 3 & -3 \end{bmatrix} \qquad d = \begin{bmatrix} d(1) \\ d(2) \\ \vdots \\ d(p) \end{bmatrix} = \begin{bmatrix} -1 \\ 1 \\ -1 \\ 1 \\ -1 \end{bmatrix}$$

Substituting this into Eq. 7.4.10 gives

$$\bar{h} = X^{\#}d = \frac{1}{29}\begin{bmatrix} 5 & 17 & 10 & 12 & 15 \\ 3 & 16 & 6 & 13 & 9 \end{bmatrix}\begin{bmatrix} -1 \\ 1 \\ -1 \\ 1 \\ -1 \end{bmatrix} = \begin{bmatrix} -\frac{1}{29} \\ \frac{11}{29} \end{bmatrix}$$

or $h_1 = -\frac{1}{29}$ and $h_0 = \frac{11}{29}$, as shown in Fig. 7.4.2. When the input sequence $x(n) = \{1, -1, 2, -2, 3, -3\}$ is applied to this filter, the output sequence from $n = 1$ to $n = 5$ is given by

$$\hat{d}(n) = \{-\tfrac{12}{29}, \tfrac{23}{29}, -\tfrac{24}{29}, \tfrac{35}{29}, -\tfrac{36}{29}\}$$

The sum of squared errors is

$$\varepsilon = \sum_{n=1}^{5} [d(n) - \hat{d}(n)]^2 = 0.51724$$

We can also use Eq. 7.4.11 to get the same sum of squared errors.

2. The second way to solve the least-squares problem is to use the orthogonality principle. This may be stated as follows.

THE ORTHOGONALITY PRINCIPLE.   Set the error orthogonal to each column in the data matrix, $x'e = 0$.
    We call X the data matrix because $x(n)$ is the input data to the filter.

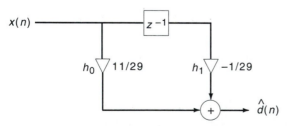

Fig. 7.4.2

A simplified statement of the orthogonality principle says to set the error orthogonal to the data. Let $\hat{d} = X\bar{h}$ and $e = d - \hat{d} = d - X\bar{h}$. The $\bar{h}$ minimizes the sum of squared errors ε if $\bar{h}$ is chosen so that $X'e = 0$. For this filter, the sum of square errors is given by $\varepsilon = d'e$.

Two vectors are *orthogonal* when their inner (dot) product is 0. Let v and w be column matrices of like dimension. Then the operation $v'w$ is the familiar inner (dot) product for column matrices. The vectors v and w are orthogonal whenever

$$v'w = 0 \qquad (7.4.12)$$

We will formalize the concept of the dot product in Section 8.2, but the important property of the dot product for any vector space is that from two vectors it produces a number. Notice in our statement of the orthogonality principle that there is one inner product $\varepsilon = d'e$ that produces the number ε. There is also an inner productlike operation $X'e = 0$ between a matrix X and a vector e that produces a column matrix 0, the zero vector. Each element in 0 is the dot product between the corresponding column of X (or the corresponding row of X') and the vector e.

We can apply the orthogonality principle to the least-squares filtering problem as follows. The principle requires that the inner product between each vector x and the error be 0, or

$$X'e = X'(d - X\bar{h}) = 0$$

Then the filter coefficient vector h that satisfies this equation is optimal. Carrying out the multiplication leads to

$$X'd = (X'X)\bar{h}$$

or

$$\bar{h} = (X'X)^{-1} X'd = X^{\#}d$$

which is Eq. 7.4.10.

You may recognize quite properly that there is essentially no difference between the two formulations of this problem. The advantage of using the orthogonality principle is convenience: It is easy to remember and easy to apply. And it applies to more than the least-squares problem, as we will see in the next chapter.

Figure 7.4.3 provides a geometric interpretation of these ideas. Two vectors x_0 and x_1 define a plane. Since $\hat{d}$ is a linear combination of the data,

$$\hat{d} = h_0 x_1 + h_1 x_0$$

then $\hat{d}$ must also lie in this plane. But d is not required to lie in this plane, so the picture will generally look like Fig. 7.4.3. Since the error is the difference $d - \hat{d}$, you can see from the diagram that the shortest error vector

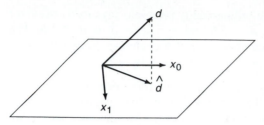

Fig. 7.4.3

occurs when the error is orthogonal to each data vector, i.e., when e is orthogonal to the plane containing x_0 and x_1. Since d is fixed, the estimate that makes the error orthogonal to the data must be optimum.

EXAMPLE 7.4.2.   Use the orthogonality principle to design a three-tap filter for the data in Example 7.4.1. Calculate the sum of squared errors.

SOLUTION: The X matrix and desired sequence are given by

$$X = \begin{bmatrix} 1 & -1 & 2 \\ -1 & 2 & -2 \\ 2 & -2 & 3 \\ -2 & 3 & -3 \end{bmatrix} \qquad d = \begin{bmatrix} 1 \\ -1 \\ 1 \\ -1 \end{bmatrix}$$

Setting the data X orthogonal to the error $e = d - X\bar{h}$ gives

$$X'e = X'(d - X\bar{h}) = 0$$

or

$$X'd = (X'X)\bar{h}$$

giving

$$\bar{h} = (X'X)^{-1} X'd = X^{\#}d$$

$$= \begin{bmatrix} -1.25 & 1.75 & 1.25 & -0.75 \\ 0.5 & 0.5 & 0.5 & 0.5 \\ 1.25 & -0.75 & -0.25 & 0.75 \end{bmatrix} \begin{bmatrix} 1 \\ -1 \\ 1 \\ -1 \end{bmatrix} = \begin{bmatrix} -1.0 \\ 0 \\ 1.0 \end{bmatrix}$$

or $h_0 = 1$, $h_1 = 1.5$, and $h_2 = -1$ as shown in Fig. 7.4.4.
    The sum of squared errors is

$$\varepsilon = d'e = d'(d - X\bar{h})$$

$$= \begin{bmatrix} 1 & -1 & 1 & -1 \end{bmatrix} \left(\begin{bmatrix} 1 \\ -1 \\ 1 \\ -1 \end{bmatrix} - \begin{bmatrix} 1 & -1 & 2 \\ -1 & 2 & -2 \\ 2 & -2 & 3 \\ -2 & 3 & -3 \end{bmatrix} \begin{bmatrix} -1 \\ 0 \\ 1 \end{bmatrix} \right) = 0$$

EXAMPLE 7.4.3. Table 7.1.2 displays 40 values of a correlated signal and 41 values of this signal are plotted in Fig. 7.1.2, illustrating the concept of least-squares prediction. Let us now design a three-tap filter for this purpose.

SOLUTION: For a three-tap filter the value of p in Fig. 7.4.1 is 2. The data matrix is given by

$$X = \begin{bmatrix} x(0) & x(1) & \cdots & x(p) \\ x(1) & x(2) & \cdots & x(p+1) \\ \vdots & \vdots & & \vdots \\ x(N) & x(N+1) & \cdots & x(p+N) \end{bmatrix} = \begin{bmatrix} 0.755 & 0.550 & 0.950 \\ 0.550 & 0.950 & 0.929 \\ 0.950 & 0.929 & 0.154 \\ \vdots & \vdots & \vdots \\ 0.074 & 1.000 & -0.461 \end{bmatrix}$$

In this matrix $N = 38$, leaving $p + N = 40$. That allows us to match up the data and desired sequences $x(n)$ and $d(n)$ so that $d(n) = x(n+1)$. Thus

$$x(n) = \{x(0), x(1), \ldots, x(40)\}$$
$$d(n) = \{x(1), x(2), \ldots, x(41)\}$$

Set the error e orthogonal to the data X by

$$X'e = 0$$

where the error vector is given by

$$e = d - \hat{d} = d - X\bar{h}$$

This gives

$$X'(d - X\bar{h}) = 0$$

or

$$X'd = X'X\bar{h}$$

Solving for $\bar{h}$ gives (see Eq. 7.4.10)

$$\bar{h} = (X'X)^{-1}X'd = X^{\#}d$$

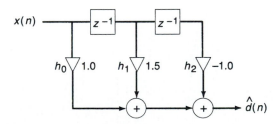

Fig. 7.4.4

When we solve this equation (by computer), we get h coefficients given by

$$h_0 = 0.7183 \qquad h_1 = -0.4032 \qquad h_2 = 0.4148$$

Equation 7.4.11 gives the sum of square errors as 7.567, which compares well with the error in Fig. 7.1.3 of 19.89. The graph in Fig. 7.4.5 shows the result of predicting $x(n)$ with this filter, as we did in Fig. 7.1.3. The closed dots represent the signal $x(n)$ and the open dots represent the predicted values of $x(n)$.

Review

Here are the steps in applying the orthogonality principle to least-squares filter design.

1. Identify p in Eq. 7.4.5 and construct the $(N + 1) \times (p + 1)$ matrix X.

$$X = \begin{bmatrix} x(0) & x(1) & \cdots & x(p) \\ x(1) & x(2) & \cdots & x(p+1) \\ \vdots & \vdots & & \vdots \\ x(N) & x(N+1) & \cdots & x(p+N) \end{bmatrix}$$

2. Set the error e orthogonal to the data X by

$$X^t e = 0$$

where the error vector is given by

$$e = d - \hat{d} = d - X\overline{h}$$

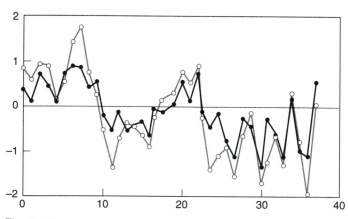

Fig. 7.4.5

This gives

$$X^t(d - X\bar{h}) = 0$$

or

$$X^t d = X^t X\bar{h}$$

Solving for $\bar{h}$ gives the same form as Eq. 7.4.10.

3. The least-squares error using the optimum filter found in step 2 is given by Eq. 7.4.11.

7.5 Problems

7.1. A particular voltage is measured by five different voltmeters, giving values 12.8, 12.4, 13.2, 12.5, and 13.0. Assume that the differences are due to instrument error instead of any fluctuation in the actual voltage, and find the least-squares estimate of the voltage. Calculate the sum of squared errors.

7.2. A noisy signal $x(t)$ is sampled at uniform intervals to obtain the sampled signal $x(n)$ given in Table 7.5.1. Suppose that we wish to model this data by a relationship of the form $x(n) = c_0 + c_1 n$.
(a) Find the least-squares estimate for c_0 and c_1.
(b) Calculate the sum of square errors.
(c) Plot the data and the least-squares line on the same graph. Does the line appear to be a reasonable fit to the data?

7.3. (a) Fit a curve $x(n) = a + bn + cn^2$ to the data in Table 7.5.1.
(b) Calculate the sum of square errors for this fit.
(c) Plot the data and the least-squares curve on the same graph. Does the curve appear to be a reasonable fit to the data?

7.4. (a) For the measurements of weight, height, and girth for eight people given in Table 7.5.2, find the regression of weight on height. That is, find the constants c_0 and c_1 that minimize the sum of squared errors for the

Table 7.5.1

n	$x(n)$	n	$x(x)$
1	26.5	6	50.0
2	36.5	7	47.0
3	44.5	8	43.0
4	50.0	9	40.0
5	52.0	10	35.0

Table 7.5.2

Subject	Weight	Height	Girth
1	170	68	38
2	220	70	42
3	180	72	36
4	150	69	32
5	170	70	36
6	200	70	40
7	160	68	36
8	190	72	34

estimate $\hat{y} = c_0 + c_1 x_1$, where x_1 is the height and y is the weight.
   (b) Find the sum of squared errors for the eight estimates.

7.5. Repeat Problem 7.4 for the girth versus weight. That is, find the regression of weight on girth, and calculate the sum of squared errors.

7.6. Find the regression of weight on both height and girth in Table 7.5.2. Here we have

$$\hat{y} = c_0 + c_1 x_1 + c_2 x_2$$

where y is the weight, x_1 is the height, and x_2 is the girth. Again calculate the sum of squared errors and see if it is less than that for Problem 7.4 or 7.5.

7.7. Solve Problem 7.2 using the pseudoinverse.

7.8. Solve Problem 7.3 using the pseudoinverse.

7.9. Solve Problem 7.4 using the pseudoinverse.

7.10. Solve Problem 7.6 using the pseudoinverse.

7.11. Given the three equations

$$6x_1 + 9x_2 = -5$$
$$3x_1 + 2x_2 = -5$$
$$-x_1 + x_2 = 5$$

   (a) Find the least-squares solution for x_1 and x_2 using the pseudoinverse.
   (b) Apply the iterative least-squares algorithm to this problem with initial guess $x = [1 \quad 1]'$. If you use hand calculation, calculate at least three steps and see if it appears to converge.

7.12. The data in Table 7.5.3 pair x and y values. These data appear to be exponential because they start with a value of y greater than 0 at $x = 0$, and increase rapidly as x increases.

Table 7.5.3

x	$y(x)$	x	$y(x)$	x	$y(x)$
0.0	1.00	0.7	1.49	1.4	2.96
0.1	1.01	0.8	1.64	1.5	3.25
0.2	1.04	0.9	1.81	1.6	3.56
0.3	1.09	1.0	2.00	1.7	3.89
0.4	1.16	1.1	2.21	1.8	4.24
0.5	1.25	1.2	2.44	1.9	4.61
0.6	1.36	1.3	2.69	2.0	5.00

(a) Fit an exponential curve $y = ae^{bx}$ to these data using least squares.
(b) Fit a polynomial $y = a + bx + cx^2$ to these data. Which model fits best?

7.13. Noisy measurements of the unit step response of a low-pass filter are given in Table 7.5.4. The measurements are taken every 0.1 s.
   (a) If this step response is of the form $y(t) = A[1 - e^{-t/\tau}]$, where the gain A is known to be 10, can you devise a method to estimate the time constant τ using regression?
   (b) Suppose the gain is unknown. Can you estimate both the gain and the time constant?

7.14. Find the line $x(n) = c_0 + c_1 n$ in Problem 7.2 by iteration.

7.15. Find the curve $x(n) = a + bn + cn^2$ in Problem 7.3 by iteration.

7.16. Can you devise a method to find the coefficients c_0, c_1, and c_2 in Problem 7.6 using iteration?

7.17. Design a two-tap filter using the orthogonality principle to provide the least-squares fit for the input sequence $x(n)$ and desired output sequence $d(n)$ shown in Table 7.5.5.

Table 7.5.4

t	$y(t)$	t	$y(t)$	t	$y(t)$
0	0.0	0.7	3.2844	1.4	4.9306
0.1	0.4662	0.8	3.4399	1.5	5.3721
0.2	0.8837	0.9	3.5281	1.6	5.4969
0.3	1.3729	1.0	4.0575	1.7	5.8116
0.4	1.9604	1.1	4.2356	1.8	6.0230
0.5	2.1537	1.2	4.5692	1.9	6.1857
0.6	2.7598	1.3	4.8651	2.0	6.3630

Table 7.5.5

$x(n)$	2	1	-1	0	2
$d(n)$	1	1	-1	-1	1

7.18. Use the orthogonality principle to design a three-tap filter for the data in Table 7.5.6. The input data $x(n)$ is a noisy sine wave, and the desired output $d(n)$ is the noise-free sinusoid. Therefore this filter should be a low-pass filter since it eliminates the high-frequency noise while retaining the low-frequency sinusoid.

7.19. Design a four-tap filter for the data in Table 7.5.6.

7.20. This problem illustrates how to use iteration on a familiar problem. Suppose we want to find $\sqrt{3}$ numerically. We must solve $x^2 = 3$ for x. Adding x^2 to both sides we get $2x^2 = x^2 + 3$. Upon dividing both sides by $2x$, we get

$$x = \frac{1}{2}\left(x + \frac{3}{x}\right)$$

which we can turn into an iterative process by

$$x_{k+1} = \frac{1}{2}\left(x_k + \frac{3}{x_k}\right)$$

(a) Start with initial guess $x_0 = 1$ and calculate the square root of 3 this way.

(b) Use this method to develop a procedure to calculate the cube root of a number. Test your formula by calculating the cube root of 8 with initial $x_0 = 1$.

Table 7.5.6

$x(n)$	$d(n)$	$x(n)$	$d(n)$
10.5011	10.0000	1.2287	0.0000
9.8553	9.8769	-1.5131	-1.5643
8.8311	9.5106	-2.5173	-3.0902
9.5674	8.0902	-3.6836	-4.5399
6.4881	7.0711	-6.9138	-5.8779
7.5587	5.8779	-6.1127	-7.0711
4.8713	4.5399	-8.1881	-8.0902
4.5215	3.0902	-8.0528	-8.9101
0.6072	1.5643	-8.6237	-9.5106

Further Reading ───────────────────────────────

1. ALEXANDER M. MOOD and FRANKLIN A. GRAYBILL, *Introduction to the Theory of Statistics,* McGraw-Hill, New York, 1963.
2. RICHARD SHIAVI, *Introduction to Applied Statistical Signal Analysis,* Asken, Homewood, Il., 1991.
3. RICHARD BRANSON, *Matrix Operations,* Schaum's Outline Series, McGraw-Hill, New York, 1989.

CHAPTER 8 _____

Optimum Filtering

8.1 Mean-Square Estimation _____

Preview

Mean-square estimation seeks to minimize the mean-square error between the estimate and the quantity being estimated. That means we need to know second-order statistics—the means, variances, and covariances of the data—because the mean-square error depends only on these quantities. The mean and autocorrelation functions contain this information, and this is often available in random signal processing.

In Chapter 7 we introduced least-squares estimation. The terms "mean-square estimation" and "least-squares estimation" are similar, so you should be careful to avoid confusing them. Least-squares techniques use samples of the data to derive good estimates. Mean-square estimation is based on statistical averages.

Linear mean-square estimation is a special case of mean-square estimation. In order to show how it fits into the overall scheme, we begin with the simplest problem in mean-square estimation and gradually add complexity until we arrive at a general estimation problem. Then we will specialize this general problem to linear estimation and show how the orthogonality principle simplifies the problem.

Let $\hat{X}$ stand for an estimate of the random variable X. The error is the difference between the value of the random variable and our estimate:

$$e = X - \hat{X}$$

The estimate $\hat{X}$ is a number; X is a random variable. Hence the error e is a random variable. We square and then average this random variable to obtain the mean-square error:

$$E(e^2) = E[(X - \hat{X})^2] \tag{8.1.1}$$

The value of $\hat{X}$ that minimizes this expression is the minimum mean-square estimate of X.

Estimating the Random Variable X by a Constant

If $\hat{X} = c$, the error is $X - c$, and the mean-square error is

$$E(e^2) = E[(X - c)^2] = \int_{-\infty}^{\infty} (\alpha - c)^2 f_X(\alpha)\, d\alpha$$

$$= c^2 - 2cE(X) + E(X^2)$$

Differentiating with respect to c and setting the derivative to 0, we have

$$2c - 2E(X) = 0 \quad \text{or} \quad c = E(X) \tag{8.1.2}$$

Therefore we should use the average $E(X)$ to estimate X when we wish to minimize the mean-square error.

> **EXAMPLE 8.1.1.**   The random variable X is defined in Fig. 8.1.1. For the time being, ignore the other random variable in the diagram; we will use it later. The experiment consists of rolling a single die, and the experimental outcome determines a value for X (and the other random variable). Find a constant c to give the minimum mean-square estimate of X and calculate the resulting mean-square error.
>
> SOLUTION: The average value of X gives the estimate.
>
> $$c = E(X) = 0.5$$
>
> The resulting mean-square error is
>
> $$E(e^2) = E[(X - 0.5)^2] \approx 2.917$$
>
> This illustrates a startling feature of mean-square estimation. The estimate need not be a possible value of X. Our estimate of 0.5 can never be correct, but the average square error will be less than that for any estimate that could possibly be correct. For example, if we fudge a little and choose as our estimate $c = 1$, we will be correct about one-sixth of the time, but the mean-square error increases to 3.167.

Here the experiment is performed with outcome ζ. This outcome ζ then determines a value of the random variable X, and it is this number we estimate. In this problem we make the estimate before performing the experiment (or

Y	4	1	0	1	4	9
X	−2	−1	0	1	2	3
ζ	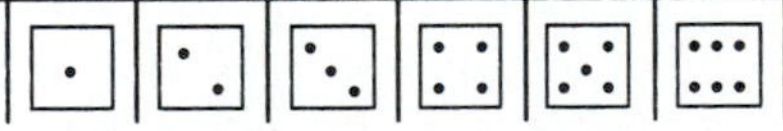 ⚀	⚁	⚂	⚃	⚄	⚅

Fig. 8.1.1. Definition of a random variable X.

at least before we have any information about the experimental outcome). In all of the following we perform the experiment first and provide partial information on which to base the estimate.

Estimating the Random Variable X given Y(ζ)

Suppose that two random variables X and Y are defined on the same sample space. The experiment is performed and you are told the value of Y. The estimate $\hat{X}$ should incorporate this additional information, so $\hat{X}$ is a function of Y, $\hat{X} = g(Y)$. The mean-square error that we wish to minimize is given by

$$E(e^2) = E\{[X - g(Y)]^2\}$$

The function that minimizes this expression is

$$g(Y) = E(X|Y) \tag{8.1.3}$$

as shown below:

$$E\{[X - g(Y)]^2\} = \int_{-\infty}^{\infty} \int_{-\infty}^{\infty} [\alpha - g(\beta)]^2 f_{XY}(\alpha, \beta) \, d\alpha \, d\beta$$

$$= \int_{-\infty}^{\infty} \int_{-\infty}^{\infty} [\alpha - g(\beta)]^2 f_X(\alpha|\beta) f_Y(\beta) \, d\alpha \, d\beta \tag{8.1.4}$$

$$= \int_{-\infty}^{\infty} f_Y(\beta) \int_{-\infty}^{\infty} [\alpha - g(\beta)]^2 f_X(\alpha|\beta) \, d\alpha \, d\beta$$

Let us define a new function $h(\beta)$ that is equal to the last integral in this expression:

$$h(\beta) = \int_{-\infty}^{\infty} [\alpha - g(\beta)]^2 f_X(\alpha|\beta) \, d\alpha \tag{8.1.5}$$

Now Eq. 8.1.4 can be written

$$E\{[X - g(Y)]^2\} = \int_{-\infty}^{\infty} f_Y(\beta) \, h(\beta) \, d\beta$$

We wish to choose $g(\beta)$ to minimize this expression. Note that $f_Y(\beta)$ is always positive because it is a probability density function. We have no control over $f_Y(\beta)$; this is the marginal pdf of the random variable Y. Hence we can operate only on $h(\beta)$. From Eq. 8.1.5 we see that $h(\beta)$ is also always positive. The product of a squared term times a density function is positive, so the integral of this positive expression is also positive. Therefore, if we choose $g(\beta)$ to minimize $h(\beta)$ for every value of β, this will minimize the mean-square error.

Expand Eq. 8.1.5 to obtain

$$h(\beta) = g^2(\beta) - 2g(\beta) \, E(X|\beta) + E(X^2|\beta)$$

Setting the derivative with respect to β equal to 0 gives

$$2g(\beta) - 2E(X|\beta) = 0$$

or

$$g(\beta) = E(X|\beta) \tag{8.1.6}$$

So to estimate X we should use the conditional expectation, the average value of the random variable X based on what we know about it. The estimate should be based on the given data $Y = \beta$, and the way to do this is to use the conditional expectation $E(X|Y = \beta)$.

This situation differs from the first (estimating X by a constant) in that the experiment is performed with outcome ζ, and we are allowed to observe the value of $Y(\zeta)$. If knowledge of Y allows us to determine ζ, then we can estimate $X(\zeta)$ exactly. That is, $E(X|\beta) = X$. Of course, this is the trivial case in which knowledge of Y gives us complete information about X. The other extreme is when knowledge of Y gives us no information about X, and $E(X|\beta) = E(X)$.

EXAMPLE 8.1.2.   In Fig. 8.1.1 the random variable Y can take on one of four different numbers, 0, 1, 4, and 9. For each possible value of Y, find the best estimate of X, and calculate the resulting mean-square error.

SOLUTION: When $Y = 4$, X must be either -2 or 2 with equal probability. Therefore

$$g(4) = E(X|Y = 4) = 0$$

Similar reasoning applies to the other values of Y, giving

$$g(0) = E(X|Y = 0) = 0$$

$$g(1) = E(X|Y = 1) = 0$$

$$g(9) = E(X|Y = 9) = 3$$

Therefore we guess $\hat{X} = 0$ for every value of Y except for $Y = 9$. For that value of Y we guess $\hat{X} = 3$. This is shown in Fig. 8.1.2. The resulting mean-square error is given by

$$E(e^2) = E[(X - \hat{X})^2]$$
$$= (\tfrac{1}{6})[(-2)^2 + (-1)^2 + 0^2 + 1^2 + 2^2 + 0^2] = 1.667$$

Notice that this is less than the error in Example 8.1.1, as it should be. Here we use more knowledge to make the estimate.

The estimate determined by Eq. 8.1.6 is a nonlinear estimate because the values of $\hat{X}$ do not lie on a straight line through the origin. The following

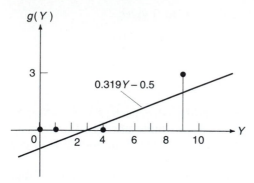

Fig. 8.1.2.   Estimates for X.

example chooses the best straight-line fit, but this is not linear estimation because the straight line does not go through the origin.

EXAMPLE 8.1.3.   Estimate the value of X given Y in Fig. 8.1.1 by points on a straight line,

$$\hat{X} = g(Y) = aY + b \qquad (8.1.7)$$

SOLUTION: The only difference between this and the previous example is that we are restricting the form of $g(Y)$. It must be a straight line, so we need to solve only for the two constants, a and b. The mean-square error that we wish to minimize is given by

$$E(e^2) = E\{[X - (aY + b)]^2\} \qquad (8.1.8)$$

Setting the partial derivative with respect to b equal to 0 gives

$$\frac{\partial}{\partial b} E(e^2) = 2b - 2E(X) + 2aE(Y) = 0$$

The value of b that minimizes Eq. 8.1.8 is therefore

$$b = m_X - am_Y \qquad (8.1.9)$$

Using this value of b, we can write Eq. 8.1.8 as

$$E(e^2) = E\{[(X - m_X) - a(Y - m_Y)]^2\}$$
$$= \sigma_X^2 - 2a\mu_{11} + a^2\sigma_Y^2$$

where μ_{11} is the covariance between X and Y (see Eq. 4.2.2). Solving for a by setting the partial derivative with respect to a equal to 0 gives

$$a = \frac{\mu_{11}}{\sigma_Y^2} \qquad (8.1.10)$$

For the random variables X and Y in Fig. 8.1.1 this gives

$$a = \frac{\mu_{11}}{\sigma_Y^2} = 0.319$$

$$b = m_X - am_Y = 0.5 - (0.319)(3.17) = -0.5$$

so that $g(Y) = 0.319Y - 0.5$. This function is plotted as the straight line in Fig. 8.1.2.

The resulting mean-square error is

$$E(e^2) = E[(X - aY - b)^2] = 1.99$$

Note that this error compares rather well with the optimum mean-square error of 1.67 from Example 8.1.2.

Review

This completes our introduction to mean-square estimation. In Section 8.3 we introduce linear mean-square estimation, which is nothing more than an estimate of the form $\hat{X} = aY$ for one data point Y. For more than one data value, the linear estimate is of the form $(\hat{X}) = \Sigma_{i=1}^{N} a_i Y_i$.

Mean-square estimation is a statistical procedure, which means that we need some statistical knowledge about the quantity we are estimating before we can derive the estimate. This is opposed to least-squares estimation, where we use the data alone to form the estimate. For mean-square estimation we need only second-order information, the means and correlations. We may be given more information, as in Example 8.1.2 where we had complete knowledge, but we use only the means and correlations in deriving our estimate.

8.2   Vector Spaces ────────────────────────────

Preview

Ask any engineer to define a vector, and she will likely answer, "a directed magnitude." This is a good answer; it is just wrong. Geometric vectors are directed magnitudes, but there are other kinds of vectors besides geometric vectors, and these other vectors don't have "magnitude and direction." Vectors, like anything else, are defined by listing properties. We list the important properties of geometric vectors in our definition below, and anything else that satisfies these properties is called a set of vectors. It turns out that many other things satisfy this definition, including random variables. This is important to us, since we will be using properties of vectors in our discussion of estimation theory.

In this section we define vector space, dot products, norms, metrics, and orthogonality. Then, in subsequent sections we will use these concepts to discuss mean-square estimation procedures. This will lead eventually to Wiener and Kalman filtering.

DEFINITION 8.2.1.   A vector space is a set $V = \{v_i\}$ together with a field of scalars $A = \{a_i\}$ that has the following two operations and seven properties.

1. We can add two vectors together and obtain a third vector. Thus we have a mechanism for combining two vectors to obtain a third.
2. We can multiply a vector by a scalar and obtain another vector. This mechanism combines a scalar with a vector to obtain another vector.

Using these two operations, vector addition and scalar multiplication, the following properties must hold for all $v_i \in V$ and all $a_i \in A$.

(1) $v_1 + v_2 = v_2 + v_1$
(2) $(v_1 + v_2) + v_3 = v_1 + (v_2 + v_3)$
(3) $a_1(v_1 + v_2) = a_1 v_1 + a_1 v_2$
(4) $(a_1 + a_2)v_i = a_1 v_i + a_2 v_i$
(5) $a_1(a_2 v_i) = (a_1 a_2)v_i$
(6) $1 \cdot v_i = v_i$
(7) There exists a unique vector v_0, called the zero vector, such that for all vectors v_i we have $0 \cdot v_i = v_0$, where 0 is the number 0.

Recall that a field is modeled after the real number system with the two operations addition and multiplication. In fact, we will have no need for any other field, so when we say "scalar" you can substitute "real number."

These are the properties that define a vector space. Any set of objects that collectively satisfies all of these properties is called a vector space. Notice that we have not defined a vector. Instead we defined a vector space, a set of vectors. A vector space is a set together with two functions, a mapping from $V \times V \rightarrow V$ that defines vector addition, and a mapping from $A \times V \rightarrow V$ that defines scalar multiplication. (The notation $V \times V$ and $A \times V$ stands for the set cross product.) Any set of objects with these two functions that satisfies the seven properties is a vector space. Furthermore, these vectors are equally as legitimate as geometric vectors, even though they may not have "magnitude" and "direction."

Ordinary geometric vectors satisfy these properties, and therefore form a vector space. To show this, let $v_1, v_2, \ldots$ be a set of ordinary geometric vectors. These are the familiar "directed magnitudes" of geometry. Let $a_1, a_2, \ldots$ be ordinary numbers (from the real number system). We will call these numbers "scalars," but they are just ordinary numbers from the field of real numbers (see Section 3.1). We can add two vectors together in any order (property 1). We can add three vectors by first adding $v_1 + v_2$ and then adding v_3, or we can add $v_2 + v_3$ before adding v_1, and we get the same result either way (property 2). Continuing in this way, you can see that geometric vectors satisfy each property listed.

But geometric vectors are not the only objects that satisfy Definition 8.2.1. All 3×2 matrices satisfy these properties. We don't usually call a matrix a vector, but all matrices of the same dimension form a vector space.

The set of real numbers satisfies these properties. Here the numbers play a dual role: They are both vector and scalar.

The set of all waveforms that can be generated in the laboratory satisfy these properties. Here, instead of $v_1, v_2, \ldots$, we have time functions $v_1(t)$, $v_2(t), \ldots$. We can add waveforms together to form another waveform. We can multiply by scalars to change the amplitude. And there is a zero vector, namely, $v_0(t) = 0$, for all time, which satisfies property 7. (These same statements apply equally to a set of discrete-time waveforms.) So you can see that waveforms (signals) are vectors.

Our interest in vectors stems from the fact that all random variables defined on the same sample space form a vector space. We can add two random variables and their sum is a third random variable. We can multiply a random variable by a scalar and the result is another random variable, and there is a zero random variable that assigns the number 0 to each experimental outcome. To illustrate, two random variables X and Y are defined on the sample space of die-toss outcomes in Fig. 8.2.1. We can add $X + Y$ and get another random variable. This means that we add the numbers for each experimental outcome. If the one-spot turns up, $X + Y = -2 + 4 = 2$. If the two-spot turns up, the sum is $1 - 1 = 0$. Continuing in this way, we can obtain a sum for each experimental outcome, which is another random variable. We can also multiply X by a scalar to obtain another random variable, and all seven properties are satisfied. So the set of all random variables that could be defined on the die-toss experiment forms a vector space. Notice that all random variables must be defined on the same sample space. Random variables defined on different sample spaces are not vectors in the same space.

So far, we have imposed little structure on our vector spaces. All we can do is multiply them by scalars and add two vectors together. We have said nothing about how to measure the length of a vector, or how to determine the distance between two vectors, or how to find the dot product of two vectors. In order to do geometry we must be able to do all three of these

Y	4	1	0	1	4	9
X	−2	−1	0	1	2	3
ζ	⚀	⚁	⚂	⚃	⚄	⚅

Fig. 8.2.1. Random variables X and Y defined on the sample space of die-toss outcomes.

things. A vector space in which we can do geometry is called an *inner product* (or *dot product*) space. That is because once we have defined the dot product, a measure of length and distance is automatically imposed. So we begin by defining an inner product.

DEFINITION 8.2.2: Inner product.   The symbol $\langle v_1|v_2\rangle$ denotes the operation of extracting a number from a pair of vectors. This is called the inner (or dot) product, and must satisfy the following four properties for all scalars and vectors.

(1) $\langle v_1|v_2\rangle = \langle v_2|v_1\rangle$
(2) $\langle v_1 + v_2|v_3\rangle \le \langle v_1|v_3\rangle + \langle v_2|v_3\rangle$
(3) $\langle a_1 v_1|v_2\rangle = a_1\langle v_1|v_2\rangle$
(4) $\langle v_1|v_1\rangle \ge 0$, and $\langle v_1|v_1\rangle = 0$ if and only if $v_1 = v_0$ (the zero vector)

If the vector space $V = \{v_i\}$ satisfies these properties, it is called an inner product space, which means that we can do geometry in this space. The length of a vector v_1, denoted by $\|v_1\|$ and called the *norm,* is defined by

$$\|v_1\| = \langle v_1|v_1\rangle^{1/2} \tag{8.2.1}$$

The distance between two vectors is given in terms of the norm by

$$d(v_1, v_2) = \|v_1 - v_2\| \tag{8.2.2}$$

This gives us a hierarchy of measures: The inner product induces a norm (a measure of length), which, in turn, induces a measure of distance between two vectors. Incidentally, we can determine the ''angle'' θ between two vectors by the formula

$$\cos \theta = \frac{\langle v_1|v_2\rangle}{\|v_1\| \, \|v_2\|} \tag{8.2.3}$$

In the space $V = R^n$ over the field R of real numbers, the standard inner product of two vectors,

$$v_1 = \begin{bmatrix} a_1 \\ \vdots \\ a_n \end{bmatrix} \quad \text{and} \quad v_2 = \begin{bmatrix} b_1 \\ \vdots \\ b_n \end{bmatrix}$$

is $\langle v_1|v_2\rangle = \sum_{i=1}^{n} a_i b_1$, which can be written in terms of matrices as $\langle v_1|v_2\rangle = v_1^t v_2$, where v_1^t is the transpose of the ($n \times 1$) matrix v_1. The norm induced by this inner product is

$$\|v_1\| = \langle v_1|v_1\rangle^{1/2} = \left(\sum_{k=1}^{n} a_k^2 \right)^{1/2}$$

and the metric (distance between two vectors) is

$$d(v_1, v_2) = \|v_1 - v_2\| = \left[\sum_{k=1}^{a} (a_k - b_k)^2 \right]^{1/2}$$

These are the usual ways to measure length and distance, and they should be familiar from Euclidean geometry.

EXAMPLE 8.2.1.   Find the inner product, the length, and the distance between the following vectors x and y. Use the usual dot product for Euclidean vectors.

$$x = \begin{bmatrix} 3 \\ -1 \\ 2 \\ 1 \end{bmatrix} \quad y = \begin{bmatrix} 2 \\ 1 \\ -1 \\ 0 \end{bmatrix}$$

SOLUTION: The dot product is $\langle x|y \rangle = x^t y = 3$. The length of each vector is

$$\|x\| = [3 \cdot 3 + (-1)(-1) + 2 \cdot 2 + 1 \cdot 1]^{1/2} = \sqrt{15}$$
$$\|y\| = [2 \cdot 2 + 1 \cdot 1 + (-1)(-1) + 0 \cdot 0]^{1/2} = \sqrt{6}$$

The distance between them is

$$d(x, y) = \|x - y\| = [1^2 + (-2)^2 + 3^2 + 1^2]^{1/2} = \sqrt{15}$$

For another example, in the space of all continuous-time energy signals, the usual inner product is defined by

$$\langle v_1(t)|v_2(t) \rangle = \int_{-\infty}^{\infty} v_1(t)\, v_2(t)\, dt$$

This induces a norm given by

$$\|v(t)\| = \left[\int_{-\infty}^{\infty} v^2(t)\, dt \right]^{1/2}$$

And the metric is

$$d(v_1, v_2) = \left[\int_{-\infty}^{\infty} [v_1(t) - v_2(t)]^2\, dt \right]^{1/2}$$

EXAMPLE 8.2.2.   Find the inner product, the norm of each waveform, and the distance between them for (see Fig. 8.2.2)

$$v_1(t) = e^{-t}u(t)$$
$$v_2(t) = u(t) - u(t - 1)$$

SOLUTION:

$$\langle v_1|v_2 \rangle = \int_0^1 e^{-t}\, dt = 1 - e^{-1} = 0.632$$

$$\|v_1\| = \left(\int_0^{\infty} e^{-2t}\, dt \right)^{1/2} = \sqrt{0.5}$$

Fig. 8.2.2

$$\|v_2\| = \left(\int_0^1 1 \, dt \right)^{1/2} = 1$$

$$d(v_1, v_2) = \left[\int_0^1 (1 - e^{-t})^2 \, dt + \int_1^\infty e^{-2t} \, dt \right]^{1/2}$$

$$= (0.168 + 0.0677)^{1/2} = 0.484$$

Similar definitions define the usual inner product, norm, and the metric for discrete time energy signals.

$$\langle v_1(n)|v_2(n)\rangle = \sum_{n=-\infty}^{\infty} v_1(n)v_2(n)$$

$$\|v_1(n)\| = \left[\sum_{n=-\infty}^{\infty} v_1^2(n) \right]^{1/2}$$

$$d(v_1, v_2) = \left[\sum_{n=-\infty}^{\infty} [v_1(n) - v_2(n)]^2 \right]^{1/2}$$

Our primary interest is in random variables. The usual dot product for random variables is simply the correlation:

$$\langle X|Y\rangle = E(XY) \tag{8.2.4}$$

The norm induced by this dot product is

$$\|X\| = [E(X^2)]^{1/2} \tag{8.2.5}$$

which is the root-mean-square (rms) value of the random variable X. This is the standard deviation if the mean of X is 0. The distance between two random variables X and Y is given by

$$d(X, Y) = \|X - Y\| = [E(X - Y)^2]^{1/2} \tag{8.2.6}$$

EXAMPLE 8.2.3.   Find the inner product, the norm of each vector, and the distance between the random variables X and Y in Fig. 8.2.1.

SOLUTION: The inner product is

$$E(XY) = \sum_{i=1}^{6} \alpha_i \beta_i P[X = \alpha_i, Y = \beta_i]$$
$$= (\tfrac{1}{6})[(-2)(4) + (-1)(1) + 0 \cdot 0 + 1 \cdot 1 + 2 \cdot 4 + 3 \cdot 9] = 4.5$$

The norm of each random variable is

$$\|X\| = [E(X^2)]^{1/2} = \left(\sum_{i=1}^{6} \alpha_i^2 P[X = \alpha_i] \right)^{1/2}$$
$$= [(\tfrac{1}{6})(4 + 1 + 0 + 1 + 4 + 9)]^{1/2} = 1.78$$
$$\|Y\| = [(\tfrac{1}{6})(16 + 1 + 0 + 1 + 16 + 81)]^{1/2} = 4.378$$

The distance between them is

$$d(X, Y) = [E(X - Y)^2]^{1/2} = \left(\sum_{i=1}^{6} (\alpha_i - \beta_i)^2 P[X = \alpha_i, Y = \beta_i] \right)^{1/2}$$
$$= \{(\tfrac{1}{6})[(-2 - 4)^2 + (-1 - 1)^2 + 0 + 0 + (2 - 4)^2 + (3 - 9)^2]\}^{1/2}$$
$$= 3.65$$

Orthogonality

Two vectors are orthogonal if their inner product is 0. We are accustomed to geometric vectors that have the additional property that orthogonal vectors are 90° apart. Since the vectors we will use do not have an apparent "direction," you should not look for "right angles" when using orthogonality. For random variables, orthogonality simply means $E(XY) = 0$.

Notice that being uncorrelated and orthogonal are not the same thing. Uncorrelated means $E(XY) = E(X) E(Y)$. Orthogonal means $E(XY) = 0$. Only if the mean of either X or Y is 0 do these two terms mean the same thing.

The Cauchy–Buniakovshy–Schwartz (CBS) Inequality

If V is an inner-product space, which means we can do geometry, then the CBS inequality holds for any valid inner product:

$$|\langle u|v \rangle|^2 \le \langle u|u \rangle \langle v|v \rangle \qquad (8.2.7)$$

The proof follows from the properties of the inner product as follows: First, if $v = v_0$, the zero vector, then we have equality by property 4. If $v \ne v_0$, then we start with

$$0 \le \langle u - \lambda v | u - \lambda v \rangle = \langle u|u \rangle - 2\lambda \langle u|v \rangle + \lambda^2 \langle v|v \rangle \qquad (8.2.8)$$

which holds for all λ. Let $\lambda = \langle u|v\rangle/\langle v|v\rangle$. Then Eq. 8.2.8 gives

$$0 \le \langle u|u\rangle - 2\frac{\langle u|v\rangle^2}{\langle v|v\rangle} + \frac{\langle u|v\rangle}{\langle v|v\rangle} = \langle u|u\rangle - \frac{\langle u|v\rangle^2}{\langle v|v\rangle}$$

which gives the CBS inequality in Eq. 8.2.7.

Here is an example of template matching using the CBS inequality. In order to store an image in computer memory, we sample the gray-scale values (how dark the image is) in small blocks called pixels. Thus, an $N \times M$ image has N rows and M columns of values, which may be stored and manipulated in a computer. Typical sizes are 128×128 or larger, but we will simplify this by using a 2×3 image array. Let

$$u = \begin{bmatrix} 3 & 1 & -2 \\ 1 & 2 & -3 \end{bmatrix} \qquad v = \begin{bmatrix} 1 & 3 & -3 \\ 2 & 1 & -2 \end{bmatrix}$$

The usual inner product for matrices multiplies corresponding pixels and sums the products. That is,

$$\langle u|v\rangle = \sum_i \sum_j u_{ij} v_{ij}$$

For the particular u and v above, the inner product is 22. We define the energy in image u as

$$E_u = \langle u|u\rangle = \|u\|^2$$

This gives

$$E_u = 28 \qquad E_v = 28$$

Thus the CBS inequality is satisfied because $|\langle u|v\rangle|^2$ is less than $\langle u|u\rangle\langle v|v\rangle$.

This example illustrates that for equal energy images, the maximum response (or correlation, or inner product) is given when the two images are identical, and this maximum response is determined by the CBS inequality.

Review

Notice that we did not define a vector in Definition 8.2.1. Instead, we defined a vector space, a set of objects that collectively satisfies the properties listed there. This means that we cannot determine that an object v by itself is or is not a vector. The entire set must satisfy all the properties of Definition 8.2.1. There are many sets besides geometric vectors that satisfy these properties, including all random variables defined on the same sample space.

The inner product, norm (length), and metric (distance) impose structure on a vector space. For random variables the usual inner product is $E(XY)$, which means the norm is the rms value, and the metric is the rms value of the difference between X and Y. See Eqs. 8.2.4, 8.2.5, and 8.2.6.

8.3 Linear Mean-Square Estimation _____

Preview

The concept of linearity applies to functions. When we say that a function is linear, we mean that the functional relationship between the domain and codomain is linear. When we say that an estimate is linear, we mean that the relationship between the data and the estimate is linear. When we say that a system is linear, we mean that the functional relationship between the input and output is linear; that is, the input–output function that describes the system is linear.

In this section we introduce the two-part test for linearity, and show how this applies to mean-square estimation. Then we introduce the idea of setting the error orthogonal to the data, which is the basic tenet in linear estimation.

A function is linear if it is additive and homogeneous. A function $f: X \rightarrow Y$ is said to be *additive* if

$$f(x_1 + x_2) = f(x_1) + f(x_2) \tag{8.3.1}$$

for all elements x_i in X. The addition on the left is in X, while the addition on the right is in Y.

We say that $f: X \rightarrow Y$ is *homogeneous* if

$$f(ax) = af(x) \tag{8.3.2}$$

for all scalars a and all elements x in X. The multiplication on the left is in X, while the multiplication on the right is in Y.

> **EXAMPLE 8.3.1.**   Let $y = 2x + 3$. Test this relationship for additivity and homogeneity.
>
> SOLUTION: Try $x_1 = 5$ and $x_2 = 6$ in Eq. 8.3.1. Then we have on the left,
>
> $$x_1 + x_2 = 5 + 6 = 11$$
>
> so
>
> $$f(x_1 + x_2) = 2 \cdot 11 + 3 = 25$$
>
> The right side of Eq. 8.3.1 gives
>
> $$f(x_1) = 2x_1 + 3 = 13$$
> $$f(x_2) = 2x_2 + 3 = 15$$
>
> giving
>
> $$f(x_1) + f(x_2) = 28$$
>
> Since we have found at least one case where Eq. 8.3.1 is not satisfied, the relationship $f(x) = 2x + 3$ is not additive (and therefore not linear).

To test for homogeneity, let $x = 5$, $a = 2$ in Eq. 8.3.2. Then on the left we have

$$ax = 10 \quad \text{and} \quad f(ax) = 2 \cdot 10 + 3 = 23$$

The right side of Eq. 8.3.2 becomes

$$af(x) = 2(2x + 3) = 2 \cdot 13 = 26$$

Now we have found one case where the function is not homogeneous, so it also cannot be linear for this reason.

As this example shows, linearity is a very restrictive condition. Not only must the relationship be a straight line, it must also pass through the origin. If in the example above the constant had been 0, giving a function of the form $y = ax$, then and only then would it be linear.

Testing a function for linearity can be reduced to an algorithm, called the *two-part test for linearity*. This algorithm, which implements Eqs. 8.3.1 and 8.3.2 in a systematic way, is illustrated in Fig. 8.3.1. Part 1 of the test accomplishes the left side of Eqs. 8.3.1 and 8.3.2, performing the addition and multiplication operations in the domain of f. Part 2 does the right side of these equations, thus performing addition and multiplication in the codomain of f. The function f is linear if $y_1 = y_2$ for all possible scalars a_1, a_2, and for all values in the domain x_1 and x_2.

EXAMPLE 8.3.2.   Test $y = 2x + 3$ for linearity using the two-part test in Fig. 8.3.1.

SOLUTION: For arbitrary a_1, a_2, x_1, and x_2, part 1 gives

$$y_1 = f(a_1 x_1 + a_2 x_2) = 2(a_1 x_1 + a_2 x_2) + 3$$

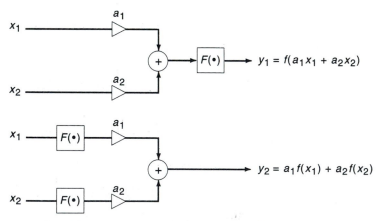

Fig. 8.3.1.   The two-part test for linearity of f.

Part 2 gives

$$y_2 = a_1 f(x_1) + a_2 f(x_2) = a_1(2x_1 + 3) + a_2(2x_2 + 3)$$

The two terms y_1 and y_2 are not equal for all values of a_1, a_2, x_1, and x_2, so the function is not linear.

An *estimator* is a function, a relation between the data and the estimate. A filter, like a function, has an input and an output, so if we identify the filter input as data and the output as the estimate, the filter serves as the estimator. If the system is linear, then it is a linear estimator. A finite-impulse response (FIR) filter has the form

$$y(n) = \sum_{i=0}^{p} h_i x(n - i) \tag{8.3.3}$$

where $x(n)$ is the input, the h_i terms are the filter taps, and $y(n)$ is the output. To show linearity, apply the two-part test as follows. Let a_1 and a_2 be arbitrary constants, and let $x_1(n)$ and $x_2(n)$ be two input sequences. Part 1 of the test gives

$$y_1(n) = \sum_{i=0}^{p} h_i[a_1 x_1(n - i) + a_2 x_2(n - i)]$$

Part 2 gives

$$y_2(n) = \sum_{i=0}^{p} h_i a_1 x_1(n - i) + \sum_{i=0}^{p} h_i a_2 x_2(n - i)$$

The two terms in $y_2(n)$ can be combined to give the form $y_1(n)$, proving linearity.

A linear estimator has the form

$$\hat{d} = h_0 x(n) + h_1 x(n - 1) + \cdots + h_p x(n - p) \tag{8.3.4}$$

where $x(i)$ is the data, the h_i's are constants, and $\hat{d}$ is our estimate of the desired quantity d. Compare Eqs. 8.3.3 and 8.3.4 to see that the FIR filter shown in Fig. 8.3.2 will serve as one type of estimator for us. The other type is an IIR filter, which we will discuss later.

The usual paradigm has the filter input $x(n)$ equal to signal plus noise:

$$x(n) = s(n) + w(n) \tag{8.3.5}$$

The parameter that we wish to estimate can be a future value of the signal, $s(n + k)$, a past value of the signal, $s(n - k)$, or the present value of the signal, $s(n)$. This gives

Extrapolation: $d(n) = s(n + k)$
Interpolation: $d(n) = s(n - k)$
Smoothing: $d(n) = s(n)$

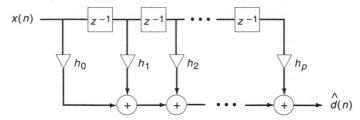

Fig. 8.3.2

where $d(n)$ is the desired parameter, the quantity to be estimated. The filter output in Fig. 8.3.2 is therefore $y(n) = \hat{d}(n)$ in each case. That is, regardless of whether we wish to predict a future value of the signal, estimate the present value, or interpolate a past value, the filter output can serve as the estimate of $d(n)$. The question to be answered in this chapter is: How do we select the filter coefficients in these filters to minimize the mean-square error? The answer lies in the orthogonality principle.

Set the Error Orthogonal to the Data

In Section 8.1 we established the general procedure for deriving a mean-square estimator: Find an expression for the squared error, take its expected value, and differentiate the expected value with respect to the parameters we seek in order to minimize the mean-square error. This leads to expressions for the unknown parameters such as Eqs. 8.1.9 and 8.1.10. We have a similar problem here, so we could follow the same procedure to arrive at expressions for the filter coefficients h_i. Less computation is required, however, by viewing random variables as vectors with the inner product between X and Y defined as $E(XY)$.

To give you an intuitive concept of minimizing error by setting the error orthogonal to the data, look at Fig. 8.3.3. The data is the vector x, the parameter to estimate is d, and the estimate is ax. You can see that the length of the error vector is minimum if the error is orthogonal to the data x. This is accomplished by choosing a so that the length of ax makes the error $d - ax$ a minimum. We will now apply this concept to several representative problems, and give a better explanation of why it is so after the examples.

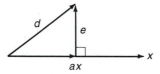

Fig. 8.3.3.   The orthogonality principle.

Single Observation

Suppose that we are given one observation $x(n)$ and wish to estimate $s(n)$, where

$$x(n) = s(n) + w(n)$$
$$d(n) = s(n)$$

Set the error $e(n) = d(n) - \hat{d}(n)$ orthogonal to the data $x(n)$.

$$E\{[d(n) - \hat{d}(n)]x(n)\} = 0 \qquad (8.3.6)$$

The estimate $\hat{d}(n)$ is $y(n) = h_0 x(n)$, giving

$$E\{[d(n) - h_0 x(n)]x(n)\} = 0$$

Since $d(n) = s(n)$, this gives

$$h_0 R_{xx}(0) = R_{sx}(0)$$

giving

$$h_0 = \frac{R_{sx}(0)}{R_{xx}(0)} \qquad (8.3.7)$$

This optimum solution agrees with Eqs. 8.1.9 and 8.1.10. Both means are 0 here, giving $b = 0$ in Eq. 8.1.9. The value of a in Eq. 8.1.10 is the same as h_0 in Eq. 8.3.7.

EXAMPLE 8.3.3.   Find the optimum h_0 and mean-square error in estimating $s(n)$ if the data is $x(n) = s(n) + w(n)$. The noise $w(n)$ is white Gaussian noise with zero mean and unit variance. The signal, which also has zero mean and is independent of the noise, has autocorrelation function given by

$$R_{ss}(n) = 0.9^{|n|}$$

SOLUTION: We need to find $R_{xx}(0)$ and $R_{sx}(0)$ to plug into Eq. 8.3.7. Since $x(n) = s(n) + w(n)$,

$$
\begin{aligned}
E[x^2(n)] = R_{xx}(0) &= E\{[s(n) + w(n)]^2\} \\
&= E[s^2(n) + E[s(n)w(n)] + E[w(n)s(n)] + E[w^2(n)] \\
&= R_{ss}(0) + R_{sw}(0) + R_{ws}(0) + R_{ww}(0)
\end{aligned}
$$

But the cross-correlation terms are 0, since $s(n)$ and $w(n)$ are zero-mean and independent. Since $w(n)$ is white with unit variance, $R_{ww}(n) = \delta(n)$. This gives

$$R_{xx}(0) = R_{ss}(0) + R_{ww}(0) = 0.9^0 + 1 = 2$$

The cross correlation $R_{SX}(0)$ is given by

$$E[s(n)x(n)] = E\{s(n)[s(n) + w(n)]\} = R_{SS}(0) = 1$$

Therefore

$$h_0 = \frac{R_{SX}(0)}{R_{XX}(0)} = \frac{1}{2}$$

The mean-square error is given by

$$
\begin{aligned}
E(e^2) &= E\{[s(n) - \hat{s}(n)]^2\} = E\{[s(n) - h_0 x(n)]^2\} \\
&= E[s^2(n) - 2h_0 s(n)x(n) + h_0^2 x^2(n)] \\
&= R_{SS}(0) - 2h_0 R_{SX}(0) + h_0^2 R_{XX}(0) = 1
\end{aligned}
$$

We will derive a simpler expression for the mean-square error in Eq. 8.3.11.

Multiple Observations

Suppose that we are given two observations, $x(n)$ and $x(n - 1)$. Then our estimate is given by

$$y(n) = h_0 x(n) + h_1 x(n - 1)$$

The error is

$$e(n) = d(n) - y(n) = d(n) - h_0 x(n) - h_1 x(n - 1)$$

Therefore, setting the error orthogonal to the data gives two equations,

$$E\{[d(n) - h_0 x(n) - h_1 x(n - 1)]x(n)\} = 0 \qquad (8.3.8a)$$
$$E\{[d(n) - h_0 x(n) - h_1 x(n - 1)]x(n - 1)\} = 0 \qquad (8.3.8b)$$

Taking expected values and transferring terms gives

$$h_0 R_{XX}(0) + h_1 R_{XX}(-1) = R_{DX}(0)$$
$$h_0 R_{XX}(1) + h_1 R_{XX}(0) = R_{DX}(1)$$

These two equations in two unknowns allow us to solve for the filter coefficients to produce the optimum linear mean-square estimate. In matrix form they are given by

$$\begin{bmatrix} R_{XX}(0) & R_{XX}(-1) \\ R_{XX}(1) & R_{XX}(0) \end{bmatrix} \begin{bmatrix} h_0 \\ h_1 \end{bmatrix} = \begin{bmatrix} R_{DX}(0) \\ R_{DX}(1) \end{bmatrix} \qquad (8.3.9)$$

EXAMPLE 8.3.4.   For the signal in Example 8.3.3, suppose that we are given two observations. Find the optimum linear estimate and the mean-square error.

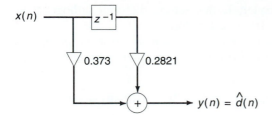

Fig. 8.3.4.   The optimum filter.

SOLUTION: The quantities that go into Eq. 8.3.9 are

$$\begin{bmatrix} 2 & 0.9 \\ 0.9 & 2 \end{bmatrix}\begin{bmatrix} h_0 \\ h_1 \end{bmatrix} = \begin{bmatrix} 1 \\ 0.9 \end{bmatrix}$$

giving $h_0 = 0.3730$ and $h_1 = 0.2821$. This first-order filter is shown in Fig. 8.3.4. The mean-square error is

$$E(e^2) = E\{[s(n) - h_0 x(n) - h_1 x(n - 1)]^2\} = 0.373$$

There is always some confusion about the sign on the arguments in the expressions for the correlation functions. For example, how do we know just where to put $R_{XX}(1)$ and $R_{XX}(-1)$ in Eq. 8.3.9? The answer is that it really does not matter as long as you are consistent. The convention adopted in this text is that $E[x(i)x(j)] = R_{XX}(i - j)$. We subtract the second argument from the first. This convention gives the following results:

$$E[x(n)y(n - 1)] = R_{XY}(1)$$
$$E[x(n - 2)y(n)] = R_{XY}(-2)$$
$$E[x(n - 2)y(n - 3)] = R_{XY}(1)$$

The Projection Theorem

For more than two data values we can write Eq. 8.3.8 as

$$E\{[d(n) - h_0 x(n) - \cdots - h_p x(n - p)]x(n - i)\} = 0, \qquad i = 0, \cdots, p \tag{8.3.10}$$

This is really $p + 1$ equations, one for each i. The *projection theorem* says that the mean-square error is minimum if the coefficients h_i are chosen to make the error orthogonal to the data $x(n)$. We previously demonstrated this with geometric vectors, but it is time for a better explanation. We can show that this is true by taking the customary approach to minimization problems, i.e., by setting the derivatives of the mean-square error with respect to the parameters equal to 0.

$$\frac{\partial}{\partial h_i} E(e^2) = E\{2[d(n) - h_0 x(n) - \cdots - h_p x(n - p)][-x(n - i)]\} = 0$$

This gives Eq. 8.3.10, so we conclude that setting the error orthogonal to the data produces the optimum linear estimator.

Here is the simpler expression for the mean-square error we promised. From Eq. 8.3.10 we can see that the error is orthogonal to any linear combination of the data, giving

$$\langle (d - \hat{d})|[a_0 x(n) + \cdots + a_p x(n - p)]\rangle = 0 \qquad \text{for any } a_0, \ldots, a_p$$

Since $\hat{d}$ is itself a linear combination of the data, this gives $\langle (d - \hat{d})|\hat{d}\rangle = 0$; that is, the error is orthogonal to the estimate. This can be used to simplify the expression for the minimum mean-square error:

$$E(e^2) = E[(d - \hat{d})^2] = E[(d - \hat{d})d] - E[(d - \hat{d})\hat{d}]$$

But this last term is 0, giving

$$E(e^2) = E[(d - \hat{d})^2] = E[(d - \hat{d})d] \qquad (8.3.11)$$

Thus, to calculate the mean-square error we find the inner product of the error with the desired quantity d. (Note that Eq. 8.3.11 is valid only if the optimum estimate is used. In other words, if some estimate is used that does not make the error orthogonal to any linear combination of the data, then our argument is not valid, and the more general formula must be used to calculate the mean-square error.)

EXAMPLE 8.3.5.   Suppose that the input signal $x(n)$ from our previous example represents the flight path of an enemy aircraft, and suppose that we wish to predict its future value, $x(n + 1)$. Thus the input signal $x(n)$ is the sum of signal $s(n)$ plus noise $x(n)$, and we wish to (a) find the optimum first-order FIR filter and resulting mean-square error; (b) find the optimum second-order FIR filter and resulting mean-square error.

SOLUTION: (a) Here $d(n) = x(n + 1)$. Setting the error orthogonal to the data $x(n)$ and $x(n - 1)$ gives

$$E\{[x(n + 1) - h_0 x(n) - h_1 x(n - 1)]x(n)\} = 0$$
$$E\{[x(n + 1) - h_0 x(n) - h_1 x(n - 1)]x(n - 1)\} = 0$$

Taking the expected values and transferring terms gives

$$h_0 R_{xx}(0) + h_1 R_{xx}(-1) = R_{xx}(1)$$
$$h_0 R_{xx}(1) + h_1 R_{xx}(0) = R_{xx}(2)$$

Putting this in matrix form and substituting values gives

$$\begin{bmatrix} 2 & 0.9 \\ 0.9 & 2 \end{bmatrix} \begin{bmatrix} h_0 \\ h_1 \end{bmatrix} = \begin{bmatrix} 0.9 \\ 0.81 \end{bmatrix}$$

Solving these two equations gives $h_0 = 0.3357$ and $h_1 = 0.2539$.

The mean-square error is given by $E[(d - \hat{d})d]$ or

$$E(e^2) = E\{[x(n + 1) - h_0 x(n) - h_1 x(n - 1)]x(n + 1)\}$$
$$= R_{xx}(0) - 0.3357 R_{xx}(1) - 0.2539 R_{xx}(2) = 1.4922$$

(b) To find the optimum second-order filter we set the error orthogonal to the data for three observations, resulting in

$$h_0 R_{xx}(0) + h_1 R_{xx}(-1) + h_2 R_{xx}(-2) = R_{xx}(1)$$
$$h_0 R_{xx}(1) + h_1 R_{xx}(0) + h_2 R_{xx}(-1) = R_{xx}(2)$$
$$h_0 R_{xx}(2) + h_1 R_{xx}(1) + h_2 R_{xx}(0) = R_{xx}(3)$$

This in matrix form with values gives

$$\begin{bmatrix} 2 & 0.9 & 0.81 \\ 0.9 & 2 & 0.9 \\ 0.81 & 0.9 & 2 \end{bmatrix} \begin{bmatrix} h_0 \\ h_1 \\ h_2 \end{bmatrix} = \begin{bmatrix} 0.9 \\ 0.81 \\ 0.729 \end{bmatrix}$$

which has the solution $h_0 = 0.297$, $h_1 = 0.202$, and $h_2 = 0.153$ as shown in Fig. 8.3.5. The mean-square error is

$$E(e^2) = E[(d - \hat{d})d]$$
$$= E\{[x(n + 1) - h_0 x(n) - h_1 x(n - 1) - h_2 x(n - 2)]x(n + 1)\}$$
$$= R_{xx}(0) - 0.297 R_{xx}(1) - 0.202 R_{xx}(2) - 0.153 R_{xx}(3)$$
$$= 1.4575$$

Notice that the addition of one more term h_2 decreases the mean-square error from 1.4922 to 1.4575.

Review

We did not discuss continuous-time systems in this section because most FIR filters are discrete-time systems, although we had an example of a FIR continuous-time system in Section 6.6. In any event, we know how to design discrete-time FIR systems, and so our discussion was confined to that subject. We design them for mean-square estimation by setting the error orthogonal to the data and solving

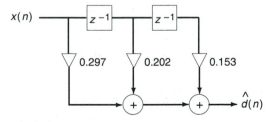

Fig. 8.3.5

the resulting simultaneous equations for the filter coefficients h_i. The mean-square error is found by taking the inner product between the error and $d(n)$, the quantity we are trying to estimate.

8.4   Spectral Factorization ──────────────

Preview

In this section we introduce the concepts of minimum-phase systems, the innovations process, the Paley–Wiener theorem, and spectral factorization. Each of these concepts applies equally to continuous-time and discrete-time systems, but we will present only the discrete-time part of these concepts. We present all of these in this section because they are connected by the following sequence of facts.

A causal minimum-phase system is stable and has a causal stable inverse. The innovations filter that we will introduce for the stochastic processes of interest are minimum-phase. Spectral factorization comes into all this because the innovations filter is one of the factors derived by spectral factorization, and the Paley–Wiener theorem defines the conditions under which this is true.

If this seems horribly complicated, do not give up, for we will use all this in presenting the causal IIR Wiener filter in the next section.

DEFINITION 8.4.1   A minimum-phase polynomial has all its zeros inside the unit circle in the Argand diagram. Conversely, a maximum-phase polynomial has all its zeros outside the unit circle. A polynomial with some zeros inside or on the unit circle, while others are outside the unit circle, is neither minimum- nor maximum-phase.

DEFINITION 8.4.2.   A minimum-phase system is a causal linear system with rational transfer function

$$H(z) = \frac{B(z)}{A(z)}$$

where both $A(z)$ and $B(z)$ are minimum-phase polynomials. A maximum-phase system is one where both $A(z)$ and $B(z)$ are maximum-phase polynomials.

EXAMPLE 8.4.1.   The pole-zero plots for two systems H_1 and H_2 are shown in Fig. 8.4.1, where

$$H_1(z) = \frac{1}{1 - 0.5z^{-1}} = \frac{z}{z - 0.5}$$

$$H_2(z) = \frac{z - 2}{z^2 - 0.3z - 0.4} = \frac{z - 2}{(z + 0.5)(z - 0.8)}$$

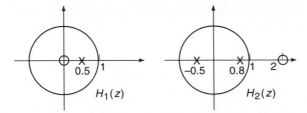

Fig. 8.4.1.   Pole-zero locations for $H_1(z)$ and $H_2(z)$.

The first system is minimum-phase because all poles and zeros are inside the unit circle. The second system is not minimum-phase because the zero is outside the unit circle. Recall that all poles must be inside the unit circle for the system to be stable. Therefore, each of the systems above is stable.

Let us pause to remind you about the relation between pole location and stability. The transfer function can be put in the form

$$H(z) = \frac{B(z)}{A(z)} = \frac{b_0 + b_1 z + b_2 z^2 + \cdots + b_m z^m}{a_0 + a_1 z + a_2 z^2 + \cdots + a_n z^n}$$

$$= C \frac{(z - z_1)(z - z_2) \cdots (z - z_m)}{(z - p_1)(z - p_2) \cdots (z - p_n)}$$

where C is a constant, $z_1, z_2, \ldots, z_m$ are the zeros, and $p_1, p_2, \ldots, p_n$ are the poles of the transfer function $H(z)$. Collectively, the poles and zeros are called *singularities*. A system is stable if all its poles are inside the unit circle because such poles give rise to exponentially decaying terms in the source-free response. The partial fraction expansion of $H(z)$ has terms of the form

$$\frac{C_i z}{z - p_i}$$

The corresponding time functions are of the form $C_i(p_i)^k$, where k is time. As time increases, these terms decay to 0 because $|p_i| < 1$. Hence the unit circle serves as the boundary between stability and instability.

We can change a nonminimum-phase system into one with minimum phase while maintaining the same magnitude response $|H(\omega)|$ as follows: If no poles or zeros are on the unit circle, move all singularities from outside the unit circle to their conjugate reciprocal locations inside the unit circle. This alters only the phase. To relocate a zero at z_0, multiply $H(z)$ by

$$\frac{1 - z_0^* z}{z - z_0} = \frac{z^{-1} - z_0^*}{1 - z_0 z^{-1}} = (-z_0^*) \frac{z - 1/z_0^*}{z - z_0} \qquad (8.4.1)$$

where z_0^* is the complex conjugate of z_0. To relocate a pole, multiply $H(z)$ by the reciprocal of this expression.

EXAMPLE 8.4.2.   If we multiply $H_2(z)$ from Example 8.4.1 by Eq. 8.4.1 to move the zero inside the unit circle, we obtain

$$H_3(z) = H_2(z) \cdot (-2) \left(\frac{z - 0.5}{z - 2} \right) = -2 \frac{z - 0.5}{(z + 0.5)(z - 0.8)}$$

The pole-zero plot for H_3 is shown in Fig. 8.4.2. Notice that $|H_2(\omega)| = |H_3(\omega)|$ for all ω. Only the phase has changed. In fact, the phase of $H_3(z)$ has magnitude less than or equal to the phase for $H_2(z)$ for all ω. This is the reason for the name "minimum phase."

Minimum-phase systems are important because they are stable and their inverses are also stable. For example, $H_1(z)$ in Example 8.4.1 represents a causal stable system with impulse response

$$h_1(n) = 0.5^n u(n)$$

The inverse system is

$$H_1^{-1}(z) = 1 - 0.5z^{-1} \leftrightarrow \delta(n) - 0.5\,\delta(n - 1)$$

which is also stable and causal.

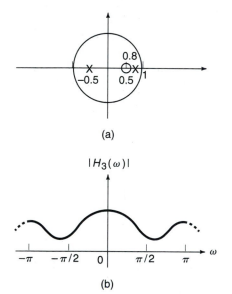

(a)

$|H_3(\omega)|$

(b)

Fig. 8.4.2. Pole-zero (a) location and (b) magnitude response for $H_3(\omega)$.

In Sections 6.4 and 6.5 we illustrated the fact that when white noise with variance σ^2 is supplied to a system with rational transfer function $H(z)$, the output has an autocorrelation function given by

$$R_{YY}(l) = \sigma^2 r_{hh}(l)$$

and a power spectrum given by

$$G_{YY}(z) = \sigma^2 |H(z)|^2 = \sigma^2 H(z)H(z^{-1})$$

If $H(z)$ is a minimum-phase causal system, then $H(z^{-1})$ is a maximum-phase anticausal system. This gives us a unique way to represent the autocorrelation function of such a signal, i.e., in terms of a minimum-phase causal system function $H(z)$. Processes that can be represented this way are called *regular*, and they satisfy the Paley–Wiener condition.

The Paley–Wiener condition is a property of Fourier transforms that allow us to state that when the power spectrum of a process y satisfies the condition

$$\int_{-\pi}^{\pi} |\ln G_{YY}(\omega)| \, d\omega < \infty \tag{8.4.2}$$

then the spectral density can be factored as

$$G_{YY}(z) = K|H(z)|^2 = K \, H_c(z) H_a(z) \tag{8.4.3}$$

where K is a constant determined by the input power, $H_c(z)$ is a minimum-phase causal system function, and $H_a(z)$ is a maximum-phase anticausal system function. The subscript on H_c stands for "causal," while the a on H_a stands for "anticausal." H_c and H_a are related to each other by

$$H_a(z) = H_c(z^{-1}) \tag{8.4.4}$$

With this notation we are now in a position to describe the innovations representation and the innovations process for a stochastic process y. Figure 8.4.3 shows two systems, $H_c(z)$ and $H_c^{-1}(z)$, which means simply that they are reciprocal functions:

$$H_c^{-1}(z) = \frac{1}{H_c(z)}$$

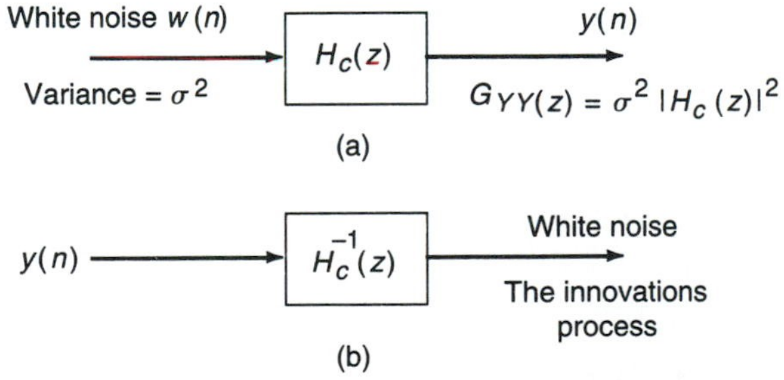

Fig. 8.4.3.   The innovations representation (a) and the innovations process (b).

The first system, $H_c(z)$, has white noise $w(n)$ with variance σ^2 as its input. The output process has the spectrum given by Eq. 8.4.3, and is called the *innovations representation*.

Since the second system is the inverse of $H_c(z)$, it "undoes" what $H_c(z)$ does. Thus, if we supply the inverse filter with a signal $y(n)$ that has spectrum $G_{YY}(z)$, the output will be white noise. This white noise derived from $y(n)$ is called the *innovations process*. (This is not just any white noise: It is derived from y.)

Wiener coined the term "innovations" because it represents the unpredictable part of $y(n)$. Notice that we have used the term "innovations" in two different (and opposite) ways here. You should not confuse the innovations representation in Fig. 8.4.3a with the innovations process that is the output of the system in Fig. 8.4.3b. Also notice that we have discussed three filters, all different. They are $H_c(z)$, $H_c^{-1}(z)$, and $H_a(z) = H_c(z^{-1})$.

Given a regular process with spectrum $G_{YY}(z)$, the procedure for finding the innovations representation is called *spectral factorization*. This is not difficult, but you must be careful to derive minimum-phase systems from $G_{YY}(z)$; otherwise the representation will not be the desired innovations representation. Here are some examples.

EXAMPLE 8.4.3.   Find the innovations representation for a stochastic process with spectrum

$$G_{XX}(z) = \frac{-2}{z - 2.5 + z^{-1}}$$

SOLUTION: Factoring the denominator, we can write

$$G_{XX}(z) = \frac{-2}{(1 - 0.5z^{-1})(z - 2)}$$

Now if we divide top and bottom by -2, we get

$$G_{XX}(z) = \frac{1}{(1 - 0.5z^{-1})(1 - 0.5z)}$$

This gives $K = 1$ in Eq. 8.4.3 and

$$H_c(z) = \frac{1}{1 - 0.5z^{-1}}$$

$$H_c^{-1}(z) = 1 - 0.5z^{-1}$$

and

$$H_a(z) = \frac{1}{1 - 0.5z}$$

EXAMPLE 8.4.4.   Find the innovations model for a process with an autocorrelation function given by

$$R_{YY}(l) = 4(0.9)^{|l|}$$

SOLUTION: A first-order impulse response $h(n) = a^n u(n)$ has a corresponding second-order impulse response found by correlating $h(n)$ with itself. The results give

$$r_{hh}(l) = \left[\frac{1}{1-a^2}\right] a^{|l|} = \frac{1}{0.19}(0.9)^{|l|} \qquad \text{since} \qquad a = 0.9$$

Since $R_{YY}(l) = \sigma^2 r_{hh}(l)$, we see that $\sigma^2 = 4(0.19) = 0.76$.

The spectrum corresponding to this type of autocorrelation function is given by

$$a^{|l|} \leftrightarrow \frac{1-a^2}{(1-az^{-1})(1-az)} \tag{8.4.5}$$

(This is a good formula to memorize.) Therefore $G_{YY}(z)$ is

$$G_{YY}(z) = \frac{4(1-0.9^2)}{(1-0.9z^{-1})(1-0.9z)}$$

$$= \frac{0.76}{(1-0.9z^{-1})(1-0.9z)}$$

Therefore $K = 0.76$ and

$$H_c(z) = \frac{1}{1-0.9z^{-1}}$$

$$H_c^{-1}(z) = 1 - 0.9z^{-1}$$

and

$$H_a(z) = \frac{1}{1-0.9z}$$

While we are preparing for the next section, there is one more topic we need to discuss. Figure 8.4.4 shows a cascade of two systems, h_1 and h_2. The input signal $x(n)$ drives h_1 to produce $v(n)$, which drives h_2 to produce $d(n)$. Thus there is a cause-and-effect relationship between these signals, so

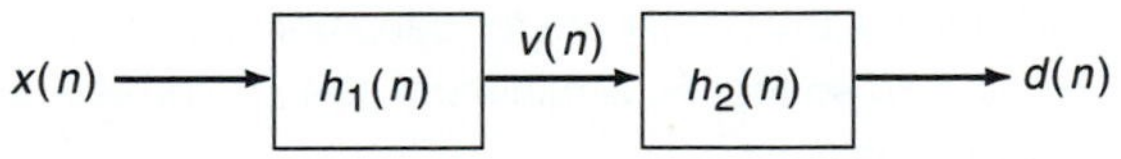

Fig. 8.4.4.   A cascade of two systems.

if we know the correlation between $x(n)$ and $d(n)$, we should be able to find the correlation between $v(n)$ and $d(n)$. This is the relationship we will need in the next section.

Since R_{DX} is known, and x and v are related by convolution, we can find R_{DV} in terms of R_{DX} as follows. Write

$$v(n) = \sum_{k=-\infty}^{\infty} h_1(n - k)x(k)$$

Multiply this expression by $d(n - l)$ and take the expected value to get

$$R_{VD}(l) = E[v(n)d(n - l)] = E\left\{\left[\sum_{k=-\infty}^{\infty} h_1(n - k)x(k)\right] d(n - l)\right\}$$

$$= \sum_{k=-\infty}^{\infty} h_1(n - k)R_{XD}(k - n + l)$$

Change the variable. Let $\lambda = n - k$.

$$R_{VD}(l) = \sum_{\lambda=-\infty}^{\infty} h_1(\lambda)R_{XD}(l - \lambda) = h_1(l) * R_{XD}(l)$$

Reversing the subscripts gives

$$R_{DV}(l) = h_1(-l) * R_{DX}(l) \tag{8.4.6}$$

This relationship between R_{DV} and R_{DX} seems rather strange because $h_1(-l)$ is the time-reversed or anticausal version of $h_1(l)$, but this is the relation we will need in the next section.

Review

A signal with a specified spectrum $G_{YY}(z)$ can be generated by supplying a unique filter $H_c(z)$ with white noise. Provided Eq. 8.4.2 is satisfied, the output signal $y(n)$ then has the spectrum

$$G_{YY}(z) = K|H_c(z)|^2 = KH_c(z)H_c(z^{-1})$$

From this we obtained three filters of interest (at least they will be of interest in the next section). They are $H_c(z)$, $H_c^{-1}(z)$, and $H_a(z) = H_c(z^{-1})$. These are the filters we will need to define the optimum IIR Wiener filter.

Physically realizable (causal) minimum-phase systems are stable and have stable inverses. Thus, if $H_c(z)$ is minimum-phase, we are assured that its inverse $H_c^{-1}(z)$ exists and is stable. This is why the concept of minimum-phase systems is important to us. Spectral factorization is the procedure for decomposing a given spectrum into components K, H_c, and H_a so that H_c is minimum phase. These are the skills we will need in the next section.

8.5   IIR Wiener Filters _____

Preview

In this section we describe how to derive the optimum IIR filter to perform mean-square estimation, so we combine the two subjects of system theory and mean-square estimation. This marriage of the two disciplines is called *Wiener filtering,* after Norbert Wiener (1894–1964), who originated the concept.

    In the preface to his book, *Statistical Theory of Communication,* Prof. Y. W. Lee explains that he used Weiner's new work, *The Extrapolation, Interpolation and Smoothing of Stationary Time Series,* to teach the first course in this discipline at MIT in 1947. His course notes evolved into his book published by Wiley in 1960. Wiener's work was often referred to as the "yellow peril" because of its yellow cover and its difficult mathematics. The main source of difficulty was the system theory in the document, not the mean-square estimation. Since that time others have discovered easier ways to explain the necessary system theory, and it is this theory that we now explore in this section.

    Wiener's original work was for continuous-time systems, since that was all they had at the time. The digital computer was essentially unknown in 1939 (the copyright date on Weiner's document), for John von Neumann did not invent the stored program concept until about 1944. We are leaving out the continuous-time system derivation for two reasons. The discrete-time system theory, which we will present, parallels Wiener's work, and most applications now are to discrete-time systems.

    A linear system output $y(n) = h(n) * x(n)$ serves as the estimate $\hat{d}(n)$ for some desired part of the input $x(n)$. The error is $e(n) = d(n) - y(n)$ and the data are $x(n)$, $-\infty < n < \infty$. We first assume that the data are available for all time, which includes the future. This will lead to physically unrealizable filters and serve as a precursor for the more complex realizable filters.

    Abiding by the orthogonality principle, we set the error orthogonal to the data, giving

$$E\{[d(n) - y(n)]x(l)\} = 0, \qquad -\infty < l < \infty$$

Substituting $y(n) = \sum_{k=-\infty}^{\infty} h(k) x(n - k)$ and taking expected values gives

$$\sum_{k=-\infty}^{\infty} h(k)R_{XX}(n - k - l) = R_{DX}(n - l), \qquad -\infty < l < \infty$$

The number of variables can be reduced if we set $\lambda = n - l$.

$$\sum_{k=-\infty}^{\infty} h(k)R_{XX}(\lambda - k) = R_{DX}(\lambda), \qquad -\infty < \lambda < \infty \qquad (8.5.1)$$

This is called the *Wiener–Hopf equation*. Since the term on the left is the convolution summation, we can write it in the form $h(l) * R_{XX}(l) = R_{DX}(l)$.

The mean-square error, found by the inner product of the error with $d(n)$, is given by

$$E(e^2) = R_{DD}(0) - \sum_{k=-\infty}^{\infty} R_{DX}(k)h(k) \tag{8.5.2}$$

Solving for the optimum filter coefficients from Eq. 8.5.1 is easy. That is the reason we begin with noncausal filters. Taking z transforms and solving for $H(z)$ gives

$$H(z)G_{XX}(z) = G_{DX}(z)$$

or

$$H(z) = \frac{G_{DX}(z)}{G_{XX}(z)} \tag{8.5.3}$$

where $G_{XX}(z)$ and $G_{DX}(z)$ are the z transforms of $R_{XX}(n)$ and $R_{DX}(n)$, respectively.

A common situation has the input signal $x(n)$ equal to the sum of signal plus zero-mean independent noise, and the desired quantity is the signal. When the signal is a first-order low-pass process and the noise is white, we have

$$x(n) = s(n) + w(n) \qquad d(n) = s(n)$$

$$R_{SS}(l) = Aa^{|l|} \qquad R_{WW}(l) = B\delta(l)$$

Solving for the spectral densities to substitute into Eq. 8.5.3 gives (see Eq. 8.4.5)

$$G_{XX}(z) = G_{SS}(z) + G_{WW}(z)$$

$$= \frac{A(1 - a^2)}{(1 - az^{-1})(1 - az)} + B, \qquad |a| < |z| < \left|\frac{1}{a}\right| \tag{8.5.4}$$

$$G_{DX}(z) = G_{SS}(z) = \frac{A(1 - a^2)}{(1 - az^{-1})(1 - az)}, \qquad |a| < |z| < \left|\frac{1}{a}\right| \tag{8.5.5}$$

We may now substitute these quantities into Eq. 8.5.3 and solve for $H(z)$ to obtain the optimum filter. From $H(z)$ we can find $h(n)$. Here is an example.

EXAMPLE 8.5.1.   Let $A = 1$, $B = 1$, and $a = 0.9$ in the specification above. Find the optimum filter $h(n)$.

SOLUTION: Substituting into Eqs. 8.5.4 and 8.5.5 gives

$$G_{XX}(z) = \frac{z^2 - (2/a)z + 1}{(z - a)(z - 1/a)}$$

$$G_{DX}(z) = \frac{z(a - 1/a)}{(z - a)(z - 1/a)}$$

Hence

$$H(z) = \frac{G_{DX}(z)}{G_{XX}(z)} = \frac{z(a - 1/a)}{z^2 - (2/a)z + 1} = \frac{-0.2111z}{z^2 - 2.222z + 1}$$

Using partial fraction expansion gives

$$\frac{H(z)}{z} = \frac{-0.21111}{(z - 0.62679)(z - 1.59543)}$$

$$= \frac{0.21794}{(z - 0.62679)} - \frac{0.21794}{(z - 1.59543)}$$

Taking the inverse z transform gives the impulse response:

$$h(n) = 0.21794[(0.62679)^n u(n) + (1.59543)^n u(-n - 1)]$$
$$= 0.21794(0.62679)^{|n|}$$

This function is plotted in Fig. 8.5.1. A system with this impulse response is physically unrealizable, meaning the output occurs before the input is applied.

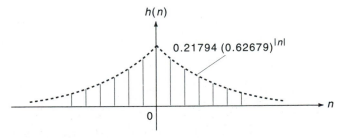

$h(n)$

$0.21794\,(0.62679)^{|n|}$

0

n

Fig. 8.5.1

The Optimum Physically Realizable Filter

Now it is time to derive the optimum realizable filter for mean-square estimation. One thought that comes to mind upon looking at Fig. 8.5.1 is why not discard the anticipatory part of the impulse response and keep the realizable part? Would that be optimum? If so, we can derive the optimum filter by first assuming that the signal is present for all time and deriving the optimum nonrealizable filter. Then we simply discard the part that gives us trouble and keep the rest, and hope this is optimum. In order to find out how well this works, we first need to go through some complicated math (the yellow peril) and compare.

First, here is what we are aiming at. We break the optimum filter $h(n)$ into two parts, $h_1(n)$ and $h_2(n)$, as shown in Fig. 8.5.2. Now we want to show that if $h_1(n)$ is the whitening filter from Section 8.4,

$$h_1(n) \leftrightarrow H_c^{-1}(z) \tag{8.5.6}$$

and if $h_2(n)$ is given by

$$h_2(n) \leftrightarrow H_2(z) = \frac{1}{k} \left[\frac{G_{DX}(z)}{H_c(z^{-1})} \right]_+ \tag{8.5.7}$$

then the optimum filter is $h(n) = h_1(n) * h_2(n)$, as shown in Fig. 8.5.2.

Recall that we use the double-headed arrow to indicate transforms (in this case, the z transform), and that the power spectral density for the data $x(n)$ is given by

$$G_{XX}(z) = KH_c(z)H_c(z^{-1}) \tag{8.5.8}$$

(see Section 8.4). The filter $H_c(z)$ is the physically realizable minimum-phase component of G_{XX}, K is a constant related to the power in the signal, and $H_c(z^{-1})$ is the maximum-phase component of the spectrum. The subscript outside the right bracket in Eq. 8.5.7 means that we will use the physically realizable part of this expression and discard the anticipatory part.

The idea behind the derivation to follow is that if $h_2(n)$ is chosen to make the error orthogonal to its input data $v(n)$, and if $h_1(n)$ is the whitening filter $H_c^{-1}(z)$, then the overall filter $h(n)$ will make the error orthogonal to the input data $x(n)$, meaning that $h(n)$ is optimum. Notice that the $v(n)$ is the innovations process for the input $x(n)$. Wiener used the term innovations because

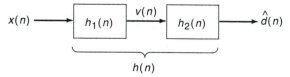

Fig. 8.5.2. The optimum filter $h(n)$ broken into the cascade of two systems, $h_1(n)$ and $h_2(n)$.

$v(n)$ contains the "new information" in $x(n)$. The signal can be split into two parts, the predictable part and the unpredictable part. (This is called the *Wold decomposition*.) The predictable part is that which can be predicted with complete accuracy based on the past history of $x(n)$. The innovations process contains the difference between the actual value of $x(n + 1)$ and the predicted value $v(n + 1)$. But before getting into the details, let us work an example to make certain we understand what each of the terms in Eqs. 8.5.6 and 8.5.7 stand for.

EXAMPLE 8.5.2.   Find the optimum physically realizable filter if

$$x(n) = s(n) + w(n) \qquad d(n) = s(n)$$
$$R_{ss}(l) = 0.9^{|l|} \qquad R_{ww}(l) = \delta(l)$$

where the noise is zero-mean and independent of the signal, as in Example 8.5.1.

SOLUTION: We need to find $G_{DX}(z)$ and $G_{XX}(z)$ and identify the terms K, $H_c(z)$, and $H_c(z^{-1})$ to substitute into Eqs. 8.5.6 and 8.5.7.

$$R_{DX}(l) = E[d(n)x(n - l)] = E\{s(n)[s(n - l) + w(n - l)]\} = R_{ss}(l)$$

since the noise and signal are independent with zero mean. Thus

$$G_{DX}(z) = G_{ss}(z) = \frac{1 - 0.9^2}{(1 - 0.9z^{-1})(1 - 0.9z)}$$

We use the same reasoning to find R_{XX}.

$$R_{XX}(l) = R_{ss}(l) + R_{ww}(l)$$

so

$$G_{XX}(l) = G_{ss}(l) + G_{ww}(l) = \frac{0.19}{(1 - 0.9z^{-1})(1 - 0.9z)} + 1$$
$$= 1.43589 \frac{(1 - 0.62679z^{-1})(1 - 0.62679z)}{(1 - 0.9z^{-1})(1 - 0.9z)}$$

From this we can identify the parameters in Eqs. 8.5.6 and 8.5.7 as

$$K = 1.43589$$
$$H_c(z) = \frac{(1 - 0.62679z^{-1})}{(1 - 0.9z^{-1})}$$
$$H_a(z) = H_c(z^{-1}) = \frac{(1 - 0.62679z)}{(1 - 0.9z)}$$

Therefore $H_2(z)$ in Eq. 8.5.7 is given by

$$H_2(z) = \frac{1}{K}\left[\frac{G_{DX}(z)}{H_c(z^{-1})}\right]_+ = 0.69643\left[\frac{0.19}{(1-0.9z^{-1})(1-0.62679z)}\right]_+$$

Expanding in partial fractions to separate the causal part gives

$$\frac{0.13232}{(1-0.9z^{-1})(1-0.62679z)} = \frac{0.30357}{1-0.9z^{-1}} + \frac{B}{1-0.62679z}$$

so

$$H_2(z) = \frac{0.30357}{1-0.9z^{-1}}$$

Multiplying by $H_1 = H_c^{-1}$ gives the final answer:

$$H(z) = \left(\frac{1-0.9z^{-1}}{1-0.62679z^{-1}}\right)\left(\frac{0.30357}{1-0.9z^{-1}}\right) = \frac{0.30357}{1-0.62679z^{-1}}$$

Hence

$$h(n) = 0.30357(0.62679)^n u(n)$$

This filter is shown in Fig. 8.5.3. Notice that the only difference between the realizable part of the optimum filter in Example 8.5.1 and this filter is in the gain. The realizable part of that filter had a gain of 0.21794, while here the gain is 0.30357. Is this important? Well, it can be. If you are trying to estimate $d(n)$, it is all-important. If you are trying to separate signal from noise, however, as in an ordinary AM radio, the gain is adjustable by the operator. It therefore assumes little importance. In presenting this theory we can only assume that the gain is very important, for we must be prepared to estimate the desired quantity $d(n)$ as accurately as possible.

Now to derive the optimum filter. We want to show that when the input signal $x(n)$ has a spectrum that can be decomposed into the components K and H_c according to Eq. 8.5.8,

$$G_{XX}(z) = KH_c(z)H_a(z) \qquad \text{(repeated) (8.5.8)}$$

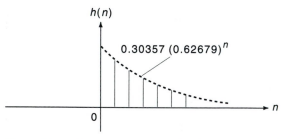

Fig. 8.5.3.   The physically realizable optimum filter.

where $H_a(z) = H_c(z^{-1})$, then the optimum filter is given in Fig. 8.5.2, where h_1 and h_2 are given in Eqs. 8.5.6 and 8.5.7.

$$h_1(n) \leftrightarrow H_c^{-1}(z) \qquad \text{(repeated) (8.5.6)}$$

$$h_2(n) \leftrightarrow H_2(z) = \frac{1}{k}\left[\frac{G_{DX}(z)}{H_c(z^{-1})}\right]_+ \qquad \text{(repeated) (8.5.7)}$$

We start with the estimate $\hat{d}(n)$ given by

$$\hat{d}(n) = \sum_{k=-\infty}^{n} h(n-k)x(k) \tag{8.5.9}$$

Notice that the upper limit on this summation is n. This is because the data $x(k)$ are available for the infinite past, but none of the future values of $x(k)$ are available. This is the only difference between the assumptions here and those for the anticipatory system in Eq. 8.5.3. The error is

$$e(n) = d(n) - \hat{d}(n) \tag{8.5.10}$$

and the data are $x(n-i)$ for $i = 0, 1, 2, \ldots$. We set the error orthogonal to the data to obtain

$$E[e(n)x(n-i)] = E\left\{\left[d(n) - \sum_{k=-\infty}^{n} h(n-k)x(k)\right]x(n-i)\right\} = 0$$

or

$$R_{DX}(i) = \sum_{k=-\infty}^{n} h(n-k)R_{XX}(k-n+i), \qquad i = 0, 1, 2, \ldots$$

We can simplify this expression if we let $l = n - k$.

$$R_{DX}(i) = \sum_{l=0}^{\infty} h(l)R_{XX}(i-l), \qquad i = 0, 1, 2, \ldots \tag{8.5.11}$$

This is the Wiener–Hopf equation. It would be the convolution of $h(l)$ with $R_{XX}(l)$ except for the troublesome restriction on the values of i. We now wish to show that $h(n)$ in Fig. 8.5.2 satisfies this equation.

Let us first consider $h_2(n)$ with white-noise input $v(n)$, where h_2 is specified by Eq. 8.5.7. From the Wiener–Hopf equation with $x = v$,

$$R_{DV}(i) = \sum_{l=0}^{\infty} R_{VV}(i-l)h_2(l), \qquad i = 0, 1, 2, \ldots$$

but since v is white noise with variance σ_V^2, we have

$$R_{VV}(l) = \sigma_V^2\delta(l)$$

so

$$h_2(l) = \frac{1}{\sigma_V^2} R_{DV}(l), \qquad l \geq 0$$

so now we need to find $R_{DV}(l)$. We derived Eq. 8.4.6 in the previous section specifically for this occasion. That formula gave the relationship as

$$R_{DV}(l) = h_1(-l) * R_{DX}(l) \qquad \text{(repeated) (8.4.6)}$$

But $h_1(-l)$ is the time-reversed or anticausal version of $h_1(l)$, so its transform is $H_1(z^{-1})$. This means that the transforms of the terms in Eq. 8.4.6 are

$$G_{DV}(z) = G_{DX}(z)H_1(z^{-1})$$

Since $H_1(z) = H_c^{-1}(z) = 1/H_c(z)$, then $H_1(z^{-1}) = 1/H_c(z^{-1})$. This gives

$$G_{DV}(z) = \frac{G_{DX}(z)}{H_c(z^{-1})}$$

since $h_2(l) = (1/\sigma_V^2)R_{DV}(l)$, $l \geq 0$,

$$H_2(z) = \frac{1}{K}[G_{DV}(z)]_+ = \frac{1}{K}\left[\frac{G_{DX}(z)}{H_c(z^{-1})}\right]_+$$

where $K = \sigma_V^2$ and the $+$ is necessary because of the condition $l \geq 0$ on h_2.

This completes our derivation of Eq. 8.6.7. Now we choose for h_1 the innovations filter h_c^{-1} and show that this, combined with h_2, makes the overall filter optimum, i.e., it makes the error orthogonal to the data. Since $v(n) = h_1(n) * x(n)$,

$$x(n) = v(n) * h_1^{-1}(n) = \sum_{k=-\infty}^{\infty} h_1^{-1}(n - k) v(k)$$

Multiply this by the error and take the inner product:

$$\langle x|e \rangle = E[x(n - i)e(n)]$$

$$= \sum_{k=-\infty}^{\infty} h_1^{-1}(n - i - k) E[v(k) e(n)], \qquad i = 0, 1, 2, \ldots$$

But $v(n)$ is orthogonal to the error because we chose h_2 to make it so. Therefore it follows that $x(n)$ is orthogonal to $e(n)$, meaning that $h(n) = h_1(n) * h_2(n)$ is optimal.

The mean-square error resulting from the filter that satisfies this equation is found by the inner product of the error with $d(n)$:

$$E[e^2] = R_{DD}(0) - \sum_{l=0}^{\infty} h(l) R_{DX}(l) \qquad (8.5.12)$$

EXAMPLE 8.5.3.   Find the mean square error in Example 8.5.2.

SOLUTION

$$R_{DD}(0) = R_{SS}(0) = 1$$
$$R_{DX}(l) = R_{SX}(l) = R_{SS}(l) = 0.9^l \qquad l \geq 0$$

Therefore

$$E(e^2) = 1 - \sum_{l=0}^{\infty} h(l) R_{SS}(l) = 1 - \sum_{l=0}^{\infty} 0.30357(0.62679)^l (0.9)^l$$

$$= 1 - 0.30357 \sum_{l=0}^{\infty} (0.56411)^l$$

From Eq. 5.1.6 we get

$$E(e^2) = 1 - 0.30357(2.29416) = 0.30357$$

Review

In order to find the filter derived in this section we need to know the input spectrum $G_{XX}(z)$ and the cross spectrum between the input and the desired signal $G_{DX}(z)$. Given these quantities (or their transforms), we first decompose $G_{XX}(z)$ into $H_c(z)$ and $H_a(z) = H_c(z^{-1})$ by

$$G_{XX}(z) = KH_c(z)H_c(z^{-1}) \qquad \text{(repeated) (8.5.8)}$$

then the optimum physically realizable filter is the cascade of $h_1(n)$ and $h_2(n)$ given by

$$h_1(n) \leftrightarrow H_c^{-1}(z) \qquad \text{(repeated) (8.5.6)}$$

$$h_2(n) \leftrightarrow H_2(z) = \frac{1}{k} \left[\frac{G_{DX}(z)}{H_c(z^{-1})} \right]_+ \qquad \text{(repeated) (8.5.7)}$$

8.6 Recursive Filtering ───────────────────

Preview

The finite-impulse response filter in Section 8.3 is optimum for a fixed-length signal. That is, if the input signal is of length p, then a filter of length p with appropriate coefficients $h_0, h_1, \ldots, h_p$ gives the optimum estimate of $d(n)$ at its output. But in real life, p increases as time increases. For example, in a radar tracking system the input signal has a beginning, and the number of terms in the input signal increases with time. Neither the fixed-length FIR filter nor the IIR filter (which also has fixed length) is appropriate for this situation. What we need is a different FIR filter at each time instant, the difference being that one more filter tap is added for each new input signal term. That is what recursive filtering accomplishes.

In this section we will consider only the simplest recursive filtering problem, which is for a first-order signal with additive white noise. We do this to illustrate the essential features of recursive filtering without too many complications. The extension of these concepts to more realistic applications should not be too difficult when you need to do so.

Here we illustrate the features of recursive filtering in the following way. Figure 8.6.1 shows a sequence of optimum filters. The minimum mean-square-error linear estimate for $d(n)$ when only one value of $x(n)$ is available for use is $\hat{d}(n) = h_0^0 x(n)$. If two data values $x(n)$ and $x(n-1)$ are available, then the optimum estimate uses a linear combination given by $\hat{d}(n) = h_0^1 x(n) + h_1^1 x(n-1)$. In general, when the present value plus p past values of the data are available, the optimum estimate is

$$\hat{d}(n) = h_0^p x(n) + h_1^p x(n-1) + \cdots + h_p^p x(n-p) \qquad (8.6.1)$$

The superscript p indicates simply that the filter tap values vary from one filter to another. That is, $h_i^p \neq h_i^{p+1}$ for every i. Of course, in all this we are assuming proper choice for the coefficients in each filter.

As the order of the filter increases, meaning the estimate is based on more data, the error should decrease. Here is an example to illustrate these ideas.

EXAMPLE 8.6.1.   Suppose we are presented with signal plus zero-mean independent noise and we wish to estimate the signal.

$$x(n) = s(n) + w(n)$$
$$d(n) = s(n)$$

where

$$R_{SS}(l) = 0.9^{|l|} \quad \text{and} \quad R_{WW}(l) = \delta(l)$$

(See Examples 8.3.3 and 8.3.4.) Find the optimum filter and the mean-square error for each case in Fig. 8.6.1.

SOLUTION: From Examples 8.3.3 and 8.3.4,

$$h_0^0 = 0.5 \qquad E[e_0^2] = 0.5$$
$$h_0^1 = 0.3730 \qquad h_1^1 = 0.2821 \qquad E[e_1^2] = 0.3730$$

For three data values,

$$\hat{d}(n) = h_0^2 x(n) + h_1^2 x(n-1) + h_2^2 x(n-2)$$

and the error is

$$e_2(n) = d(n) - h_0^2 x(n) - h_1^2 x(n-1) - h_2^2 x(n-2)$$

Setting the error $e_2(n)$ orthogonal to the data, which are $x(n)$, $x(n-1)$,

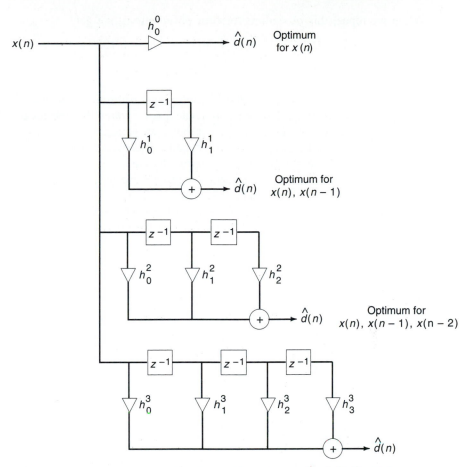

Fig. 8.6.1. Optimum filters for data of length exactly 1, 2, 3, or 4.

and $x(n - 2)$, gives three equations in three unknowns:

$$\begin{bmatrix} R_{XX}(0) & R_{XX}(-1) & R_{XX}(-2) \\ R_{XX}(1) & R_{XX}(0) & R_{XX}(-1) \\ R_{XX}(2) & R_{XX}(1) & R_{XX}(0) \end{bmatrix} \begin{bmatrix} h_0^2 \\ h_1^2 \\ h_2^2 \end{bmatrix} = \begin{bmatrix} R_{DX}(0) \\ R_{DX}(1) \\ R_{DX}(2) \end{bmatrix}$$

or

$$\begin{bmatrix} 2 & 0.9 & 0.81 \\ 0.9 & 2 & 0.9 \\ 0.81 & 0.9 & 2 \end{bmatrix} \begin{bmatrix} h_0^2 \\ h_1^2 \\ h_2^2 \end{bmatrix} = \begin{bmatrix} 1.0 \\ 0.9 \\ 0.81 \end{bmatrix}$$

This gives

$$h_0^2 = 0.3298 \qquad h_1^2 = 0.2250 \qquad h_2^2 = 0.1702 \qquad E[e_2^2] = 0.3298$$

When we repeat this procedure for four observations, we get

$$h_0^3 = 0.3137 \qquad h_1^3 = 0.2037 \qquad h_2^3 = 0.1389$$
$$h_3^3 = 0.1051 \qquad E[e_3^2] = 0.3137$$

The results of this example plus the h^4 coefficients, along with the associated mean-square error, are displayed in Table 8.6.1. Notice that the mean-square error (mse) decreases as we use more data (longer filters) in our estimates. This is called *block processing,* because successive blocks of data are processed by successively larger systems. As each new data value is received, the processor must be redesigned with new coefficients h_i^p and one new delay added to the system. This begs the question of whether we can design an equivalent recursive filter, one with time-varying parameters, that produces the same estimate at each step. If so, it would need to be of the form

$$\hat{d}(n) = A_n \, \hat{d}(n-1) + K_n x(n) \tag{8.6.2}$$

where A_n and K_n are the time-varying coefficients to be determined. For this to be equivalent to Eq. 8.6.1, we start at $n = 0$ and set

$$\hat{d}(0) = h_0^0 x(0)$$

Then successive steps give

$$
\begin{aligned}
\hat{d}(1) &= A_1 \, \hat{d}(0) + K_1 x(1) \\
&= A_1[h_0^0 x(0)] + K_1 x(1) \\
\hat{d}(2) &= A_2 \, \hat{d}(1) + K_2 x(2) \\
&= A_2[A_1 h_0^0 x(0) + K_1 x(1)] + K_2 x(2)
\end{aligned}
$$
$$\text{etc.}$$

If we know the block-processing parameters h_i^p, we can use these equations to solve for the A_i and K_i parameters. But this is not what we want. We need to be able to find the A_i's and K_i's directly.

Table 8.6.1

	h_0	h_1	h_2	h_3	h_4	mse
h^0	0.5					0.5
h^1	0.373	0.2821				0.373
h^2	0.3298	0.2250	0.1702			0.3298
h^3	0.3137	0.2037	0.1389	0.1051		0.3137
h^4	0.3075	0.1955	0.1269	0.0866	0.0655	0.3075

The approach taken by R. E. Kalman in 1960 began with a general model for the signal based on the same idea as the innovations representation. If the signal could be generated by a system with known characteristics, this would allow him to find the optimum recursive estimates. These estimates turn out to be related to the parameters of the system that generated the signal in the first place. For instance, the signal in Example 8.6.1 has an exponential correlation function, and we know that the innovations representation for this signal is a first-order system described by any of the three descriptions,

$$\text{Impulse response:} \qquad h(n) = \alpha^n u(n)$$

$$\text{Transfer function:} \qquad H(z) = \frac{1}{1 - \alpha z^{-1}} \qquad (8.6.3)$$

$$\text{Difference equation:} \qquad s(n) = \alpha s(n-1) + \eta(n)$$

where the difference equation description has white-noise input $\eta(n)$ and output signal $s(n)$. The observations are in the form

$$x(n) = s(n) + w(n) \qquad (8.6.4)$$

The system diagram in Fig. 8.6.2 combines these last two equations to give the signal model. The system driven by white noise $\eta(n)$ produces the signal $s(n)$. The noise $w(n)$, which is independent of the driving force noise $\eta(n)$, is added to $s(n)$ to produce the observation $x(n)$. [We are using two different noise terms here. The $\eta(n)$ noise is part of the innovations model. The $w(n)$ noise is our usual additive noise.]

The mean-square linear recursive estimate for $s(n)$ is given by Eq. 8.6.2. As we will show later, A_n is given in terms of the parameter α by

$$A_n = (1 - K_n)\alpha \qquad (8.6.5)$$

giving

$$\hat{d}(n) = \alpha \hat{d}(n-1) + K_n[x(n) - \alpha \hat{d}(n-1)] \qquad (8.6.6)$$

This allows us to express the recursive estimate as the sum of two terms given by

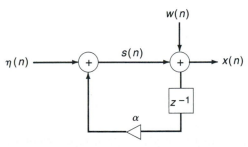

Fig. 8.6.2.   The signal model.

Term 1: $\hat{d}_1(n) = \alpha\,\hat{d}(n-1)$ The forward prediction term

Term 2: $\hat{d}_2(n) = K_n[x(n) - \alpha\,\hat{d}(n-1)]$ The residual or correction term

where α is the parameter in Eq. 8.6.3, and K_n is a time-varying gain to be determined. We can derive a system diagram from this equation, just as we derived Fig. 8.6.2 from Eq. 8.6.3. This is the recursive filter shown in Fig. 8.6.3. For proper choice of K_n and initial estimate $\hat{s}(0)$, it will produce the same estimates as the sequence of block-processing units in Fig. 8.6.1.

The way to derive these proper values of K_n is to set the error orthogonal to the data and solve the resulting equations. That is a rather involved procedure, which we will get to shortly. In the meantime, let us present the procedure for evaluating the parameters and do an example.

The changing mean-square error determines the time-varying gain K_n as follows: Label this mean-square error as $\varepsilon(n)$, where

$$\varepsilon(n) = E[e^2(n)] = E\{[d(n) - A_n\,\hat{d}(n-1) - K_n x(n)]^2\} \quad (8.6.7a)$$

Since the estimate we are using is linear, this is also given by the inner product of the error with $d(n)$,

$$\varepsilon(n) = E[e(n)d(n)] \quad (8.6.7b)$$

Then K_n is given in terms of $\varepsilon(n)$ by

$$K_n = \frac{\varepsilon(n)}{\sigma_W^2} \quad (8.6.8)$$

You may notice we have traded one problem for another. Since K_n depends on $\varepsilon(n)$ in Eq. 8.6.8, and $\varepsilon(n)$ depends on K_n from Eq. 8.6.7a, it seems we are going around in circles. And we are, but we can escape from this loop by the fact that $\varepsilon(n)$ depends on the previous mean-square error $\varepsilon(n-1)$.

$$\varepsilon(n) = \left[\frac{\sigma_\eta^2 + \alpha^2\varepsilon(n-1)}{\sigma_\eta^2 + \sigma_W^2 + \alpha^2\varepsilon(n-1)}\right]\sigma_W^2 \quad (8.6.9)$$

with initial value given by

$$\varepsilon(0) = \frac{\sigma_S^2\sigma_W^2}{\sigma_S^2 + \sigma_W^2} \quad (8.6.10)$$

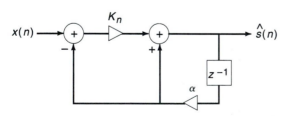

Fig. 8.6.3.   The recursive filter. For proper choice of K_n and initial estimate $\hat{s}(0)$ it is exactly equivalent to the sequence of block processors in Fig. 8.6.1.

We will derive these equations later.

EXAMPLE 8.6.2.   For the conditions in Example 8.6.1, calculate the first three Kalman estimates.

SOLUTION: The initial error term is given by

$$\varepsilon(0) = \frac{\sigma_S^2 \sigma_W^2}{\sigma_S^2 + \sigma_W^2} = \frac{1 \cdot 1}{1 + 1} = 0.5$$

Equation 8.6.8 gives

$$K_0 = \frac{0.5}{1} = 0.5$$

so

$$\hat{s}(0) = K_0 \, x(0) = 0.5 \, x(0)$$

For $n = 1$, the variance σ_η^2 is related to the variance σ_S^2 of the signal by

$$\sigma_\eta^2 = (1 - \alpha^2)\sigma_S^2 = (1 - \alpha^2)R_{SS}(0) = 1 - 0.81 = 0.19$$

Plugging this into Eq. 8.6.9 with $n = 1$ gives

$$\varepsilon(n) = \left[\frac{\sigma_\eta^2 + \alpha^2\varepsilon(n-1)}{\sigma_\eta^2 + \sigma_W^2 + \alpha^2\varepsilon(n-1)}\right]\sigma_W^2$$

$$= \left[\frac{0.19 + 0.81(0.5)}{0.19 + 1 + 0.81(0.5)}\right] = 0.3730$$

Then, from Eq. 8.6.8, $K_1 = 0.3730$. Therefore,

$$\hat{s}(1) = \alpha\hat{s}(0) + K_1[x(1) - \alpha\hat{s}(0)] = 0.373 \, x(1) + 0.2821 \, x(0)$$

which agrees with the coefficients in Example 8.6.1.
    For $n = 2$,

$$\varepsilon(2) = \left[\frac{0.19 + 0.81(0.373)}{0.19 + 1 + 0.81(0.373)}\right] = 0.3298$$

$$K_2 = \frac{\varepsilon(2)}{\sigma_W^2} = 0.3298$$

$$\hat{s}(2) = \alpha\hat{s}(1) + K_2[x(2) - \alpha\hat{s}(1)]$$
$$= 0.3298 \, x(2) + 0.225 \, x(1) + 0.1702 \, x(0)$$

which agrees again with the values in Table 8.6.1. As you can see, these recursive estimates are equal to those obtained by block processing.

Derivation of the Kalman Filter

We wish to show now that the recursion formulas 8.6.6, 8.6.8, and 8.6.9 can be derived by setting the error orthogonal to the data. The data are

$$x(n) = s(n) + w(n) \qquad n = 0, 1, 2, \ldots$$

The estimate is

$$\hat{s}(n) = A_n \hat{s}(n-1) + K_n x(n)$$

and the error is

$$\varepsilon(n) = s(n) - \hat{s}(n)$$

We break the following analysis into two parts, $l = 0$ and $l > 0$.

$l = 0$.   First, for $l = 0$, setting the error orthogonal to the data gives

$$E[e(n)x(n)] = E\{e(n)[s(n) + w(n)]\}$$
$$= E[e(n)s(n)] + E[e(n)w(n)] = 0$$

The first term is $\varepsilon(n)$, the mean-square error, because this term is the inner product of the error with the term to be estimated. Substituting this term and expanding the second term gives

$$\varepsilon(n) + E[e(n)w(n)] = \varepsilon(n) + E\{[s(n) - A_n\hat{s}(n-1) - K_n x(n)]w(n)\}$$
$$= \varepsilon(n) + E[s(n)w(n)] - A_n E[\hat{s}(n-1)w(n)]$$
$$- K_n E[x(n)w(n)] = 0$$

Since the noise is uncorrelated with $s(n)$, and also uncorrelated with past values of $x(n)$, the second and third terms are zero. This gives

$$\varepsilon(n) - K_n \sigma_w^2 = 0$$

or

$$K_n = \frac{\varepsilon(n)}{\sigma_w^2} \qquad \text{(repeated) (8.6.8)}$$

This gives a formula for calculating K_n if we can calculate $\varepsilon(n)$. This is provided by the second part of the derivation, for $l > 0$.

$l > 0$.   Now when we set the error orthogonal to the data we get

$$E[e(n)x(n-l)] = E\{[s(n) - A_n\hat{s}(n-1) - K_n x(n)]x(n-l)\}$$
$$= E[s(n)x(n-l)] - A_n E[\hat{s}(n-1)x(n-l)]$$
$$- K_n E[x(n)x(n-l)] = 0$$

The first term is $R_{SS}(l)$. So is the last term, though it may not be so obvious. The last term is $R_{XX}(l) = R_{SS}(l) + R_{WW}(l)$, but for $l > 0$ we have $R_{WW}(l) = 0$ [because $w(n)$ is white noise]. This gives

$$(1 - K_n)R_{SS}(l) - A_n\{[s(n - 1) - e(n - 1)]x(n - l)\} = 0$$

But $E[e(n - 1)x(n - l)] = 0$ for $l > 0$, giving

$$(1 - K_n)R_{SS}(l) - A_n[s(n - 1)x(n - l)]$$
$$= (1 - K_n)R_{SS}(l) - A_n R_{SS}(l - 1)$$
$$= 0$$

or

$$R_{SS}(l) = \frac{A_n}{1 - K_n} R_{SS}(l - 1) \qquad (8.6.11)$$

This first-order homogeneous difference equation has the solution

$$R_{SS}(l) = R_{SS}(0)\alpha^l, \qquad l > 0 \qquad (8.6.12)$$

where

$$\alpha = \frac{A_n}{1 - K_n} \qquad (8.6.13)$$

which gives us Eq. 8.6.5.

We have succeeded in deriving Eqs. 8.6.5 and 8.6.6. Now we need to show that the recursive error equation 8.6.9 is valid. The mean-square error at each step is the inner product of the error with the quantity we are estimating.

$$\varepsilon(n) = \langle s(n)|e(n)\rangle = E\{s(n)[s(n) - \alpha(1 - K_n)\hat{s}(n - 1) - K_n x(n)]\}$$
$$= E[s^2(n)] - \alpha(1 - K_n)E[s(n)\hat{s}(n - 1)] - K_n E[s(n)x(n)]$$

The first and last expected values are $R_{ss}(0) = \sigma_s^2$. This gives

$$\varepsilon(n) = \sigma_s^2(1 - K_n) - \alpha(1 - K_n)E[s(n)\hat{s}(n - 1)]$$
$$= (1 - K_n)(\sigma_s^2 - \alpha E\{[\alpha s(n - 1) + \eta(n)]\hat{s}(n - 1)\})$$

The input noise $\eta(n)$ is uncorrelated, implying that it is uncorrelated with past values of the signal. Therefore,

$$\varepsilon(n) = (1 - K_n)[\sigma_s^2 - \alpha^2 E\{s(n - 1)\hat{s}(n - 1)\}] \qquad (8.6.14)$$

From this we may now derive Eqs. 8.6.9 and 8.6.10. Initially, with $n = 0$, if we assume that there is no prior estimate $\hat{s}(n - 1)$, we get

$$\varepsilon(n) = (1 - K_n)\sigma_s^2$$

Substituting $K_n = \varepsilon(n)/\sigma_W^2$ from Eq. 8.6.8 and performing a few manipulations gives Eq. 8.6.10.

For $n > 0$ we can use Eq. 8.6.14 to derive Eq. 8.6.9, but first let us display an expression for $\varepsilon(n - 1)$:

$$\varepsilon(n - 1) = E[s(n - 1)e(n - 1)] = E\{s(n - 1)[s(n - 1) - \hat{s}(n - 1)]\}$$
$$= \sigma_S^2 - E[s(n - 1)\hat{s}(n - 1)]$$

Solving for $E[s(n - 1)\hat{s}(n - 1)]$ and substituting into Eq. 8.6.14 gives

$$\varepsilon(n) = (1 - K_n)\{\sigma_S^2 - \alpha^2[\sigma_S^2 - \varepsilon(n - 1)]\} \qquad (8.6.15)$$

This provides a recursive relationship between $\varepsilon(n)$ and $\varepsilon(n - 1)$. Recall that $\sigma_\eta^2 = (1 - \alpha^2)\sigma_S^2$ and that $K_n = \varepsilon(n)/\sigma_W^2$. Substituting these two quantities into Eq. 8.6.15 gives Eq. 8.6.9.

Review

In order to use the Kalman filter derived in this section, we need to know:

1. The signal has exponential autocorrelation function, meaning that a signal with identical second-order statistics can be generated by the system in Fig. 8.6.2. The parameters α and σ_η^2 must be known.
2. The additive noise $w(n)$ is white with known variance σ_W^2.

Then we perform the following steps.

Step 1: Set $n = 0$ and calculate the initial mean-square error:

$$\varepsilon(0) = \frac{\sigma_S^2 \sigma_W^2}{\sigma_S^2 + \sigma_W^2}$$

Step 2: Calculate $K_n = \varepsilon(n)/\sigma_W^2$.
Step 3: Input the data $x(n)$ and calculate the estimate:

$$\hat{s}(n) = \alpha\hat{s}(n - 1) + K_n[x(n) - \alpha\hat{s}(n - 1)]$$

[For $n = 0$ assume $\hat{s}(n - 1) = 0$, so $\hat{s}(0) = K_0 x(0)$.]
Step 4: Let $n = n + 1$.
Step 5: Update the error:

$$\varepsilon(n) = \left[\frac{\sigma_\eta^2 + \alpha^2\varepsilon(n - 1)}{\sigma_\eta^2 + \sigma_W^2 + \alpha^2\varepsilon(n - 1)}\right]\sigma_W^2$$

where $\sigma_\eta^2 = (1 - \alpha^2)\sigma_S^2$.
Step 6: Go to step 2.

8.7   Problems

8.1.   Let X and Y be random variables defined on the $\{H, T\}$ sample space. Define vector addition and scalar multiplication by the rules

$$\begin{bmatrix} x_1 \\ x_2 \end{bmatrix} + \begin{bmatrix} y_1 \\ y_2 \end{bmatrix} = \begin{bmatrix} x_1 + y_1 \\ x_2 + y_2 \end{bmatrix} \qquad a\begin{bmatrix} x_1 \\ x_2 \end{bmatrix} = \begin{bmatrix} ax_1 \\ 0 \end{bmatrix}$$

Is this a vector space? Show why or why not.

8.2. Let A be the ternary field as in Problem 2.8b. Let $X = \{x, y, z\}$. Fill in the blanks in the following table for vector addition to satisfy the properties listed in Definition 8.2.1.

+	x	y	z
x	x	y	z
y	y		
z	z		

Example: $x + z = z$.
Also fill in the following table to define scalar multiplication.

A \ X	x	y	z
0			
1			
2			

8.3. Two random variables X and Y are defined on the sample space given in Fig. 8.7.1. There are three possible experimental outcomes ζ_1, ζ_2, and ζ_3, and they occur with equal probability.
(a) Find the mean-square estimate of X and the resulting mean-square error.
(b) Find the mean-square estimate of X given Y and the resulting mean-square error.

X	-2	0	1
Y	1	0	1
ζ	ζ_1	ζ_2	ζ_3

Fig. 8.7.1

8.4. Repeat Problem 8.3 for the probabilities

$$P(\zeta_1) = P(\zeta_2) = \tfrac{1}{4} \qquad P(\zeta_3) = \tfrac{1}{2}$$

8.5. Let X be a random variable that assumes the values $-2, -1, 0, 1, 2$ with equal probability. Let $Y = X + N$, where N is another random variable, statistically independent from X, that assumes any of the values $-1, 0, 1$ with equal probability. Find the linear estimate of X given Y and the resulting mean-square error.

8.6. Let X and N be as in Problem 8.5, but now we perform the experiment as follows: A number is selected for X and recorded. A number N_1 is chosen and added to X to form $Y_1 = X + N_1$. A second number N_2, independent of N_1, is chosen and added to the same X to form $Y_2 = X + N_2$. If you are shown the two numbers Y_1 and Y_2, how can you estimate X with a linear estimator? Also find the resulting mean-square error and see that it is smaller than in Problem 8.5.

8.7. Repeat Problem 8.6 with three observations Y_1, Y_2, and Y_3.

8.8. Let N be a Gaussian random variable with zero mean and unit variance. Repeat Problem 8.5.

8.9. A continuous-time, stationary stochastic process $X(t)$ with mean 0 has correlation

$$R_{XX}(\tau) = \frac{\sin(\pi\tau)}{\pi\tau}$$

(a) One sample is taken at time t. It is $x(t) = 1.5$. Find the linear estimate of $x(t + \tfrac{1}{2})$ and the resulting mean-square error.
(b) Now suppose that you are given two values of x: $x(t) = 1.5$ and $x(t - \tfrac{1}{2}) = -0.5$. Repeat part (a).

8.10. Which of the following discrete-time systems are minimum-phase systems?

(a) $H_1(z) = \dfrac{z^2 + 1}{z^2 + 0.6z - 0.16}$

(b) $H_2(z) = \dfrac{z(z - 0.5)}{z^2 + 0.6z - 0.16}$

(c) $H_3(z) = \dfrac{z^2 + 1.7z + 0.6}{z^2 + 0.6z - 0.16}$

8.11. For each minimum-phase system in Problem 8.10, do the following:
(a) Find the impulse response of the system.
(b) Find the inverse system and its impulse response.
(c) Convolve the two impulse responses and show that the result is a delta function.

8.12. Factor the following complex spectral density functions into minimum- and maximum-phase components.

(a) $G_1(z) = \dfrac{-2.5}{z - 2.9 + z^{-1}}$

(b) $G_2(z) = \dfrac{z - 2.5 + z^{-1}}{z - 2.05 + z^{-1}}$

8.13. Find the innovations representation for each process in Problem 8.12.

8.14. Factor the following complex spectral density functions into minimum- and maximum-phase components.

(a) $G_1(z) = \dfrac{-4}{z - 4.25 + z^{-1}}$

(b) $G_2(z) = \dfrac{-5z^{-2}}{1 - 2.5z^{-2} + z^{-4}}$

8.15. Find the innovations representation for each process in Problem 8.14.

8.16. Suppose that we are presented with signal plus zero-mean independent noise and we wish to estimate the signal.

$$x(n) = s(n) + w(n)$$
$$d(n) = s(n)$$

where

$$R_{ss}(l) = 2(0.8)^{|l|} \qquad R_{ww}(l) = 0.5\delta(l)$$

Find the optimum realizable IIR Wiener filter and the resulting mean-square error.

8.17. For the conditions in Problem 8.16, let $d(n) = s(n + 1)$ (prediction) and repeat the problem.

8.18. For the conditions in Problem 8.16, let $d(n) = s(n - 1)$ (smoothing) and repeat the problem.

8.19. Suppose that we are presented with signal plus zero-mean independent noise and we wish to estimate the signal.

$$x(n) = s(n) + w(n)$$
$$d(n) = s(n)$$

where

$$R_{ss}(l) = 1.2(0.7)^{|l|} \qquad R_{ww}(l) = 0.2\delta(l)$$

Find the optimum realizable IIR Wiener filter and the resulting mean-square error.

8.20. For the conditions in Problem 8.19, let $d(n) = s(n + 1)$ (prediction) and repeat the problem.

8.21.   For the conditions in Problem 8.19, let $d(n) = s(n - 1)$ (smoothing) and repeat the problem.

8.22.   Suppose that we are presented with signal plus zero-mean independent noise and we wish to estimate the signal.

$$x(n) = s(n) + w(n)$$
$$d(n) = s(n)$$

where

$$R_{ss}(l) = 2(0.8)^{|l|} \qquad R_{ww}(l) = 0.5\delta(l)$$

Find the recursive (Kalman) estimate and the resulting mean-square error for $n = 1, 2, 3,$ and 4.

8.23.   For the conditions in Problem 8.22, let $d(n) = s(n + 1)$ (prediction) and repeat the problem.

8.24.   For the conditions in Problem 8.22, let $d(n) = s(n - 1)$ (smoothing) and repeat the problem.

8.25.   Suppose that we are presented with signal plus zero-mean independent noise and we wish to estimate the signal.

$$x(n) = s(n) + w(n)$$
$$d(n) = s(n)$$

where

$$R_{ss}(l) = 1.2(0.7)^{|l|} \qquad R_{ww}(l) = 0.2\delta(l)$$

Find the recursive (Kalman) estimate and the resulting mean-square error for $n = 1, 2, 3,$ and 4.

8.26.   For the conditions in Problem 8.16, let $d(n) = s(n + 1)$ (prediction) and repeat the problem.

8.27.   For the conditions in Problem 8.16, let $d(n) = s(n - 1)$ (smoothing) and repeat the problem.

Further Reading _____

1. ATHANASIOS PAPOULIS, *Probability, Random Variables, and Stochastic Processes,* 3rd ed., McGraw-Hill, New York, 1991.
2. CHARLES W. THERRIEN, *Discrete Random Signals and Statistical Signal Processing,* Prentice Hall, Englewood Cliffs, NJ, 1992.
3. JAMES A. CADZOW, *Foundations of Digital Signal Processing and Data Analysis,* Macmillan, New York, 1987.

CHAPTER 9 _____

Template Matching

9.1 Eigenvectors _____

Preview

Few would argue against calling Carl Friedrich Gauss (1777–1855) the world's greatest mathematician. It seems there is a Gauss law, Gauss rule, or Gauss formula in every discipline that uses mathematics. Unlike Einstein, Gauss showed early signs of genius. At the age of three he watched his father incorrectly add the wages of the bricklaying crew he supervised. When the sum was complete, the young Carl said, "The tally is wrong, father."

In grade school his teacher asked the class to sum the integers from 1 to 100. Gauss wrote down the numbers in the following order:

1	100
2	99
3	98
⋮	⋮
50	51

He noticed that the sum of each row is 101, and there are 50 such rows. Therefore the sum is 5050. He had discovered an instance of the famous formula $n(n + 1)/2$ for the sum of the first n integers.

For his Ph.D. thesis, completed at the age of 19, Gauss proved what we now call the fundamental theorem of algebra. Every polynomial of the form

$$a_n x^n + a_{n-1} x^{n-1} + \cdots + a_1 x + a_0 = 0$$

has a solution. We will solve such a formula for the eigenvalues of a given $n \times n$ matrix in the following.

The German term *eigen* means characteristic. An eigenvector is a characteristic vector, and an eigenvalue is a characteristic value. These characteristic quantities describe important properties of a matrix, and since a correlation matrix describes important properties of a multivariate distribution, it follows that eigenvalues and eigenvectors can be used to describe the distribution of random variables. Of course, the best description is the n-dimensional cdf or pdf, but this may not be available to us, while the covariance matrix often is available or easily estimated.

In this section we introduce the concept of eigenvectors, describe their use in one application to data analysis, and describe Hotelling's algorithm for finding the eigenvalues and vectors of a symmetric matrix.

Consider what happens when we multiply a matrix A times a vector x. An $n \times n$ matrix A operating on an $n \times 1$ vector x produces an $n \times 1$ vector y:

$$y = Ax \qquad (9.1.1)$$

Usually, y points in a different direction from x. For example, let A and x_1 be given by

$$A = \begin{bmatrix} 3 & 2 \\ -1 & 0 \end{bmatrix} \qquad x_1 = \begin{bmatrix} 1 \\ 1 \end{bmatrix}$$

Then

$$y_1 = Ax_1 = \begin{bmatrix} 5 \\ -1 \end{bmatrix}$$

The vectors x_1 and y_1 are shown in Fig. 9.1.1, and as you can clearly see, they point in different directions.

An eigenvector of the matrix A is a vector x such that Ax points in the same or the opposite direction from x. Thus

$$Ax = \lambda x \qquad (9.1.2)$$

where λ is a scalar. Any vector x that satisfies this equation is called an *eigenvector* of the matrix A, and the corresponding constant λ is called an

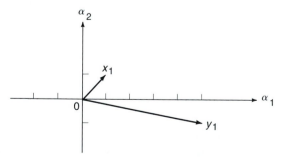

Fig. 9.1.1. x_1 is not an eigenvector of A.

eigenvalue. Using the A matrix above, if x_2 is given by

$$x_2 = \begin{bmatrix} -2 \\ 1 \end{bmatrix}$$

then

$$y_2 = Ax_2 = \begin{bmatrix} -4 \\ 2 \end{bmatrix}$$

which means $\lambda = 2$. This is shown in Fig. 9.1.2.

The eigenvector is not unique. For example, for $\lambda = 2$, there are an infinite number of eigenvectors, some of which are

$$\begin{bmatrix} -2 \\ 1 \end{bmatrix} \quad \begin{bmatrix} -4 \\ 2 \end{bmatrix} \quad \begin{bmatrix} 2 \\ -1 \end{bmatrix} \quad \cdots$$

So when we specify eigenvectors we are really specifying the direction, not a unique vector. The eigenvalues, however, are unique numbers.

Eigenvalues and eigenvectors are found for a given matrix as follows. From the defining equation $Ax = \lambda x$ we have

$$(\lambda I - A)x = O_v \qquad (9.1.3)$$

where O_v stands for the zero vector. Therefore, for nonzero x we have

$$|\lambda I - A| = 0 \qquad (9.1.4)$$

which is called the *characteristic equation.* Carl Gauss proved that such an equation has at least one root, and this fact is now called the *fundamental theorem of algebra.* Thus we know that the characteristic equation has at least one root λ, where this root is a complex or real number. This means that there may be any number of distinct eigenvalues from 1 to n.

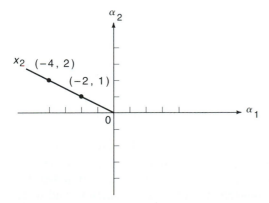

Fig. 9.1.2. x_2 is an eigenvector of A.

The procedure for finding eigenvalues and eigenvectors uses Eqs. 9.1.4 and 9.1.3. First find the distinct eigenvalues with Eq. 9.1.4, then substitute these into Eq. 9.1.3 to find the eigenvectors. The following example illustrates the procedure.

EXAMPLE 9.1.1.   Find the eigenvalues and eigenvectors for the A matrix above.

SOLUTION: The roots of the characteristic equation are found from

$$|\lambda I - A| = \begin{bmatrix} \lambda - 3 & -2 \\ 1 & \lambda \end{bmatrix} = \lambda^2 - 3\lambda + 2 = 0$$

which gives $\lambda_1 = 1$ and $\lambda_2 = 2$. Next the eigenvectors are found by substituting each eigenvalue into Eq. 9.1.3. For $\lambda = 1$,

$$\begin{bmatrix} \lambda - 3 & -2 \\ 1 & \lambda \end{bmatrix}\begin{bmatrix} x_1 \\ x_2 \end{bmatrix} = \begin{bmatrix} -2 & -2 \\ 1 & 1 \end{bmatrix}\begin{bmatrix} x_1 \\ x_2 \end{bmatrix} = \begin{bmatrix} 0 \\ 0 \end{bmatrix}$$

This gives two identical equations, each of which say that $x_1 + x_2 = 0$, or $x_1 = -x_2$. Therefore, for $\lambda = 1$, any of the following are eigenvectors:

$$\begin{bmatrix} -1 \\ 1 \end{bmatrix} \qquad \begin{bmatrix} -2 \\ 2 \end{bmatrix} \qquad \begin{bmatrix} 1 \\ -1 \end{bmatrix} \qquad \cdots$$

For $\lambda = 2$ we use Eq. 9.1.3 to obtain

$$(2I - A)x = \begin{bmatrix} 2 - 3 & -2 \\ 1 & 2 \end{bmatrix}\begin{bmatrix} x_1 \\ x_2 \end{bmatrix} = \begin{bmatrix} -1 & -2 \\ 1 & 2 \end{bmatrix} = \begin{bmatrix} 0 \\ 0 \end{bmatrix}$$

Again we have two identical equations, each of which say that

$$x_1 = -2x_2$$

giving the eigenvector $[-2 \quad 1]'$, which agrees with the eigenvector we "accidentally" found before. Notice that the eigenvalues and eigenvectors are paired. For each vector there is a corresponding value, and although the eigenvectors specify only a direction, the eigenvalues are unique.

Our interest is limited to the eigenvectors of symmetric positive semi-definite matrices because these describe covariance matrices. Therefore let us pause to explain what this means, and to list some properties of these matrices that we will use. A square $n \times n$ matrix A is said to be positive definite if the quadratic form $x'Ax$ is greater than 0 for all nonzero vectors x. Notice that $x'Ax$ is a scalar quantity (a number), and it is this number that must be greater than 0 for any nonzero vector x. The matrix is said to be positive semidefinite if the quadratic product is greater than or equal to 0 for all nonzero vectors x. The quadratic product can be expressed in terms of the elements a_{ij} of A and the components x_i of x as

$$x'Ax = \sum_{i=1}^{n} \sum_{j=1}^{n} a_{ij} x_i x_j \tag{9.1.5}$$

A matrix cannot be positive definite if its determinant is 0.

EXAMPLE 9.1.2.   Determine whether the matrix

$$A = \begin{bmatrix} 2 & -1 \\ -1 & 1 \end{bmatrix}$$

is positive definite.

SOLUTION: Form the quadratic product to get

$$x'Ax = 2x_1^2 - 2x_1 x_2 + x_2^2$$
$$= (x_1 - x_2)^2 + x_1^2$$

This quantity is positive for any choice of nonzero x_1 and x_2, so the matrix A is positive definite.

EXAMPLE 9.1.3.   Test the matrix B to see if it is positive definite, where

$$B = \begin{bmatrix} 2 & 1 & 3 \\ -1 & 1 & 2 \\ 0 & 1 & 1 \end{bmatrix}$$

SOLUTION: Form the quadratic product to get

$$x'Bx = 2x_1^2 + x_2^2 + x_3^2 + 3x_1 x_3 + 3x_2 x_3$$

Notice that if $x_1 = 0$, the quadratic form is

$$x_2^2 + x_3^2 + 3x_2 x_3$$

which is negative for $x_2 = 1$ and $x_3 = -1$. Since we have found one set of values for x_1, x_2, and x_3 that make the quadratic form negative, the matrix A is not positive definite.

A matrix A is symmetric if it equals its own transpose, $A = A'$. The first matrix in the examples above was symmetric and the second was not. All covariance and correlation matrices are symmetric.

Properties of Symmetric Positive Semidefinite Matrices

Let A be a nonnegative (i.e., positive definite or positive semidefinite) square symmetric matrix of size $n \times n$. Then

1. The eigenvalues, which are roots of the characteristic equation $|\lambda I - A| = 0$, are nonnegative real numbers. These roots may be ordered so that $\lambda_1 \geq \lambda_2 \geq \cdots \geq \lambda_n$.
2. Corresponding to each root λ_i there exists an eigenvector q_i such that $Aq_i = \lambda_i q_i$. The set of eigenvectors can be chosen to be orthonormal, so that

$$q_i^t q_i = 1 \quad \text{and} \quad q_i^t q_j = 0 \qquad \text{for } i \neq j$$

3. The original matrix A can be constructed from the eigenvalues and orthonormal eigenvectors by

$$A = \lambda_1 q_1 q_1^t + \lambda_2 q_2 q_2^t + \cdots + \lambda_n q_n q_n^t \tag{9.1.6}$$

and the identity matrix can be constructed from

$$I = q_1 q_1^t + q_2 q_2^t + \cdots + q_n q_n^t \tag{9.1.7}$$

EXAMPLE 9.1.4.   Show that Eqs. 9.1.6 and 9.1.7 hold for the following matrix:

$$A = \begin{bmatrix} 1 & \frac{1}{2} & 0 \\ \frac{1}{2} & 1 & \frac{1}{2} \\ 0 & \frac{1}{2} & 1 \end{bmatrix}$$

SOLUTION: The characteristic equation is given by

$$|\lambda I - A| = \lambda^3 - 3\lambda^2 + 2.5\lambda - 0.5 = 0$$

which gives the three eigenvalues, $\lambda_1 = 1.70711$, $\lambda_2 = 1$, $\lambda_3 = 0.29289$. The three normalized eigenvectors corresponding to each eigenvalue are given by

$$q_1 = \begin{bmatrix} \frac{1}{2} \\ \frac{1}{\sqrt{2}} \\ \frac{1}{2} \end{bmatrix} \qquad q_2 = \begin{bmatrix} \frac{1}{\sqrt{2}} \\ 0 \\ -\frac{1}{\sqrt{2}} \end{bmatrix} \qquad q_3 = \begin{bmatrix} \frac{1}{2} \\ -\frac{1}{\sqrt{2}} \\ \frac{1}{2} \end{bmatrix}$$

Calculating $q_i q_i^t$ for each i gives

$$q_1 q_1^t = \begin{bmatrix} \dfrac{1}{4} & \dfrac{1}{\sqrt{8}} & \dfrac{1}{4} \\[3mm] \dfrac{1}{\sqrt{8}} & \dfrac{1}{2} & \dfrac{1}{\sqrt{8}} \\[3mm] \dfrac{1}{4} & \dfrac{1}{\sqrt{8}} & \dfrac{1}{4} \end{bmatrix}$$

$$q_2 q_2^t = \begin{bmatrix} \dfrac{1}{2} & 0 & -\dfrac{1}{2} \\[3mm] 0 & 0 & 0 \\[3mm] -\dfrac{1}{2} & 0 & \dfrac{1}{2} \end{bmatrix}$$

$$q_3 q_3^t = \begin{bmatrix} \dfrac{1}{4} & -\dfrac{1}{\sqrt{8}} & \dfrac{1}{4} \\[3mm] -\dfrac{1}{\sqrt{8}} & \dfrac{1}{2} & -\dfrac{1}{\sqrt{8}} \\[3mm] \dfrac{1}{4} & -\dfrac{1}{\sqrt{8}} & \dfrac{1}{4} \end{bmatrix}$$

Adding these three matrices gives the identity matrix, in accordance with Eq. 9.1.7. Multiplying by the eigenvalues and then adding gives the matrix A in accordance with Eq. 9.1.6.

Data Structures

Now that we have discussed concentration ellipses in Section 4.4, and eigenvectors in this section, we can tie the two together and increase our understanding of each topic. Suppose that we have an n-dimensional Gaussian random variable

$$X = \begin{bmatrix} X_1 \\ X_2 \\ \vdots \\ X_n \end{bmatrix}$$

with mean m and covariance matrix Λ_X. Now think about selecting samples from this distribution. As we select more and more samples, they will form a cluster about the mean, somewhat like a cloud in the sky. This cloud will be thick near the mean, and thin out as the distance from the mean becomes large. The shape of this cloud will be elliptical, in the sense that contours of constant density will be ellipsoids. In three dimensions this is like a flat football. The major axis through the mean and in the direction of maximum spread is the eigenvector paired with the largest eigenvalue of Λ_X. The next axis, formed by constructing a line through the mean and in the direction of maximum spread perpendicular to the first axis, is the eigenvector paired with the next largest eigenvalue. This process continues until all coordinates are accounted for.

Recall that for a symmetric nonnegative matrix (a covariance matrix) the eigenvalues may be ranked:

$$\lambda_1 \geq \lambda_2 \geq \cdots \geq \lambda_n$$

with corresponding normalized eigenvectors

$$q_1, q_2, \ldots, q_n$$

If Λ_X is positive definite, then none of the eigenvalues is repeated and the eigenvectors are mutually orthogonal. If Λ_X is positive semidefinite, then some of the eigenvalues will be repeated, but it is still possible to find a set of mutually orthogonal eigenvectors. Repeated eigenvalues mean simply that the spread in two different directions is the same. In other words, the football is fully inflated.

Ellipsoids of constant density are described by

$$(x - m)^t \Lambda_X^{-1} (x - m) = d^2 \qquad (9.1.8)$$

where d is the Mahalanobis distance from the mean m to the point x (see Section 4.4). The projections of these samples on the n eigenvectors have a maximum variance on each vector, with the largest variance of λ_1 on q_1, the next largest spread of λ_2 on q_2, etc. In other words, the eigenvalues λ_1, $\lambda_2, \ldots, \lambda_n$ provide a measure of the spread or variance of the distribution in the direction of $q_1, q_2, \ldots, q_n$.

The transformation

$$y = Q^t(x - m) \qquad (9.1.9)$$

where $Q = [q_1 \quad q_2 \quad \cdots \quad q_n]$ is the matrix of eigenvectors, changes Eq. 9.1.8 of an ellipsoid in the x coordinates into the simple form

$$\frac{y_1^2}{\lambda_1} + \frac{y_2^2}{\lambda_2} + \cdots + \frac{y_n^2}{\lambda_n} = d^2 \qquad (9.1.10)$$

where the λ_i's are the eigenvalues of Λ_x. The surface of this ellipsoid intersects the y-coordinate axes at the points

$$y_i = \pm d \sqrt{\lambda_i} \qquad (9.1.11)$$

Thus, if $d = 1$, the distance to the elliptical boundary is the standard deviation in that direction.

From this we see that the eigenvalues and vectors of a covariance matrix give a geometric description of the distribution of a random vector. If this random vector is Gaussian, the description is complete in the sense that the mean and covariance matrix completely describe a Gaussian random vector. If the vector has some other distribution the eigenvalue–vector description is not as complete, nevertheless it may provide valuable information.

EXAMPLE 9.1.5.   Suppose that X is a two-dimensional Gaussian random vector with mean and covariance matrix given by

$$m = \begin{bmatrix} 0 \\ 2 \end{bmatrix} \qquad \Lambda_x = \begin{bmatrix} 2 & 1 \\ 1 & 2 \end{bmatrix}$$

(a) Plot the eigenvectors and concentration ellipse for $d = 1$ on the same graph.

(b) Translate and rotate the axes by the transformation in Eq. 9.1.9 and plot the result.

SOLUTION: (a) The eigenvalues are $\lambda_1 = 3$, $\lambda_2 = 1$ with corresponding eigenvectors

$$q_1 = \begin{bmatrix} \dfrac{1}{\sqrt{2}} \\ \dfrac{1}{\sqrt{2}} \end{bmatrix} \qquad q_2 = \begin{bmatrix} \dfrac{1}{\sqrt{2}} \\ -\dfrac{1}{\sqrt{2}} \end{bmatrix}$$

These vectors along with the concentration ellipse are plotted in Fig. 9.1.3a.

(b) The translation and the rotation are accomplished by Eq. 9.1.9.

$$\begin{bmatrix} y_1 \\ y_2 \end{bmatrix} = \begin{bmatrix} \dfrac{1}{\sqrt{2}} & \dfrac{1}{\sqrt{2}} \\ \dfrac{1}{\sqrt{2}} & -\dfrac{1}{\sqrt{2}} \end{bmatrix} \begin{bmatrix} x_1 \\ x_2 - 2 \end{bmatrix}$$

Substituting this into Eq. 9.1.8 with the proper covariance matrix gives

$$\frac{y_1^2}{\sqrt{3}} + \frac{y_2^2}{1} = 1$$

This ellipse is shown in Fig. 9.1.3b.

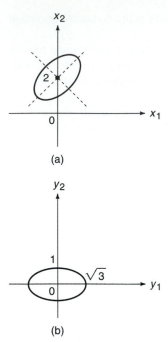

(a)

(b)

Fig. 9.1.3. The vectors and the concentration ellipse for Example 9.1.5.

Hotelling's Iterative Method

In his 1933 paper on principle component analysis, Hotelling developed an iterative procedure for calculating the eigenvalues and vectors of any symmetric matrix. Since correlation matrices are symmetrical, this method may be used in our application. Let A be the given $n \times n$ symmetric matrix whose eigenvalues and vectors are sought. The following is *Hotelling's algorithm.*

Step 1: Take any n-dimensional column vector v_0 as the initial trial vector.

Step 2: Form the product $Av_k = u_k$. (Initially, $k = 0$.)

Step 3: Divide u_k by its largest (in magnitude) element, and denote the result as v_{k+1}. This becomes our new trial vector.

Step 4: Increment k and repeat steps 2 and 3 until two successive trial vectors v_k and v_{k+1} differ by less than a specified amount. (Hotelling proved that this convergence will eventually occur.)

Step 5: The largest element of u_k (that is, the number by which $Av_k = u_k$ was divided to get v_{k+1}) is the largest eigenvalue λ_1 of A.

Step 6: Normalize the last trial vector v_k to obtain q_1, the normalized eigenvector corresponding to λ_1.

Step 7: Form the outer product $q_1 q_1^t$, multiply it by λ_1, and subtract the result from A. The resulting difference matrix is called the first residual matrix A_1.

$$A_1 = A - \lambda_1 q_1 q_1^t \tag{9.1.12}$$

Step 8: Carry out steps 1–7 using A_1 in place of A. The result will be the largest eigenvalue and associated vector for A_1. Because of Eq. 9.1.6, these will be the second largest eigenvalue and its vector for the original matrix A.

This procedure can be repeated until all the eigenvalues and their vectors from Eq. 9.1.6 are found. If the matrix A is positive definite there will be n distinct positive real eigenvalues and their vectors will be mutually orthogonal.

EXAMPLE 9.1.6. Let us apply Hotelling's method to find the first eigenvalue and vector for the matrix A in Example 9.1.3, where

$$A = \begin{bmatrix} 1 & \frac{1}{2} & 0 \\ \frac{1}{2} & 1 & \frac{1}{2} \\ 0 & \frac{1}{2} & 1 \end{bmatrix}$$

SOLUTION:

Step 1: Use initial trail vector $v_0 = [1 \quad 0 \quad 0]^t$.
Step 2: $Av_0 = u_0 = [1 \quad \frac{1}{2} \quad 0]^t$.
Step 3: The largest element in u_0 is 1, so $v_1 = u_0$.
Step 4: Go to step 2 and let $k = 1$. $Av_1 = u_1 = [1.25 \quad 1 \quad 0.25]^t$.
Step 5: Step 3, again: Dividing each element by 1.25 gives $v_2 = [1 \quad 0.8 \quad 0.2]^t$.

Continuing in this way, the sequence of v_i vectors converges to the answer $\lambda_1 = 1.70711$ and $v_k = [0.707 \quad 1 \quad 0.707]^t$. Normalizing v_k gives

$$q_1 = \begin{bmatrix} \dfrac{1}{2} \\[2mm] \dfrac{1}{\sqrt{2}} \\[2mm] \dfrac{1}{2} \end{bmatrix}$$

Here is pseudocode to implement Hotelling's algorithm on the computer.

```
procedure hotel(a) /* find eigenvalues and vectors of a */
define a(N,N), v(N), u(N), old(N)
define λ(N), x(N,N) /* store eigenvalues in λ,   */
    e = 0.0001    /* eigenvectors in x          */
```

```
    k = 1
    while k ≤ N
    begin
        i = 1
        while i ≤ N /* initialize v, u, and old */
        begin
            v(i) = 0
            u(i) = 0
            old(i) = 0
            i = i + 1
        end for i
        v(1) = 1
        d = 10
        while (d > e)
        begin
            call a_times_v(a,v) /*  u = a*v        */
            call compare(u,old) /* compare u to old u */
            maxu = −1000
            i = 1
            while i ≤ N /* find max element of u */
            begin
                old(i) = u(i)
                b = u(i)
                if(b < 0) b = −b
                if(b > maxu) maxu = b
                i = i + 1
            end for i
            i = 1              /* scale elements in u */
            while i ≤ N    /* and store in v     */
            begin
                v(i) = u(i)/maxu
                i = i + 1
            end for i
        end for (d > e)
        λ(k) = maxu
        call normal(v)  /* normalize v to unit length */
        call newa(a,v,k) /* a = a − λ*v*v-transpose    */
        call store(v)  /* store v in x              */
        k = k + 1
    end for k
end hotel

procedure a_times_v(a,v) /* multiply a*v  */
define a(N,N), v(N), u(N)
```

```
    i = 1
    while i ≤ N
    begin
        sum = 0
        j = 1
        while j ≤ N
        begin
            sum = sum + a(i,j)*v(j)
            j = j + 1
        end for j
        u(i) = sum
        i = i + 1
    end for i
    return(u)
end a_times_v

procedure compare(u,old)    /* compare u to old u and */
define u(N), old(N)         /* calculate d        */
    d = 0
    i = 1
    while i ≤ N
    begin
        dif = old(i) − u(i)
        if(dif < 0) dif = −dif
        d = d + dif
        i = i + 1
    end for i
    return(d)
end for compare

procedure normal(v)    /* normalize v */
define v(N)
    sum = 0
    i = 1
    while i ≤ N
    begin
        sum = sum + v(i)*v(i)
        i = i + 1
    end for i
    b = sqrt(sum)
    i = 1
    while i ≤ N
    begin
        v(i) = v(i)/b
```

```
            i = i + 1
      end for i
      return(v)
end normal

procedure newa(a,v,k)  /* a = a - λ*v*v-transpose */
define a(N,N), v(N)
      i = 1
      while i ≤ N
      begin
            j = 1
            while j ≤ N
            begin
                  a(i,j) = a(i,j) - λ(k)*v(i)*v(j)
                  j = j + 1
            end for j
            i = i + 1
      end for i
      return(a)
end newa

procedure store(v,x,k)  /* stores eigenvectors in x */
define v(N), x(N,N)
      i = 1
      while i ≤ N
      begin
            x(k,i) = v(i)
            i = 1 + 1
      end for i
      return(x)
end store
```

Review

The eigenvector of a square matrix is a vector that points in the same (or opposite) direction after being multiplied by the matrix. This leads to the eigenvector equation, Eq. 9.1.2.

$$Ax = \lambda x \qquad \text{(repeated) (9.1.2)}$$

We use this equation in different forms to solve for the eigenvalues and vectors. If

$$(\lambda I - A)x = O_v \qquad \text{(repeated) (9.1.3)}$$

for a nonzero vector x, then we must have

$$|\lambda I - A| = 0 \qquad \text{(repeated) (9.1.4)}$$

which is the characteristic equation. From this we solve for the eigenvalues λ_i, and then for each λ_i we use Eq. 9.1.3 to find the associated vector.

Correlation and covariance matrices are symmetric positive semidefinite, which means that they have several useful properties. We have listed several of these in this section, and have demonstrated one application to data analysis. For these matrices the eigenvalues are nonnegative real numbers, and the eigenvectors can be chosen mutually orthogonal.

Hotelling's algorithm for calculating the eigenvalues and vectors for a given symmetric matrix provides us with a step-by-step procedure. If you do much of this, or if you have a large matrix to work with, you may want to write a computer program to implement the algorithm.

9.2 Matched Filters

Preview

Matched filters and frequency-selective filters form the two basic ways to detect signals. They differ in their objectives because frequency-selective filters detect any signal in the frequency band of interest. Matched filters look for a particular waveshape, hence they are more selective.

Of the several ways to introduce this topic, we have selected to use the eigenvector–Lagrange multiplier approach. We formulate the problem by deriving an eigenvector equation and treating it as a problem in constrained maximization. Perhaps this is not the most direct approach, but it gives us a chance to use the ideas presented in the last section, and it adds to the diversity of tools that you can use.

Given a deterministic signal $s(n)$ of length p in noise $w(n)$, let

$$x(n) = s(n) + w(n) \qquad (9.2.1)$$

be the received signal. The purpose of a matched filter is to determine if the known signal $s(n)$ is present in the sequence $x(n)$. We process the signal $x(n)$ by a causal linear FIR filter. Our goal is to design the filter $h(n)$ to produce a maximum response when $s(n)$ is received. Figure 9.2.1 shows the situation with $s(n)$ a sinusoidal pulse beginning at time n_0 and ending at time $n_p = n_0 + p$. A matched filter produces a peak response when the input sequence $x(n)$ contains the signal $s(n)$. There is no such peak when the input sequence $x(n)$ contains only noise.

Figure 9.2.2 shows the filter $h(n)$ with input $x(n) = s(n) + w(n)$. The

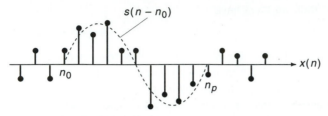

Fig. 9.2.1.   The filter input is $x(n) = s(n) + w(n)$.

response $y(n)$ can be expressed as two terms: $y_s(n)$, the response to signal alone, and $y_w(n)$, which is everything else. By this we mean that $y_s(n)$ is the response when $w(n) = 0$, and $y_w(n)$ is the difference $y(n) - y_s(n)$ when $w(n)$ is present. Our purpose is to design a filter $h(n)$ so that the signal-to-noise ratio (SNR) defined below is maximum at $n = n_p$.

$$\text{SNR} = \frac{y_s^2(n_p)}{E[y_w^2(n_p)]} \tag{9.2.2}$$

Notice that this is not what you might have expected for a "signal-to-noise" ratio. The numerator is a voltage squared, not an average power term. The denominator is an average power, so the ratio is really a voltage squared over an average power term. Despite its shortcomings, this measure of goodness will serve us well, for we can solve for the filter $h(n)$ to maximize this ratio.

Figure 9.2.3 shows the form of the filter with $p + 1$ taps $h(0), h(1), \ldots, h(p)$. Define the four vectors

$$h = \begin{bmatrix} h(0) \\ h(1) \\ \vdots \\ h(p) \end{bmatrix} \quad x = \begin{bmatrix} x(n_0) \\ x(n_0 + 1) \\ \vdots \\ x(n_p) \end{bmatrix} \quad s = \begin{bmatrix} s(n_0) \\ s(n_0 + 1) \\ \vdots \\ s(n_p) \end{bmatrix} \quad w = \begin{bmatrix} w(n_0) \\ w(n_0 + 1) \\ \vdots \\ w(n_p) \end{bmatrix}$$

Then

$$y(n_p) = \sum_{k=0}^{p} h(k) x(n_p - k) = h'\bar{x} \tag{9.2.3}$$

where $\bar{x}$ is the reversal of x. (We introduced the reversal operation in Section 7.4.) We can also express the signal output term y_s and noise output y_w as

$$y_s(n_p) = h'\bar{s} \qquad y_w(n_p) = h'\bar{w}$$

$x(n) = s(n) + w(n) \longrightarrow \boxed{h(n)} \longrightarrow y(n) = y_s(n) + y_w(n)$

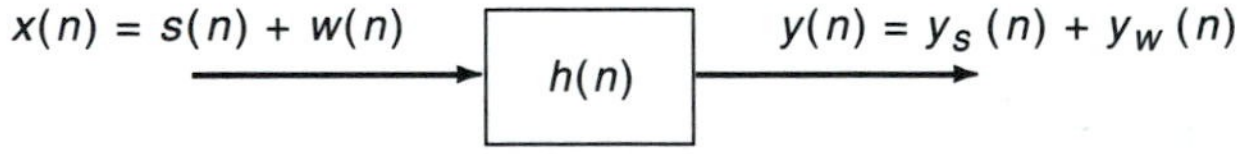

Fig. 9.2.2.   The matched filter $h(n)$.

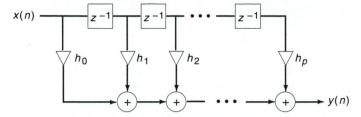

Fig. 9.2.3.   The form of the filter with $p + 1$ taps.

With this notation we can express the SNR in Eq. 9.2.2 in terms of the filter by first writing

$$|y_S(n_p)|^2 = (h^t \bar{s})(\bar{s}^t h)$$

and

$$E[y_W^2(n_p)] = E[h^t \bar{w} \bar{w}^t h] = h^t \bar{R}_{ww} h$$

but $\bar{R}_{ww} = R_{ww}$, since R_{ww} is symmetric. Therefore the SNR is

$$\text{SNR} = \frac{h^t \bar{s} \bar{s}^t h}{h^t R_{ww} h} \tag{9.2.4}$$

Now we must choose h to maximize SNR.

In the following derivation you must understand that for given signal, noise, and filter, the ratio in Eq. 9.2.4 is just a number. It is this number we maximize by choosing h.

EXAMPLE 9.2.1.   Calculate the SNR in Eq. 9.2.4 for the two filters h_1 and h_2 when the signal and noise correlation matrix are given by

$$h_1 = \begin{bmatrix} 1 \\ 2 \\ 1 \end{bmatrix} \qquad h_2 = \begin{bmatrix} 0 \\ 1 \\ 1 \end{bmatrix} \qquad s_1 = \begin{bmatrix} 1 \\ 1 \\ 0 \end{bmatrix} \qquad R_{ww} = \begin{bmatrix} 3 & 2 & 1 \\ 2 & 3 & 2 \\ 1 & 2 & 3 \end{bmatrix}$$

SOLUTION: First, for h_1 the numerator is

$$h_1^t \bar{s} \bar{s}^t h_1 = \begin{bmatrix} 1 & 2 & 1 \end{bmatrix} \begin{bmatrix} 0 \\ 1 \\ 1 \end{bmatrix} \bar{s}^t h = 3^2 = 9$$

and

$$h_1^t R_{ww} h_1 = \begin{bmatrix} 1 & 2 & 1 \end{bmatrix} \begin{bmatrix} 3 & 2 & 1 \\ 2 & 3 & 2 \\ 1 & 2 & 3 \end{bmatrix} \begin{bmatrix} 1 \\ 2 \\ 1 \end{bmatrix} = 36$$

Hence

$$SNR_1 = \frac{9}{36} = \frac{1}{4}$$

For h_2, similar calculations give $SNR_2 = \frac{4}{10}$. Therefore h_2 gives a larger SNR than h_1 and is thus better according to our criterion.

Now we wish to choose h to maximize the SNR in Eq. 9.2.4. As you can see from this example, the choice may not be obvious. The first filter had a larger norm than the second, yet the second gave a larger SNR. We can, however, solve for the optimum filter by changing the unconstrained maximization problem into a constrained problem and using Lagrange multipliers. To do this we constrain the denominator by setting

$$h^t R_{ww} h = 1 \qquad (9.2.5)$$

With this constraint we wish to maximize the numerator by the method of Lagrange. Thus we write

$$L = h^t \bar{s} \bar{s}^t h + \lambda(1 - h^t R_{ww} h) \qquad (9.2.6)$$

Review of Lagrange Multipliers

Let us pause to remind you of the ideas behind Lagrange multipliers. If this does not jog your memory sufficiently, you may need to review this concept in your calculus book. The problem of finding maxima and minima of a function of several variables without added conditions is called a problem in free maxima and minima. With added conditions this problem is called constrained. The method of Lagrange changes a problem in constrained maxima and minima into a problem in free maxima and minima by adding variables. Here is a simple example to illustrate these ideas.

EXAMPLE 9.2.2.   Find the minimum of

$$f(x, y) = x^2 + y^2 \qquad (9.2.7)$$

subject to the constraint that (x, y) is a point on the line

$$2x - y + 1 = 0 \qquad (9.2.8)$$

SOLUTION: Introduce a new variable λ and combine Eq. 9.2.7 with its constraint to obtain

$$L(x, y, \lambda) = (x^2 + y^2) + \lambda(2x - y + 1) \qquad (9.2.9)$$

Finding critical points of Eq. 9.2.7 subject to 9.2.8 is equivalent to finding critical points of 9.2.9 considered as a function of three variables, x, y, and λ. To proceed, calculate the partial derivatives of L and set them to 0.

$$L_x = 2x + 2\lambda = 0$$
$$L_y = 2y - \lambda = 0$$
$$L_\lambda = 2x - y + 1 = 0$$

Note that $L_\lambda = 0$ is the side condition, Eq. 9.2.8. Solving these equations gives $x = -\frac{1}{3}$, $y = \frac{1}{3}$, and $\lambda = \frac{2}{3}$. Therefore the minimum of $f(x, y)$ subject to the constraint is

$$f(x, y) = (-\tfrac{1}{3})^2 + (\tfrac{1}{3})^2 = \tfrac{2}{9}$$

Statement of the Method.   In order to find the critical points of $f(x, y)$ subject to the side condition

$$g(x, y) = 0$$

form the function

$$L(x, y, \lambda) = f(x, y) + \lambda g(x, y)$$

and find the critical points of L considered as a function of three variables, x, y, λ. The method is general and may include several side conditions.

$$L(x, y, \lambda_1, \lambda_2) = f(x, y) + \lambda_1 g_1(x, y) + \lambda_2 g_2(x, y)$$

You can see that Eq. 9.2.6 is formulated as a free problem in two parameters, h and λ. For a given signal vector s and noise correlation matrix R_{ww}, we can use the method of Lagrange directly and find the optimum filter h. We can also take another step without knowing the specific values of s and R_{ww}, which will lead to a more general solution. To this end, take the gradient of L in Eq. 9.2.6 with respect to h:

$$\nabla_h L = \bar{s}\bar{s}^t h - \lambda R_{ww} h = 0$$

or

$$\bar{s}\bar{s}^t h = \lambda R_{ww} h \tag{9.2.10}$$

This is a generalized eigenvector equation. To see why, assume that R_{ww}^{-1} exists (i.e., R_{ww} is positive definite) and premultiply both sides to get

$$(R_{ww}^{-1}\bar{s}\bar{s}^t)h = \lambda h \tag{9.2.11}$$

Now compare this equation to the eigenvector form Eq. 9.1.2. You can see that h is an eigenvector of the matrix in the bracket, and λ is an eigenvalue. All but one of the eigenvalues is 0, because $\bar{s}\bar{s}^t$ has rank 1. The nonzero eigenvalue is the signal-to-noise ratio. To see this, multiply both sides of Eq. 9.2.10 by h^t and solve for λ. This gives the SNR from Eq. 9.2.4.

The optimum h is given by

$$h = R_{ww}^{-1}\bar{s} \tag{9.2.12}$$

To see this, substitute into Eq. 9.2.11 to get

$$R_{WW}^{-1} \bar{s} \bar{s}' R_{WW}^{-1} \bar{s} = \lambda R_{WW}^{-1} \bar{s}$$

or

$$\bar{s}(\bar{s}' R_{WW}^{-1} \bar{s}) = \lambda \bar{s}$$

This equation is valid with $\lambda = \bar{s}' R_{WW}^{-1} \bar{s}$. Thus the signal-to-noise ratio is

$$\text{SNR} = \bar{s}' R_{WW}^{-1} \bar{s} \tag{9.2.13}$$

and the optimum filter is given by

$$h = R_{WW}^{-1} \bar{s} \tag{9.2.14}$$

EXAMPLE 9.2.3.   Find the optimum filter for Example 9.2.1. There the signal and noise correlation matrix were given by

$$s = \begin{bmatrix} 1 \\ 1 \\ 0 \end{bmatrix} \qquad R_{WW} = \begin{bmatrix} 3 & 2 & 1 \\ 2 & 3 & 2 \\ 1 & 2 & 3 \end{bmatrix}$$

SOLUTION: The inverse R_{WW}^{-1} is given by

$$R_{WW}^{-1} = \begin{bmatrix} 0.625 & -0.5 & 0.125 \\ -0.5 & 1.0 & -0.5 \\ 0.125 & -0.5 & 0.625 \end{bmatrix}$$

The optimum filter is given by

$$h = R_{WW}^{-1} \bar{s} = \begin{bmatrix} 0.625 & -0.5 & 0.125 \\ -0.5 & 1.0 & -0.5 \\ 0.125 & -0.5 & 0.625 \end{bmatrix} \begin{bmatrix} 0 \\ 1 \\ 1 \end{bmatrix} = \begin{bmatrix} -0.375 \\ 0.5 \\ 0.125 \end{bmatrix}$$

From Eq. 9.2.4, the steps to find the signal-to-noise ratio are

$$h' \bar{s} \bar{s}' h = \begin{bmatrix} -0.375 & 0.5 & 0.125 \end{bmatrix} \begin{bmatrix} 0 \\ 1 \\ 1 \end{bmatrix} \quad \bar{s}' h = (0.625)^2$$

and

$$h' R_{WW} h = \begin{bmatrix} -0.375 & 0.5 & 0.125 \end{bmatrix} \begin{bmatrix} 3 & 2 & 1 \\ 2 & 3 & 2 \\ 1 & 2 & 3 \end{bmatrix} h$$

$$= \begin{bmatrix} 0 & 1 & 1 \end{bmatrix} \begin{bmatrix} -0.375 \\ 0.5 \\ 0.125 \end{bmatrix} = 0.625$$

Hence,

$$SNR = 0.625$$

Notice that if $w(n)$ is white, then $R_{ww} = \sigma^2 I$ and the optimum filter is

$$h = R_{ww}^{-1} \bar{s} = \frac{1}{\sigma^2} \bar{s} \qquad (9.2.15)$$

This is the origin of the label "matched." If the noise is white, then the filter has the same shape as the signal to which it is matched (except for reversal).

Review

The purpose of a matched filter differs from that of a frequency-selective filter. The matched filter looks for a particular shape in the time domain, while the frequency-selective filter accepts any frequency components in the passband of the filter. Therefore a matched filter is more selective: It will respond only to input signals that have waveshapes similar to the desired signal.

The criterion used to derive the optimum matched filter is to maximize a signal-to-noise ratio. We chose to use the eigenvector–Lagrange multiplier approach, but the most common derivation of the filter uses the Schwartz (CBS) inequality, so you may see this other approach in other books. The end result is the same, however. Equation 9.2.15 depicts the form of the filter for white noise, and Eq. 9.2.14 shows the form for nonwhite noise.

9.3   Basis and Dimension

Preview

The purpose of this section is to introduce you to some general notions about vector spaces, in preparation for the Gram–Schmidt orthogonalization procedure. We will need the concepts of basis, dimension, orthogonality, and spanning set. We have already introduced the idea of a vector space and dot (inner) product in Section 8.2.

Recall that the definition for a vector space is derived from the properties of geometric vectors, and that an inner product imposes considerable structure on a vector space. In particular, an inner product $\langle v_1 | v_2 \rangle$ induces a norm by the formula

$$\|v\| = \langle v | v \rangle^{1/2}$$

This, in turn, induces a metric or distance measure between two vectors given by

$$d(v_1, v_2) = \|v_1 - v_2\|$$

This means that we can do geometry. If we can define a valid inner product for a vector space, then we can measure angles, length, and distance. We can find the angle θ between v_1 and v_2 from

$$\cos \theta = \frac{\langle v_1 | v_2 \rangle}{\| v_1 \| \| v_2 \|}$$

In this section we will expand our knowledge to other aspects of vector spaces, the ideas associated with basis and dimension. You should be warned right away that the dimension of a vector space may not be what you thought it was. For example, the following vector may have any dimension from 1 to N:

$$v = \begin{bmatrix} a_1 \\ \vdots \\ a_N \end{bmatrix}$$

The concept of dimension applies not to a single vector, but to a set of vectors. All vectors lying in a plane have dimension 2, even though they may have N components.

Another muddled concept is that of basis. We usually think of the orthonormal coordinate axes in N-dimensional Euclidean space as a basis, and they are. But we will encounter some strange vector spaces where the concept of basis may not be so simple. For these spaces we must understand that a basis is a set of vectors with two properties: They are linearly independent, and we can express every vector in the space as a linear combination of the basis vectors. So we have several concepts to introduce in this section. To begin at the beginning, we must first define linear independence.

DEFINITION 9.3.1.   A set of vectors $\{v_1, v_2, \ldots, v_n\}$ in the vector space V over the field A is said to be linearly independent if the only scalars $\{a_i\}$ such that

$$\sum_{i=1}^{n} a_i v_i = 0 \tag{9.3.1}$$

are all 0. The set is linearly dependent if it is not independent.

Since only a warped mind can understand this definition upon first reading, you may find it better to use the definition for dependence. A set is linearly dependent if we can express any one of the vectors as a linear combination of the others. A set is either dependent or independent, with no in-between, so we can use either definition.

EXAMPLE 9.3.1.   Test the following three Euclidean vectors for dependence:

$$v_1 = \begin{bmatrix} 2 \\ 1 \\ -1 \end{bmatrix} \qquad v_2 = \begin{bmatrix} -3 \\ 0 \\ 2 \end{bmatrix} \qquad v_3 = \begin{bmatrix} -7 \\ -2 \\ 4 \end{bmatrix}$$

SOLUTION: To show dependence we must be able to write one of these vectors as a linear combination of the other two. For example,

$$v_1 = av_2 + bv_3$$

or

$$\begin{bmatrix} 2 \\ 1 \\ -1 \end{bmatrix} = a \begin{bmatrix} -3 \\ 0 \\ 2 \end{bmatrix} + b \begin{bmatrix} -7 \\ -2 \\ 4 \end{bmatrix}$$

which gives $a = 0.5$ and $b = -0.5$. Therefore the vectors are linearly dependent.

EXAMPLE 9.3.2.   Let P_n denote the space of all polynomials of degree n or less. Test the following polynomials in P_2 for either dependence or independence:

$$p_1(x) = 1 + 2x$$
$$p_2(x) = 2x + 3x^2$$
$$p_3(x) = 1 + 2x + 3x^2$$

SOLUTION: To test for independence we write

$$a_1 p_1(x) + a_2 p_2(x) + a_3 p_3(x) = 0$$

and see if we can find nonzero values of a_i to make this equation true. If so, the vectors are dependent. If not, they are independent. Plugging into this equation gives

$$a_1(1 + 2x) + a_2(2x + 3x^2) + a_3(1 + 2x + 3x^2) = 0$$

Equating like coefficients gives three equations:

$$\begin{bmatrix} 1 & 0 & 1 \\ 2 & 2 & 2 \\ 0 & 3 & 3 \end{bmatrix} \begin{bmatrix} a_1 \\ a_2 \\ a_3 \end{bmatrix} = \begin{bmatrix} 0 \\ 0 \\ 0 \end{bmatrix} \tag{9.3.2}$$

The determinant of the coefficient matrix is 6, meaning that the vectors are independent. Since the determinant of the coefficient matrix is not 0, Eq. 9.3.2 is true only if all the a_i's are 0, which is the condition for independence.

EXAMPLE 9.3.3.   Figure 9.3.1 shows three random variables X_1, X_2, and X_3 defined on the sample space of the die-toss experiment. Determine if these vectors are independent.

SOLUTION: We begin with Eq. 9.3.1 and write

$$a_1 X_1 + a_2 X_2 + a_3 X_3 = 0 \tag{9.3.3}$$

X_1	-2	-1	0	1	2	3
X_2	4	1	0	1	4	9
X_3	2	1	0	1	2	3
ζ	1	2	3	4	5	6

Fig. 9.3.1.   Three random variables defined on the sample space of the die-toss experiment.

We can interpret this equation as follows: Perform the experiment and obtain the outcome; for example, suppose the five-spot turned up. The values of X_1, X_2, and X_3 for this outcome are 2, 4, and 2, respectively. Then we must be able to find scalars a_i to make Eq. 9.3.3 true for this experimental outcome (which is easy to do). Now the hard part comes in repeating the experiment with a different outcome, say, the two-spot. We must use the same scalars a_i to make Eq. 9.3.3 true for these values of the random variables, namely, -1, 1, and 1. If the three random variables are dependent, we can do this for all outcomes. If not, then they are independent.

We can express all this compactly if we write the values of the three random variables as column vectors and express Eq. 9.3.3 as

$$a_1 \begin{bmatrix} -2 \\ -1 \\ 0 \\ 1 \\ 2 \\ 3 \end{bmatrix} + a_2 \begin{bmatrix} 4 \\ 1 \\ 0 \\ 1 \\ 4 \\ 9 \end{bmatrix} + a_3 \begin{bmatrix} 2 \\ 1 \\ 0 \\ 1 \\ 2 \\ 3 \end{bmatrix} = \begin{bmatrix} 0 \\ 0 \\ 0 \\ 0 \\ 0 \\ 0 \end{bmatrix}$$

Writing these equations in matrix form gives

$$\begin{bmatrix} -2 & 4 & 2 \\ -1 & 1 & 1 \\ 0 & 0 & 0 \\ 1 & 1 & 1 \\ 2 & 4 & 2 \\ 3 & 9 & 3 \end{bmatrix} \begin{bmatrix} a_1 \\ a_2 \\ a_3 \end{bmatrix} = \begin{bmatrix} 0 \\ 0 \\ 0 \\ 0 \\ 0 \\ 0 \end{bmatrix} \qquad (9.3.4)$$

This is an overdetermined set of equations, meaning that there are more equations than unknowns. For this situation we need to invoke a theorem from linear algebra. (You can probably find this theorem in your calculus book. Look in the chapter on linear algebra.)

THEOREM.   A homogeneous system of m equations in n unknowns (as in Eq. 9.3.4) has a solution $a_1, a_2, \ldots, a_n$ in which not all the a_i's are 0 if and only if the rank r of the coefficient matrix is less than n.

The rank of a matrix is the largest-dimension square submatrix with nonzero determinant. The first, second, and fourth rows of the coefficient matrix in Eq. 9.3.4 form a 3×3 matrix with nonzero determinant, so the rank of this matrix is 3. The only possible set of a_i's that satisfy the equation are all 0, meaning that the vectors are independent.

EXAMPLE 9.3.4.   Repeat the example above for the three random variables shown in Fig. 9.3.2.

SOLUTION: Repeating the procedure above gives

$$a_1 \begin{bmatrix} -2 \\ -1 \\ 0 \\ 1 \\ 2 \\ 3 \end{bmatrix} + a_2 \begin{bmatrix} 4 \\ 1 \\ 0 \\ 1 \\ 4 \\ 9 \end{bmatrix} + a_3 \begin{bmatrix} 0 \\ -1 \\ 0 \\ 3 \\ 8 \\ 15 \end{bmatrix} = \begin{bmatrix} 0 \\ 0 \\ 0 \\ 0 \\ 0 \\ 0 \end{bmatrix}$$

These vectors are dependent. To show this, write the equations above in matrix form to give

$$\begin{bmatrix} -2 & 4 & 0 \\ -1 & 1 & -1 \\ 0 & 0 & 0 \\ 1 & 1 & 3 \\ 2 & 4 & 8 \\ 3 & 9 & 15 \end{bmatrix} \begin{bmatrix} a_1 \\ a_2 \\ a_3 \end{bmatrix} = \begin{bmatrix} 0 \\ 0 \\ 0 \\ 0 \\ 0 \\ 0 \end{bmatrix} \tag{9.3.5}$$

This coefficient matrix has rank 2, meaning that the vectors are dependent. In fact, we can write

$$2X_1 + X_2 - X_3 = 0$$

X_1	-2	-1	0	1	2	3
X_2	4	1	0	1	4	9
X_3	0	-1	0	3	8	15
ζ	1	2	3	4	5	6

Fig. 9.3.2

This brings us to the point where we can introduce the idea of a basis. A basis is a set of vectors with two properties: We can express every vector as a linear combination of the basis vectors, and this representation is unique. The following definition assures us that these two conditions are met.

DEFINITION 9.3.2.   Given a vector space V, a subset $\{v_1, v_2, \ldots, v_n\}$ of V forms a basis for V if the vectors in the subset are linearly independent, and if the addition of any other nonzero vector from V makes the set dependent.

Let us now consider carefully what this means. Figure 9.3.3 shows vectors in two-dimensional space. Let e_1 and e_2 be candidates for basis vectors. Then we can express v_1 as a linear combination of the two, but not v_2. There are not enough basis vectors because e_1 and e_2 are dependent. Figure 9.3.4 shows three candidate basis vectors, e_1, e_2, and e_3. Here there are enough vectors to express v_1, but there is more than one way to do this. We can express v_1 in terms of e_1 and e_2, or e_1 and e_3, or e_2 and e_3. The problem here is that the vectors are again dependent.

In order to have a basis we must have enough, but not too many, vectors. Definition 9.3.2 specifies just how many this is. Figure 9.3.5 shows just the right amount for a two-dimensional space. Now we can express any vector, such as v_1, as a linear combination of e_1 and e_2 in one and only one way.

The vectors in Example 9.3.1 cannot form a basis for three-dimensional Euclidean space, because they are dependent. The vectors in Example 9.3.2 can be used as a basis for P_2 because they satisfy Definition 9.3.2. The vectors in Example 9.3.3 do not form a basis for that vector space. The three vectors there are independent, but we can add three more vectors before they are forced to be dependent. Therefore a basis for that space contains more than three vectors (in fact, six).

EXAMPLE 9.3.5.   The vectors $\beta = \{p_1, p_2, p_3\}$ in Example 9.3.2 form a basis for P_2. This means that we can express any other polynomial from P_2 in only one way in terms of these three polynomials. Express

$$p_4(x) = 5 + 2x - x^2$$

in terms of the three polynomials from that example.

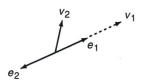

Fig. 9.3.3.   Vectors in two-dimensional space.

Fig. 9.3.4.   Three candidate basis vectors.

SOLUTION: Write $p_4(x) = a_1 p_1(x) + a_2 p_2(x) + a_3 p_3(x)$, or

$$5 + 2x - x^2 = a_1(1 + 2x) + a_2(2x + 3x^2) + a_3(1 + 2x + 3x^2)$$

Equating like coefficients and solving, we get $a_1 = \frac{4}{3}$, $a_2 = -4$, and $a_3 = \frac{11}{3}$. Therefore we can express $p_4(x)$ as

$$p_4(x) = \tfrac{4}{3} p_1(x) - 4 p_2(x) + \tfrac{11}{3} p_3(x)$$

Note that this representation is unique. No other combination of $p_1(x)$, $p_2(x)$, and $p_3(x)$ equals $p_4(x)$.

Another useful term is the coefficient matrix for the vector p_4 with respect to the basis β. This is given by

$$[p_4(x)]_\beta = \begin{bmatrix} \frac{4}{3} \\ -4 \\ \frac{11}{3} \end{bmatrix} \tag{9.3.6}$$

EXAMPLE 9.3.6.   The usual basis for E^3 (three-dimensional Euclidean space) is

$$e_1 = \begin{bmatrix} 1 \\ 0 \\ 0 \end{bmatrix} \qquad e_2 = \begin{bmatrix} 0 \\ 1 \\ 0 \end{bmatrix} \qquad e_3 = \begin{bmatrix} 0 \\ 0 \\ 1 \end{bmatrix}$$

However, an equally valid basis is

$$v_1 = \begin{bmatrix} 1 \\ 1 \\ 0 \end{bmatrix} \qquad v_2 = \begin{bmatrix} 0 \\ 1 \\ 1 \end{bmatrix} \qquad v_3 = \begin{bmatrix} 1 \\ 0 \\ 1 \end{bmatrix}$$

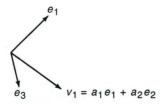

Fig. 9.3.5

The vectors $\{v_1, v_2, v_3\}$ are independent, and if we add any other vector that is different from these (besides the zero vector) the set is no longer independent.

The dimension of a vector space is the number of vectors in the basis. That bears repeating. The dimension is not the number of coordinates, or the number of axes in the space. The dimension is the number of basis vectors. (Of course, the number of coordinates, or the number of axes, are equal to the dimension. But that is not what determines the dimension of a space. It is the number of basis vectors.)

And now one final term. Whenever each vector of a vector space V can be expressed as a linear combination of a set of vectors $\{v_1, v_2, \ldots, v_n\}$ of V, we say the space is spanned by $\{v_1, v_2, \ldots, v_n\}$. This is written

$$V = \text{span}\{v_1, v_2, \ldots, v_n\}$$

Note that this is not the same thing as a basis. There is no requirement that the representation be unique. In other words, a spanning set can contain more vectors than the dimension of the space. Every basis is a spanning set, but so is any other set that contains a basis. For example, all the vectors e_i plus the v_i vectors in Example 9.3.6 form a spanning set for that space.

The vectors in Example 9.3.1 are dependent. Therefore they span a space of dimension less than 3. In fact, they span a space of two dimensions, which is a plane.

Review

A basis provides a unique representation of any vector as a linear combination of the basis vectors. It does this by being independent and maximal. By maximal, we mean that the addition of any other nonzero vector makes the set dependent.

The dimension of a vector space is the number of vectors in the basis set. A spanning set contains a basis for the space spanned.

9.4   Gram–Schmidt Orthogonalization ⎯⎯⎯⎯⎯

Preview

Breaking a problem into independent parts makes it easier to solve. The key word here is "independent": One must carve the bird at its joints. The Gram–Schmidt orthogonalization procedure breaks a set of vectors into independent parts. Although this is not the same as breaking a problem into independent parts, we will show that the two can go together. The purpose of this section is to introduce the Gram–Schmidt procedure and illustrate how it applies to a variety of vector spaces. Then we will use this knowledge in applications to matched filter design and template matching in pattern recognition.

The Gram–Schmidt procedure starts with a given set of vectors (which necessarily span some space) and derives a set of orthonormal vectors from this set. These can then serve as an orthonormal basis for the space spanned by the set.

There are many types of vectors, Euclidean n-dimensional vectors, continuous-time waveforms, discrete-time signals, $M \times N$ images, and random variables, to name a few. We will phrase our presentation in terms of arbitrary vectors v and then indicate through examples how the notation can be changed to account for specific types of vectors.

We begin by assuming that we have a given set of vectors $\{v_1, v_2, \ldots, v_M\}$, where M is the number of vectors in the set. From this set we wish to derive an orthonormal set of vectors $\{e_1, e_2, \ldots, e_D\}$, where D is the dimension of the space spanned by the set $\{v_i\}$. If the vectors $\{v_i\}$ are independent, there will be M orthonormal vectors $\{e_i\}$, and $D = M$. If the $\{v_i\}$ are dependent, then there will be less than M vectors in the orthonormal set, and $D < M$. The Gram–Schmidt procedure accomplishes this by the following steps:

Define

$$e_1 = \frac{v_1}{\|v_1\|} \tag{9.4.1}$$

where

$$\|v_1\| = \langle v_1 | v_1 \rangle^{1/2} \tag{9.4.2}$$

For $j \geq 2$,

$$e_j = \frac{z_j}{\|z_j\|} \tag{9.4.3}$$

where

$$z_j = v_j - \sum_{k=1}^{j-1} \langle e_k | v_j \rangle e_k \tag{9.4.4}$$

This yields a set of basis vectors with the properties

(normal) $\|e_j\| = 1$ for all j (9.4.5)

(orthogonal) $\langle e_k | e_j \rangle = 0$ for $k \neq j$ (9.4.6)

Here is a pictorial guide to help in understanding these formulas. Figure 9.4.1a shows three geometric vectors that all point in different directions, meaning they span a space of three dimensions. We arbitrarily label them v_1, v_2, and v_3, and v_1 and v_2 lie in the plane. The Gram–Schmidt procedure first picks e_1 as a unit-length vector in the direction of v_1, as directed by Eq. 9.4.1. Next we ignore v_3 and concentrate on e_1 and v_2, as shown in Fig. 9.4.1b. We find the component of v_2 that is orthogonal to e_1, change its

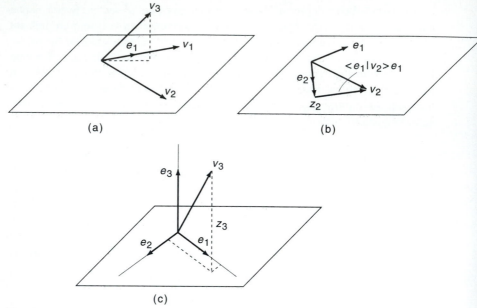

Fig. 9.4.1

length to 1, and call it e_2. The following formulas do this. They are Eqs. 9.4.4 and 9.4.3 with $j = 2$.

$$z_2 = v_2 - \langle e_1 | v_2 \rangle e_1$$

$$e_2 = \frac{z_2}{\|z_2\|}$$

Finally, we construct e_3 by finding the component of v_3 that is orthogonal to the plane containing e_1 and e_2. To do this we set $j = 3$ in Eqs. 9.4.4 and 9.4.3 and get

$$z_3 = v_3 - \langle e_1 | v_3 \rangle e_1 - \langle e_2 | v_3 \rangle e_2 \qquad (9.4.7)$$

as shown in Fig. 9.4.1c. Then we normalize z_3 to get e_3.

Having found an orthonormal basis for the three-dimensional space, we can express an arbitrary vector v in terms of its components with respect to the basis vectors as

$$v = k_1 e_1 + k_2 e_2 + k_3 e_3 \qquad (9.4.8)$$

EXAMPLE 9.4.1.   (a) Given the column vectors x_1, x_2, and x_3 shown below, use the Gram–Schmidt procedure to construct an orthonormal basis for E^3.

$$x_1 = \begin{bmatrix} 1 \\ 1 \\ 0 \end{bmatrix} \qquad x_2 = \begin{bmatrix} 0 \\ 1 \\ 1 \end{bmatrix} \qquad x_3 = \begin{bmatrix} 1 \\ 0 \\ 1 \end{bmatrix}$$

(b) Express the vector $y = [2 \quad -1 \quad 3]'$ in terms of its coordinates with respect to the orthonormal basis $\{e_1, e_2, e_3\}$. (See Eq. 9.3.6.)

SOLUTION: (a) First calculate e_1:

$$\|x_1\| = \langle x_1 | x_1 \rangle^{1/2} = [1 \quad 1 \quad 0] \begin{bmatrix} 1 \\ 1 \\ 0 \end{bmatrix}^{1/2} = \sqrt{2}$$

Therefore

$$e_1 = \frac{x_1}{\sqrt{2}} = \begin{bmatrix} \dfrac{1}{\sqrt{2}} \\ \dfrac{1}{\sqrt{2}} \\ 0 \end{bmatrix}$$

Next, to calculate e_2 we first find z_2, given by

$$z_2 = x_2 - \langle e_1 | x_2 \rangle e_1 = \begin{bmatrix} 0 \\ 1 \\ 1 \end{bmatrix} - \frac{1}{\sqrt{2}} \begin{bmatrix} \dfrac{1}{\sqrt{2}} \\ \dfrac{1}{\sqrt{2}} \\ 0 \end{bmatrix} = \begin{bmatrix} -\dfrac{1}{2} \\ \dfrac{1}{2} \\ 1 \end{bmatrix}$$

The norm of z_2 is $(\frac{3}{2})^{1/2}$, giving

$$e_2 = \frac{z_2}{\|z_2\|} = \begin{bmatrix} -\dfrac{1}{\sqrt{6}} \\ \dfrac{1}{\sqrt{6}} \\ \dfrac{2}{\sqrt{6}} \end{bmatrix}$$

(Here we should check that $\langle e_2 | e_2 \rangle = 1$ and $\langle e_1 | e_2 \rangle = 0$.) Next, z_3 is given by

$$z_3 = x_3 - \langle e_1 | x_3 \rangle e_1 - \langle e_2 | x_3 \rangle e_2 = \begin{bmatrix} \dfrac{2}{3} \\ -\dfrac{2}{3} \\ \dfrac{2}{3} \end{bmatrix}$$

giving

$$e_3 = \begin{bmatrix} \dfrac{1}{\sqrt{3}} \\[2mm] -\dfrac{1}{\sqrt{3}} \\[2mm] \dfrac{1}{\sqrt{3}} \end{bmatrix}$$

(Again, check that $\|e_3\| = 1$, $\langle e_1 | e_3 \rangle = 0$, and $\langle e_2 | e_3 \rangle = 0$.)

(b) In order to express the vector y in terms of $\{e_1, e_2, e_3\}$, we write $y = k_1 e_1 + k_2 e_2 + k_3 e_3$, or

$$\begin{bmatrix} 2 \\ -1 \\ 3 \end{bmatrix} = k_1 \begin{bmatrix} \dfrac{1}{\sqrt{2}} \\[2mm] \dfrac{1}{\sqrt{2}} \\[2mm] 0 \end{bmatrix} + k_2 \begin{bmatrix} -\dfrac{1}{\sqrt{6}} \\[2mm] \dfrac{1}{\sqrt{6}} \\[2mm] \dfrac{2}{\sqrt{6}} \end{bmatrix} + k_3 \begin{bmatrix} \dfrac{1}{\sqrt{3}} \\[2mm] -\dfrac{1}{\sqrt{3}} \\[2mm] \dfrac{1}{\sqrt{3}} \end{bmatrix} \qquad (9.4.9)$$

which gives three equations in three unknowns. We have two ways to solve these equations. First, proceeding without thought we write

$$2 = k_1 \left(\frac{1}{\sqrt{2}} \right) - k_2 \left(\frac{1}{\sqrt{6}} \right) + k_3 \left(\frac{1}{\sqrt{3}} \right)$$

$$-1 = k_1 \left(\frac{1}{\sqrt{2}} \right) + k_2 \left(\frac{1}{\sqrt{6}} \right) - k_3 \left(\frac{1}{\sqrt{3}} \right)$$

$$3 = 0 + k_2 \left(\frac{1}{\sqrt{6}} \right) + k_3 \left(\frac{1}{\sqrt{3}} \right)$$

Solving these three equations gives $k_1 = 1/\sqrt{2}$, $k_2 = 3/\sqrt{6}$, and $k_3 = 1/\sqrt{12}$, or

$$y = \begin{bmatrix} 2 \\ -1 \\ 3 \end{bmatrix} = \frac{1}{\sqrt{2}} \begin{bmatrix} \dfrac{1}{\sqrt{2}} \\[2mm] \dfrac{1}{\sqrt{2}} \\[2mm] 0 \end{bmatrix} + \frac{3}{\sqrt{6}} \begin{bmatrix} -\dfrac{1}{\sqrt{6}} \\[2mm] \dfrac{1}{\sqrt{6}} \\[2mm] \dfrac{2}{\sqrt{6}} \end{bmatrix} + \sqrt{12} \begin{bmatrix} \dfrac{1}{\sqrt{3}} \\[2mm] -\dfrac{1}{\sqrt{3}} \\[2mm] \dfrac{1}{\sqrt{3}} \end{bmatrix}$$

We say that the coordinates of y with respect to the basis $\beta = \{e_1, e_2, e_3\}$ are

$$[y]_\beta = \begin{bmatrix} \dfrac{1}{\sqrt{2}} \\[2mm] \dfrac{3}{\sqrt{6}} \\[2mm] \sqrt{12} \end{bmatrix}$$

A second way to solve these equations is to use the orthogonality of the basis vectors e_1, e_2, and e_3. If we multiply Eq. 9.4.9 by e_1' we get

$$e_1'y = k_1 e_1' e_1 + k_2 e_1' e_2 + k_3 e_1' e_3$$

but the last two terms are 0 because of orthogonality. This gives

$$e_1'y = k_1 = \frac{1}{\sqrt{2}}$$

Of course, we can use similar procedures to find k_2 and k_3.

For this example, the three vectors are independent. Therefore the Gram–Schmidt procedure yields three orthonormal vectors. The vectors v_1, v_2, v_3 in Example 9.3.1 are dependent. Therefore, the Gram–Schmidt procedure should yield only two orthonormal vectors. To see what happens there, let

$$v_1 = \begin{bmatrix} 2 \\ 1 \\ -1 \end{bmatrix} \qquad v_2 = \begin{bmatrix} -3 \\ 0 \\ 2 \end{bmatrix} \qquad v_3 = \begin{bmatrix} -7 \\ -2 \\ 4 \end{bmatrix}$$

as defined in Example 9.3.1. Applying the Gram–Schmidt procedure, we obtain

$$e_1 = \frac{v_1}{\|v_1\|} = \begin{bmatrix} \dfrac{2}{\sqrt{6}} \\[2mm] \dfrac{1}{\sqrt{6}} \\[2mm] -\dfrac{1}{\sqrt{6}} \end{bmatrix}$$

$$z_2 = v_2 - \langle e_1|v_2\rangle e_1 = \begin{bmatrix} -\dfrac{1}{3} \\[2mm] \dfrac{4}{3} \\[2mm] \dfrac{2}{3} \end{bmatrix}$$

$$e_2 = \frac{z_2}{\|z_2\|} = \begin{bmatrix} -\dfrac{1}{\sqrt{21}} \\[2mm] \dfrac{4}{\sqrt{21}} \\[2mm] \dfrac{2}{\sqrt{21}} \end{bmatrix}$$

But when we calculate e_3 we obtain

$$e_3 = v_3 - \langle e_1 | v_3 \rangle e_1 - \langle e_2 | v_3 \rangle e_2 = \begin{bmatrix} 0 \\ 0 \\ 0 \end{bmatrix}$$

which illustrates that the Gram–Schmidt procedure yields only as many orthonormal vectors as there are independent vectors in the set $\{v_i\}$.

Now let us expand these ideas to include other sets of vectors besides geometric vectors and column matrices. A set of waveforms forms a vector space, and once we have defined an inner product we can apply the Gram–Schmidt procedure. Here is an example.

EXAMPLE 9.4.2.   Let P_2 be the set of all polynomials of degree 2 or less. Define the inner product between two polynomials $p_i(t)$ and $p_j(t)$ by

$$\langle p_i(t) | p_j(t) \rangle = \int_0^1 p_i(t)\, p_j(t)\, dt \qquad (9.4.10)$$

Let $p_1(t) = 1$, $p_2(t) = t$, and $p_3(t) = t^2$.
(a) Show that $\{p_1, p_2, p_3\}$ is a basis for P_2.
(b) Construct an orthonormal basis for P_2.

SOLUTION: (a) To show that $\{p_1, p_2, p_3\}$ is a basis for P_2, we must show that the vectors are independent and maximal. To show independence we write

$$a_1 p_1(t) + a_2 p_2(t) + a_3 p_3(t) = 0$$

or

$$a_1 + a_2 t + a_3 t^2 = 0$$

The only coefficients that make this true are all 0, so the set is independent. The set is maximal because we know that the dimension of P_2 is 3, and there are three vectors in the set.

(b) Since $e_1(t)$ is just $p_1(t)$ normalized to 1, we need the formula for the norm. Substituting the inner product from Eq. 9.4.10 into Eq. 8.2.1 gives the norm,

$$\|p_1(t)\| = \left[\int_0^1 p_1^2(t)\, dt \right]^{1/2}$$

Therefore

$$e_1(t) = 1, \qquad 0 < t < 1$$

Next,

$$z_2(t) = p_2(t) - \langle e_1 | p_2 \rangle e_1(t)$$

where

$$\langle e_1 | p_2 \rangle = \int_0^1 t\, dt = 0.5$$

Therefore $z_2(t) = t - 0.5$. When it is normalized, this becomes

$$e_2(t) = \sqrt{12}\,(t - 0.5)$$

Next,

$$z_3(t) = p_3(t) - \langle e_1 | p_3 \rangle e_1(t) - \langle e_2 | p_3 \rangle e_2(t)$$

$$\langle e_1 | p_3 \rangle = \frac{1}{3}$$

$$\langle e_2 | p_3 \rangle = \frac{1}{\sqrt{12}}$$

After normalizing, this gives $e_3(t) = 6\sqrt{5}\,(t^2 - t + \tfrac{1}{6})$.

EXAMPLE 9.4.3.   Given the three waveforms shown in Fig. 9.4.2, derive an orthonormal set of vectors by the Gram–Schmidt procedure.

SOLUTION: We first normalize $s_1(n)$ to obtain $e_1(n)$, shown in Fig. 9.4.3a. Then, to calculate $z_2(n) = s_2(n) - \langle e_1 | s_2 \rangle e_1(n)$, we find

$$\langle e_1(n) | s_2(n) \rangle = 1$$

Therefore $z_2 = s_2 - e_1$, as shown in Fig. 9.4.3b. When this is normalized we obtain $e_2(n)$.

Now, to calculate $z_3(n) = s_3(n) - \langle e_1 | s_3 \rangle e_1(n) - \langle e_2 | s_3 \rangle e_2(n)$, we find

$$\langle e_1 | s_3 \rangle = 0$$

$$\langle e_2 | s_3 \rangle = -\frac{1}{\sqrt{3}}$$

This gives z_3 as shown in Fig. 9.4.3c, and normalizing gives $e_3(n)$.

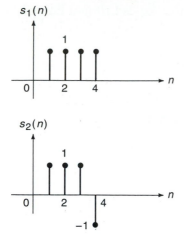

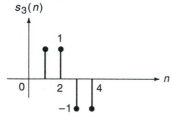

Fig. 9.4.2.   Three waveforms.

Review

The Gram–Schmidt procedure derives a set of orthonormal vectors from an arbitrary set of vectors. If the given set is independent, the number of derived vectors will equal the number in the original set. If not, the number of derived vectors will be less. But in any event, the orthonormal set serves as a basis for the space spanned by the original set.

We have outlined the Gram–Schmidt procedure and demonstrated its use in several settings. The next section implements the Gram–Schmidt procedure in a computer program.

9.5   Gram–Schmidt Computer Program _____

Preview

In this section we will show you how to write a computer program to find the orthonormal basis vectors $\{e_i\}$ and the dot (inner) products $\langle e_i | s_j \rangle$ for a given set of signals $\{s_1(n), s_2(n), \ldots, s_{NS}(n)\}$. You should easily be able to extend this to

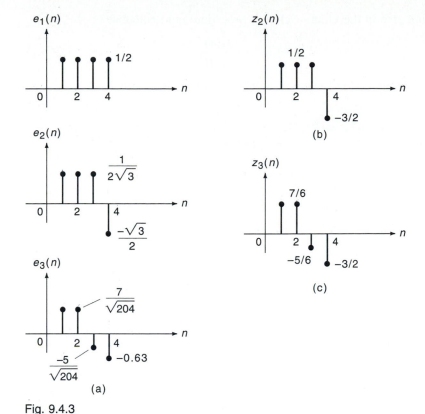

Fig. 9.4.3

other vector spaces besides waveforms. We will use the pseudocode introduced in Section 1.3. We also list FORTRAN and C versions of this code.

Suppose we are given a set of signals $\langle s_1(n), s_2(n), \ldots, s_{NS}(n)\rangle$, where NS stands for the number of signals in the set. Each signal is of length N. This means that we can think of each signal as a vector with N components, or we can think of each as a waveform (time function) of length N. We wish to calculate the set of orthonormal vectors $e_1(n), e_2(n), \ldots, e_{NS}(n)$, in the space spanned by the set of signals. Note that the dimension of this space is at most NS. (Since there are NS vectors, they can point in at most NS different directions.) Therefore we assume that the signals are independent and that $NS \leq N$ in the following. Let us define the variables

$$dp(NS, NS) = \text{dot products between } e_i \text{ and } s_j$$
$$s(NS, N) = NS \text{ signals, each of length } N$$
$$e(NS, N) = \text{orthonormal vectors to be derived}$$

The first step in the Gram–Schmidt procedure is to calculate

$$e_1(n) = \frac{s_1(n)}{\|s_1(n)\|} \tag{9.5.1}$$

The following code segment accomplishes this calculation.

```
procedure gram(s)
define s(NS,N), e(NS,N), dp(NS,NS)
    sum = 0          /* calculate norm of s(1,n) */
    n = 1
    while n ≤ N
    begin
        sum = sum + s(1,n)*s(1,n)
        n = n + 1
    end for n
    norm = sqrt(sum)
    dp(1,1) = norm    /* dot product of e1·s1 */
    n = 1
    while n ≤ N       /* calculate e(1,n)   */
    begin
        e(1,n) = s(1,n)/norm
        n = n + 1
    end for n
```

The first part calculates the denominator of Eq. 9.5.1, and the last part divides each term in $s_1(n)$ by the norm to produce $e_1(n)$. We do not need the dot product between e_1 and s_1 for the Gram–Schmidt procedure, but its calculation is included here because we will need this in future applications. It so happens that $dp(1, 1)$ is always the norm of $s_1(n)$, so this is a convenient place to define it.

The next part of our program calculates $z_j(n)$ and $e_j(n)$ for $j \geq 2$ from Eqs. 9.4.4 and 9.4.3, respectively. Notice that the dot products in Eq. 9.4.4, given by

$$dp(k, j) = \langle e_k(n)|s_j(n)\rangle$$

are always for $k < j$. Here is the pseudocode. As a bonus, the last part of this code also calculates $dp(j, j)$. This is not needed in the Gram–Schmidt procedure, but it is often needed in other applications.

```
    j = 2
    while j ≤ NS
    begin
        n = 1
        while n ≤ N    /* initialize z(n) */
        begin
```

```
            z(n)  =  s(j,n)
            n = n + 1
      end for n
      k = 1
      while k ≤ (j - 1)  /* calculate dp(k,j) */
      begin
            dp(k,j) = 0
            n = 1
            while n ≤ N
            begin
                  dp(k,j)  =  dp(k,j)  +  e(k,n)*s(j,n)
                  n = n + 1
            end for n
            n = 1
            while n ≤ N      /* calculate z */
            begin
                  z(n)  =  z(n)  -  dp(k,j)*e(k,n)
                  n = n + 1
            end for n
      end for k
      sum = 0
      n = 1
      while n ≤ N     /* find norm of z */
      begin
            sum  =  sum  +  z(n)*z(n)
            n = n + 1
      end for n
      norm  =  sqrt(sum)
      n = 1
      while n ≤ N     /* calculate e(j,n) */
      begin
            e(j,n)  =  z(n)/norm
            n = n + 1
      end for n
      dp(j,j) = 0    /*  now for the extra */
      n = 1          /*  added bonus       */
      while n ≤ N    /*  calculate dp(j,j) */
      begin
            dp(j,j)  =  dp(j,j)  +  e(j,n)*s(j,n)
            n = n + 1
      end for n
   end for j
end gram
```

The outer do loop from $j = 2$ to $j = NS$ allows us to calculate all dot products and unit vectors. That is, for each j we calculate all the dot products from $i = 1$ to $i = (j - 1)$, followed by the desired vector $e_j(n)$. There are five second-level do loops. The first sets $z(n) = s(j, n)$ in preparation for subtracting the e_j terms multiplied by the dot products $dp(i, j)$. The second loop accomplishes this subtraction after calculating the appropriate dot product in the first third-level loop. The second third-level loop does the subtracting. The next two second-level loops calculate the norm and divides this into $z(n)$ to obtain $e(j, n)$. The last do loop is not needed for the Gram–Schmidt procedure, but we will have use for the $dp(j, j)$ term it calculates later.

This completes the code for calculating the dot products and the orthonormal vectors e_j. Here is the FORTRAN version, which is provided to make sure you understand how to translate this pseudocode into FORTRAN. The program is in the form of a subroutine to which the main program must supply the parameters NS, N, and the signal set $S(NS, N)$. The routine returns the two arrays $E(NS, N)$ and $DP(NS, NS)$.

```
      SUBROUTINE GRAM(NS, N, S, E, DP)
      DIMENSION S(NS,N), E(NS,N), DP(NS,NS)
      SUM = 0
      DO 10 I=1,N
10    SUM = SUM + S(1,I)*S(1,I)
      NORM = SQRT(SUM)
C
C     CALCULATE DP(1,1)
C
      DP(1,1) = NORM
C
C     CALCULATE E(1,N)
C
      DO 11 I = 1,N
      E(1,I) = S(1,I)/NORM
11    CONTINUE
C
C     SECOND PART OF PROGRAM
C
      DO 12 J = 2, NS
C
C     INITIALIZE Z(N)
C
      DO 13 I = 1, N
      Z(I) = S(J,I)
```

```
13      CONTINUE
C
C       CALCULATE DP AND Z
C
        DO 14 K = 1, J - 1
        DP(K,J) = 0
        DO 15 I = 1, N
        DP(K,J) = DP(K,J) + E(K,I)*S(J,I)
15      CONTINUE
        DO 16 I = 1, N
        Z(I) = Z(I) - DP(K,J)*E(K,I)
16      CONTINUE
14      CONTINUE
C
C       CALCULATE NORM OF Z
C
        SUM = 0
        DO 17 I = 1, N
        SUM = SUM + Z(I)*Z(I)
17      CONTINUE
        NORM = SQRT(SUM)
C
C       CALCULATE E(J,N)
C
        DO 18 I = 1, N
        E(J,I) = Z(I)/NORM
18      CONTINUE
C
C       CALCULATE DP(J,J)
C
        DP(J,J) = 0
        DO 19 I = 1, N
        DP(J,J) = DP(J,J) + E(J,I)*S(J,I)
19      CONTINUE
12      CONTINUE
        RETURN
        END
```

So as not to leave you C programmers out, here is the translation of our pseudocode into C. We assume that the variables NS, N, and the arrays $s[NS][N]$, $e[NS][N]$, and $dp[NS][NS]$ are globally defined, so we do not have to worry with passing pointers to arrays. Otherwise the following routine is self-contained.

```
gram()
{
    int i,j,n,k:
    float sum, z[N], norm;

    sum = 0;        /* find norm of s[1][n] */
    for(n = 0; n < N; n++)
        sum += s[0][n]*s[0][n];
    norm = sqrt( (double) sum );

    dp[0][0] = norm;  /* define de[0][0] for future use */

    for(n = 0; n < N; n++) /* calculate e[0][n]  */
        e[0][n] = s[0][n] / norm;

    for(j = 1; j < NS; j++) /* second part of the program */
    {
        for(n = 0; n < N; n++) /* initialize z[n] */
            z[n] = s[j][n];
        for(k = 0; k < j; k++)
        {
            dp[k][j] = 0; /* calculate dp[k][j]  */
            for(n = 0; n < N; n++)
                dp[k][j] += e[k][n]*s[j][n];

            for(n = 0, n < N; n++) /* calculate z */
                z[n] -= dp[k][j]*e[k][n];
        }

        sum = 0;     /* calculate norm of z */
        for(n = 0; n < N; n++)
            sum += z[n]*z[n];
        norm = sqrt( (double) sum);

        for(n = 0; n < N; n++)    /* calculate e[j][n] */
            e[j][n] = z[n] / norm;

        dp[j][j] = 0; /* not needed for gram-schmidt */
        for(n = 0; n < N; n++)
            dp[j][j] += e[j][n]*s[j][n];
    }
}
```

Review

The program described in this section calculates the orthonormal basis vectors $\{e_i(n)\}$ and the inner products $\langle e_i | s_j \rangle$ from a given set of signals $\{s_1(n), s_2(n), \ldots, s_{NS}(n)\}$. We will use this in the next section to calculate robust filters and to design templates for pattern recognition.

9.6   Template Matching

Preview

One problem with matched filters is that they are too "sharp." They respond well to the signal to which they are matched, but their response falls off sharply to other signals. Many applications would be better served if we could somehow reduce the selectivity of matched filters. Some applications would profit from filters designed to respond equally to more than one signal.

Another problem occurs in template matching. Templates designed for one class may or may not respond well to other classes. Typically, the design for one class ignores the similarities and differences that may exist between that class and other classes. Thus template design should account for these similarities and differences, instead of accounting only for the patterns in one class.

In this section we present a template (or filter) design procedure based on Gram–Schmidt orthogonalization that can be used to accomplish these goals.

Given a finite set of classes $i = 1, 2, \ldots, c$, our goal is to design a template for class i with two properties:

1. When class i is the input, the response is large.
2. When another class is present, the response is small.

Let us assume that the design is based on a training set $S = \{s_1, s_2, \ldots, s_{NS}\}$, where NS is the total number of patterns in the training set. We allow any number of patterns from each class in this set. There may be N_1 patterns from class 1, N_2 patterns from class 2, etc., of course with $N_1 + N_2 + \cdots + N_c = NS$.

These signals are imbedded in a space of dimension no more than NS. This is an important point. The space spanned by a finite set of signals has finite dimension, even though the space of all signals may be of infinite dimension. You can see this because the Gram–Schmidt procedure derives a finite set of basis vectors for the space spanned by the NS signals. (And remember that the dimension of a vector space is the number of basis vectors.)

Mathematically, matched filtering and template matching are the same operation. The common link is the inner product between the input signal (pattern) and the filter (template) response. This inner product takes different forms depending on the type of vector we are dealing with.

If the vectors are continuous-time waveforms, then the filter performs the operation

$$\langle x(t)|s_i(t)\rangle = \int_0^T x(t)\,s_i(t)\,dt \qquad (9.6.1)$$

where the signal $s_i(t)$ has length T, $x(t)$ is the input to the filter, and $s_i(t) = h_i(T - t)$ is the time-reversed impulse response of the filter. Figure 9.6.1a pictures this operation. A bank of c such filters, one for each class, constitutes the receiver. We decide which of the c classes is present based on the largest output.

By the Gram–Schmidt procedure we can derive a set of basis vectors for the space spanned by the set $S = \{s_1(t), s_2(t), \ldots, s_{NS}(t)\}$. This gives up to NS orthonormal basis vectors $E = \{e_1(t), e_2(t), \ldots, e_{NS}(t)\}$. This basis can be used to express each vector $s_i(t)$ in the set S.

$$s_i(t) = k_{i1}e_1(t) + k_{i2}e_2(t) + \cdots + k_{iNS}e_{NS}(t) \qquad (9.6.2)$$

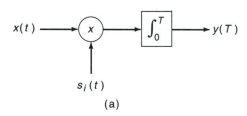

(a)

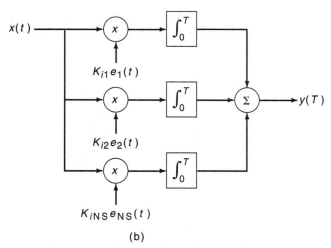

(b)

Fig. 9.6.1.   Two versions of a matched filter.

Figure 9.6.1b shows a structure for the filter in terms of this basis set. The two filter representations are equivalent because of Eq. 9.6.2. A glance at Fig. 9.6.1b illustrates the potential for modifying filter or template design procedures. When we modify the coefficients k_{ij}, we change the filter characteristics. Furthermore, a slight change in any coefficient should cause a slight change in the overall filter response, and a larger change should elicit a larger change in the response. Our purpose in this section is to illustrate a systematic and logical method of effecting these changes to accomplish desired goals. But first here is an example to illustrate the operations in expressing matched filters in terms of their Gram–Schmidt components.

EXAMPLE 9.6.1.   For the three signals shown in Fig. 9.6.2, find (a) a set of orthonormal basis vectors for the space spanned by these signals, and (b) express the three signals in terms of their components with respect to these basis vectors.

SOLUTION: (a) With the inner product defined by

$$\langle v_i(t) | v_j(t) \rangle = \int_0^3 v_i(t)\, v_j(t)\, dt$$

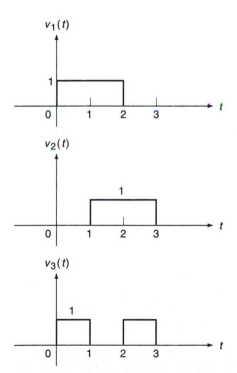

Fig. 9.6.2.   Three signals for Example 9.6.1.

the norm of v_1 is given by (see Eqs. 9.4.1–9.4.4)

$$\|v_1\| = \langle v_1|v_1\rangle^{1/2} = \left[\int_0^2 1 \, dt \right]^{1/2} = \sqrt{2}$$

Therefore

$$e_1(t) = \frac{v_1(t)}{\|v_1(t)\|} = \frac{1}{\sqrt{2}} v_1(t)$$

as shown in Fig. 9.6.3a. Calculating $z_2(t)$ from Eq. 9.4.4 gives

$$z_2(t) = v_2(t) - \langle e_1|v_2\rangle e_1(t)$$

$$= v_2(t) - \frac{1}{\sqrt{2}} e_1(t)$$

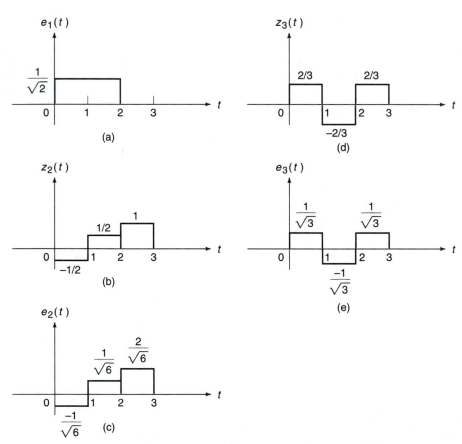

Fig. 9.6.3.   The orthonormal basis vectors for the signal set in Fig. 9.6.2.

The function $z_2(t)$ is shown in Fig. 9.6.3b. Working from the figure, we calculate

$$\|z_2(t)\| = \left[\int_0^1 \left(-\frac{1}{2}\right)^2 dt + \int_1^2 \left(\frac{1}{2}\right)^2 dt + \int_2^3 (1)^2 dt \right]^{1/2} = \left(\frac{3}{2}\right)^{1/2}$$

Therefore

$$e_2(t) = \frac{z_2(t)}{\|z_2(t)\|}$$

is the function shown in Fig. 9.6.3c.

With

$$z_3(t) = v_3(t) - \langle e_1|v_3\rangle e_1(t) - \langle e_2|v_3\rangle e_2(t)$$

we calculate

$$\langle e_1|v_3\rangle = \frac{1}{\sqrt{2}}$$

$$\langle e_2|v_3\rangle = \frac{1}{\sqrt{6}}$$

This gives

$$z_3(t) = v_3(t) - \frac{1}{3}v_1(t) - \frac{1}{3}v_2(t)$$

which is shown in Fig. 9.6.3d. Since $\|z_3\| = 2/\sqrt{3}$, we calculate $e_3(t)$ as shown in Fig. 9.6.3e.

(b) The three waveforms are expressed in terms of these basis vectors as

$$v_i(t) = \sum_{j=1}^{3} k_{ij}e_j(t) = k_{i1}e_1(t) + k_{i2}e_2(t) + k_{i3}e_3(t) \qquad (9.6.3)$$

The advantage of orthonormal basis vectors now comes into play in solving for the k_{ij} coefficients. Multiply both sides of Eq. 9.6.3 by $e_j(t)$ and find the inner product to obtain

$$\langle e_j|v_i\rangle = k_{i1}\langle e_j|e_1\rangle + k_{i2}\langle e_j|e_2\rangle + k_{i3}\langle e_j|e_3\rangle \qquad (9.6.4)$$

Since the e_j basis vectors are orthonormal, all but one of the terms on the right will be 0. The nonzero inner product will be 1, giving

$$k_{ij} = \langle v_i|e_j\rangle \qquad (9.6.5)$$

Therefore, for the nine coefficients in this problem we get

$$k_{11} = \sqrt{2} \qquad k_{12} = 0 \qquad k_{13} = 0$$

$$k_{21} = \frac{1}{\sqrt{2}} \qquad k_{22} = \frac{3}{\sqrt{6}} \qquad k_{23} = 0$$

$$k_{31} = \frac{1}{\sqrt{2}} \qquad k_{32} = \frac{1}{\sqrt{6}} \qquad k_{33} = \frac{2}{\sqrt{3}}$$

If the vectors are discrete-time signals or $N \times 1$ vectors, the inner product operation is given by

$$\langle x(n)|s_i(n)\rangle = \sum_{n=1}^{N} x(n)\, s_i(n) = x^t s_i \tag{9.6.6}$$

Figure 9.6.4a shows this operation. Again we have one such filter for each class, and the filter with the largest response determines our guess. Expressing $s_i(n)$ with respect to the basis generated by the Gram–Schmidt procedure gives

$$s_i(n) = k_{i1}e_1(n) + k_{i2}e_2(n) + \cdots + k_{iNS}e_{NS}(n) \tag{9.6.7}$$

This is pictured in Fig. 9.6.4b.

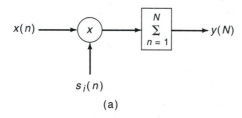

(a)

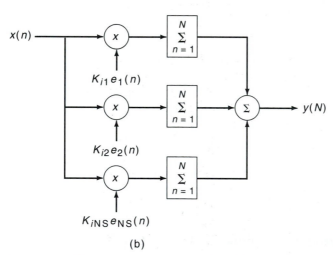

(b)

Fig. 9.6.4.   Two versions of a matched filter.

If the vectors are $M \times N$ images, we represent these images by an $M \times N$ matrix whose entries are the gray-level values. The inner product is (usually) the sum of the products of corresponding pixels.

$$\langle x|s_i \rangle = \sum_{m=1}^{M} \sum_{n=1}^{N} x(m, n) \, s_i(m, n)$$

Figure 9.6.4 can, with the modification of using double sums, once again be used to picture this operation.

We are trying to establish the important point in this introductory material that Figs. 9.6.1 and 9.6.4 picture the operations in template matching no matter what vectors are involved. The inner product can be expressed as in part (a) of these figures, or in terms of the orthonormal basis vectors of part (b).

Matched Filter and Template Design

Templates (or matched filters) can be designed with or without regard to the similarities and differences that exist between classes. We will call templates designed without regard to these similarities and differences "type A" templates, and those designed by taking these into account "type B" templates. Type A templates are usually chosen to be similar to typical patterns in each class.

To illustrate what we mean by type B, consider the design of templates to distinguish between the numerals 0 and 1. The first two parts of Fig. 9.6.5 show typical patterns, while the third part shows the two patterns superimposed. The black pixels are common to each other, and so are the white pixels. The gray pixels are where the differences exist. Since the common pixels cannot be used to distinguish between the two classes (assuming binary images), their assigned values in a template are unimportant. Thus we should concentrate on the gray areas.

Figure 9.6.6 shows type A and type B templates for the two classes. The type A templates are just copies of the patterns in Fig. 9.6.5. The bottom part of Fig. 9.6.6 shows the type B templates. The black, white, and gray shades are supposed to imply different values for those portions of the image. For the class 0 template (bottom left), we place positive values in the black pixels and negative values in the gray pixels. The values in the white pixels,

Fig. 9.6.5

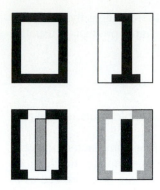

Fig. 9.6.6

which are common to both images, are immaterial if they are the same in both images. Likewise, for the class 1 template shown in the bottom right of the figure, we place positive values in the black pixels, negative values in the gray pixels, and identical values as before in the white pixels.

Let $S = \{s_1(t), s_2(t), \ldots, s_{NS}(t)\}$ be the training set on which we base the design. Let $E = \{e_1(t), e_2(t), \ldots, e_{NS}(t)\}$ be the basis vectors derived by the Gram–Schmidt procedure from the set S. The following design procedure chooses the k_{ij} coefficients in Fig. 9.6.1b or 9.6.4b for the class i template.

Step 1: Normalize all training set patterns so they contain equal energy $\|s_i\|$, and choose a set of desired responses. Let r_j be the desired response to input pattern s_j. Thus $r_j = y(N)$ when the input pattern is s_j. (To design the template for class i we would normally choose all responses to patterns in class i large, and all other responses small.)

Step 2: Let s_1 be the input in Fig. 9.6.4b with desired response r_1, and calculate k_{i1} from the relationship

$$k_{i1}\langle e_1|s_1\rangle = r_1 \qquad (9.6.8)$$

Note that when s_1 is present, only the first branch will respond. This is because all other branch multipliers, e_2 through e_{NS}, are orthogonal to s_1 and their branch responses are 0.

Step 3: To find k_{i2} we use a similar procedure. The desired response of the filter to pattern s_2 is given by

$$k_{i1}\langle e_1|s_2\rangle + k_{i2}\langle e_2|s_2\rangle = r_2 \qquad (9.6.9)$$

The multipliers e_3 through e_{NS} are orthogonal to s_2 and their response to s_2 is 0. Having previously found k_{i1}, we can solve Eq. 9.6.9 for k_{i2}.

Step 4: The remaining gains $k_{ij}, j = 3, 4, \ldots, NS$, are found in a similar manner, adding one more term in the relationship such as Eq. 9.6.9 at each stage. A closed-form expression for the gain at each stage j is given by

$$k_{ij} = \frac{r_j - \sum_{l=1}^{j-1} k_{il}\langle e_l | s_j \rangle}{\langle e_j | s_j \rangle} \qquad (9.6.10)$$

Step 5: The template for class i is then given by

$$\Im_i = \sum_{j=1}^{NS} k_{ij} e_j \qquad (9.6.11)$$

We have one template for each class, $\{\Im_1, \Im_2, \ldots, \Im_c\}$, where c is the number of classes. To design $\Im_i$, the template for class i, select a set of desired responses that makes the response of patterns in class i large and the response of all other classes small. To design $\Im_j$, with $j \neq i$, choose a different set of responses, the set appropriate for this template, and repeat the procedure. Therefore we must repeat the procedure once for each class.

EXAMPLE 9.6.2.   Here is a simple example to illustrate some of the features of template design. The two patterns s_1 and s_2 shown below have in common the four corners and the center pixel. The other four pixels are different in each pattern. Type A templates are just copies of s_1 and s_2. Let us design type B templates to (a) have a response equal to 1 for their class and a response of 0 for the other class, and (b) have a response of 1 for their class and a response of -1 for the other class.

$$s_1 = \begin{bmatrix} 0 & 1 & 0 \\ 0 & 1 & 0 \\ 0 & 1 & 0 \end{bmatrix} \qquad s_2 = \begin{bmatrix} 0 & 0 & 0 \\ 1 & 1 & 1 \\ 0 & 0 & 0 \end{bmatrix}$$

SOLUTION: First we derive the basis vectors, since they are common to both parts (a) and (b) of this example.

$$e_1 = \frac{s_1}{\|s_1\|} = \begin{bmatrix} 0 & \frac{1}{\sqrt{3}} & 0 \\ 0 & \frac{1}{\sqrt{3}} & 0 \\ 0 & \frac{1}{\sqrt{3}} & 0 \end{bmatrix}$$

$$z_2 = s_2 - \langle e_1 | s_2 \rangle e_1$$

where $\langle e_1 | s_2 \rangle = 1/\sqrt{3}$, giving

$$z_2 = \begin{bmatrix} 0 & -\dfrac{1}{3} & 0 \\[2mm] 1 & \dfrac{2}{3} & 1 \\[2mm] 0 & -\dfrac{1}{3} & 0 \end{bmatrix}$$

$$e_2 = \frac{z_2}{\|z_2\|} = \begin{bmatrix} 0 & -0.2041 & 0 \\ 0.61237 & 0.4082 & 0.61237 \\ 0 & -0.2041 & 0 \end{bmatrix}$$

Having derived the two basis vectors e_1 and e_2, let us now design the filters.

(a) The template for class 1, shown in Fig. 9.6.7a, should have response 1 to s_1. Since $\langle e_2 | s_1 \rangle = 0$, this gives

$$k_{11}\langle e_1 | s_1 \rangle = 1$$

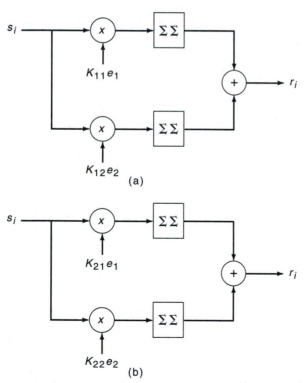

(a)

(b)

Fig. 9.6.7.   (a) Template for class 1 and (b) template for class 2 in Example 9.6.2.

Since $\langle e_1|s_1\rangle = \sqrt{3}$,

$$k_{11} = \frac{1}{\sqrt{3}}$$

We chose 0 for the response to s_2, giving

$$\frac{1}{\sqrt{3}}\langle e_1|s_2\rangle + k_{12}\langle e_2|s_2\rangle = 0$$

Since $\langle e_1|s_2\rangle = 1/\sqrt{3}$ and $\langle e_2|s_2\rangle = 1.63294$, $k_{12} = -0.20413$. This gives the template

$$\Im_1 = k_{11}e_1 + k_{12}e_2 = \begin{bmatrix} 0 & 0.375 & 0 \\ -0.125 & 0.25 & -0.125 \\ 0 & 0.375 & 0 \end{bmatrix}$$

The template for class 2, shown in Fig. 9.6.7b, should have response 0 to s_1. Since the response of the second branch, $k_{22}\langle s_1|e_2\rangle$, is always 0, this gives $k_{21} = 0$. The response to s_2 should be 1, giving

$$k_{22}\langle s_2|e_2\rangle = 1 \quad \text{or} \quad k_{22} = \frac{1}{1.63294} = 0.61239$$

This gives the second template,

$$\Im_2 = k_{21}e_1 + k_{22}e_2 = \begin{bmatrix} 0 & -0.125 & 0 \\ 0.375 & 0.25 & 0.375 \\ 0 & -0.125 & 0 \end{bmatrix}$$

Check:

$$\langle s_1|\Im_1\rangle = 1 \qquad \langle s_1|\Im_2\rangle = 0$$
$$\langle s_2|\Im_1\rangle = 0 \qquad \langle s_2|\Im_2\rangle = 1$$

(b) Here we repeat the procedure above, but with a different set of desired responses. The template for class 1, shown in Fig. 9.6.7a, should have response 1 to s_1. With $\langle e_2|s_1\rangle = 0$ this gives

$$k_{11}\langle e_1|s_1\rangle = 1$$

Since $\langle e_1|s_1\rangle = \sqrt{3}$,

$$k_{11} = \frac{1}{\sqrt{3}}$$

The response to s_2 is now -1, giving

$$\frac{1}{\sqrt{3}}\langle e_1|s_2\rangle + k_{12}\langle e_2|s_2\rangle = -1$$

Since $\langle e_1|s_2\rangle = 1/\sqrt{3}$ and $\langle e_2|s_2\rangle = 1.63294$, $k_{12} = -0.81652$. This gives the template

$$\mathfrak{I}_1 = k_{11}e_1 + k_{12}e_2 = \begin{bmatrix} 0 & 0.5 & 0 \\ -0.5 & 0 & -0.5 \\ 0 & 0.5 & 0 \end{bmatrix}$$

The template for class 2, shown in Fig. 9.6.7b, should have response -1 to s_1. Since the response of the second branch, $k_{22}\langle s_1|e_2\rangle$, is 0 to s_1, this gives

$$k_{21}\langle s_1|e_1\rangle = -1 \qquad \text{or} \qquad k_{21} = -\frac{1}{\sqrt{3}}$$

The response to s_2 should be 1, giving

$$k_{21}\langle s_2|e_1\rangle + k_{22}\langle s_2|e_2\rangle = 1 \qquad \text{or} \qquad k_{22} = \frac{\frac{4}{3}}{1.63294} = 0.81652$$

This gives the second template,

$$\mathfrak{I}_2 = k_{21}e_1 + k_{22}e_2 = \begin{bmatrix} 0 & -0.5 & 0 \\ 0.5 & 0 & 0.5 \\ 0 & -0.5 & 0 \end{bmatrix}$$

Check:

$$\langle s_1|\mathfrak{I}_1\rangle = 1 \qquad \langle s_1|\mathfrak{I}_2\rangle = -1$$
$$\langle s_2|\mathfrak{I}_1\rangle = -1 \qquad \langle s_2|\mathfrak{I}_2\rangle = 1$$

Pseudocode for Template Design

Let us assume that we have a training set consisting of *NS* samples of known classification. We first apply the Gram–Schmidt procedure from Sections 9.4 and 9.5 to derive the orthonormal vectors plus the dot products $dp(i, j) = \langle e_i|s_j\rangle$. Therefore we can supply to the following routine the *NS* samples $\{s(j, n), j = 1, 2, \ldots, NS\}$, the *NS* orthonormal vectors $\{e(j, n), j = 1, 2, \ldots, NS\}$, the dot products, and the correct class for each sample. Here are some of the terms we will use:

NS = number of samples in the training set
N = dimension of each sample
NC = number of classes
$s(NS, N)$ = the *NS* samples (signals), each of length *N*
$e(NS, N)$ = the orthonormal vectors derived from the samples
$dp(NS, NS)$ = dot product between *e* and *s*
$class(NS)$ = the correct classification for each sample

```
procedure template_designs(s,e,dp,class)
define s(NS,N),  e(NS,N),  dp(NS,NS),  class(NS)
define k(NS)          /* filter gains  */
define T(NC,N)        /* the NC templates to be designed */
    cl = 1            /* cl = 1 for the first template */
    while cl ≤ NC    /* calculate gains and design */
    begin             /* template for each class */
        call gains(cl, class)    /* gains k(j) for class cl */
        call template(cl,s,e,k)  /* design template for */
        cl = cl + 1              /* class cl            */
    end for cl
end template_design

procedure gains(cl, class)
define r(NS), k(NS) /* r(j) is the desired response */
    j = 1            /* to signal j                 */
    while j ≤ NS    /* initialize responses         */
    begin
        r(j) = 0
        if(cl = class(j)) r(j) = 1
        j = j + 1
    end for j
    j = 1
    while j ≤ NS            /* calculate gains k   */
    begin                   /* from Eq. 9.6.10     */
        k(j) = r(j)
        i = 1
        while i < j
        begin
            k(j) = k(j) - k(i)*dp(i,j)
            i = i + 1
        end for i
        k(j) = k(j) / dp(j,j)
        j = j + 1
    end for j
    return(k)
end gains

procedure template(cl,s,e,k)
define s(NS,N),  e(NS,N),  k(NS)
define z(N)           /* temporary template storage */
    n = 1
    while n ≤ N      /* initialize z  */
    begin
```

```
            z(n) = 0
            n = n + 1
      end for n
      i = 1
      while i ≤ NS        /* z holds template   */
      begin               /* before normalization */
            n = 1
            while n ≤ N
            begin
                z(n) = z(n) + k(i)*e(i,n)
                n = n + 1
            end for n
            i = i + 1
      end for i
      sum = 0
      n = 1
      while n ≤ N         /* sum = energy in z */
      begin
            sum = sum + z(n)*z(n)
            n = n + 1
      end for n
      n = 1
      while n ≤ N         /* normalize z    */
      begin               /* and store in T */
            T(cl,n) = z(n)/sqrt(sum)
            n = n + 1
      end for n
      return(T)
end template
```

Review

Type A templates are matched to their class without regard to which patterns are in other classes. Type B templates take these other patterns into consideration and achieve both high response to their class and low response to other classes. When all patterns in one class are similar, this works well. When the diversity in each class is wide, however, type B templates do not perform well.

In this section we have developed design procedures for type B templates based on Gram–Schmidt orthogonalization. Together with the previous two sections, this should provide you with a workable design package. The pseudocode presented here has been implemented and tested, so it should be reliable.

9.7 Problems

9.1. Determine which of the following matrices are positive definite, or positive semidefinite:

$$A = \begin{bmatrix} 2 & -1 \\ 1 & 2 \end{bmatrix} \qquad B = \begin{bmatrix} 2 & 1 & 0 \\ 1 & 2 & 1 \\ 0 & 1 & 2 \end{bmatrix} \qquad C = \begin{bmatrix} 3 & 1 & 1 \\ 1 & -1 & 1 \\ 1 & 1 & 3 \end{bmatrix}$$

9.2. Find the eigenvalues and eigenvectors of matrix A in Problem 9.1. Note that these will be complex valued. Check that Eq. 9.1.3 is satisfied for each eigenvalue–eigenvector pair.

9.3. Find the eigenvalues and eigenvectors of matrix B in Problem 9.1. Check that Eqs. 9.1.6 and 9.1.7 are satisfied.

9.4. Find the eigenvalues and eigenvectors for matrix A below. Check that Eqs. 9.1.6 and 9.1.7 are satisfied.

$$A = \begin{bmatrix} 3 & 1 \\ 1 & 3 \end{bmatrix}$$

9.5. If v is an eigenvector of A corresponding to the eigenvalue λ, then v is an eigenvector of A^n for $n \geq 1$. The eigenvalues are different, however, because an eigenvalue of A is not necessarily an eigenvalue of A^n. Determine the relationship of the eigenvalues by finding A^2 and its eigenvalues for the matrix in Problem 9.4.

9.6. Find at least two matrices that have $v_1 = [1 \quad 1]'$ and $v_2 = [1 \quad -1]'$ as their eigenvectors. What are the eigenvectors of

$$A = \begin{bmatrix} 1 & 0 \\ 0 & 1 \end{bmatrix}$$

9.7. The purpose of this problem is to illustrate the relation between the concentration ellipse (Section 4.4) and the eigenvalues of a covariance matrix. Let X be a two-dimensional Gaussian random variable with mean 0 and covariance matrix given by

$$\Lambda_X = \begin{bmatrix} 1 & \rho \\ \rho & 1 \end{bmatrix}$$

(a) Sketch the contours of Mahalanobis distance 1 from the origin if $\rho = 0$, 0.7, and -0.7. (See Example 4.4.5.)

(b) Find the eigenvalues and eigenvectors of the covariance matrix, and sketch the vectors on the same plot. Each eigenvalue is the variance of X in the direction of the corresponding eigenvector. Mark the square root of the eigenvalues on the eigenvectors and check that they are proportional to the ellipse boundary.

9.8.   Given a FIR filter with impulse response h, suppose that the received signal has noise autocorrelation function $R_{ww}(l)$, where

$$h = \begin{bmatrix} h_0 \\ h_1 \\ h_2 \end{bmatrix} = \begin{bmatrix} 1 \\ 2 \\ 1 \end{bmatrix} \qquad R_{ww}(l) = \begin{bmatrix} 2 & 1 & 0 \\ 1 & 2 & 1 \\ 0 & 1 & 2 \end{bmatrix}$$

Two signals are received:

$$s_1 = \begin{bmatrix} 1 \\ 2 \\ 1 \end{bmatrix} \qquad s_2 = \begin{bmatrix} 2 \\ 1 \\ 1 \end{bmatrix}$$

Which signal produces the largest signal-to-noise ratio?

9.9.   Find the values of x, y, z that minimize the function

$$f(x, y, z) = x^2 + y^2 + z^2$$

subject to the constraints

$$x + 2y + z - 1 = 0$$
$$2x - y - 3z - 4 = 0$$

9.10.   (a) Suppose that we wish to detect the signal $s = [1 \quad 2]'$ in white Gaussian noise with variance 0.5. Find the optimum filter coefficients and calculate the signal-to-noise ratio.
(b) Now suppose that the noise is colored with autocorrelation function

$$R_{ww}(l) = 0.8^{|l|} \qquad \text{for all } l$$

Repeat part (a).

9.11.   Repeat Problem 9.10 if the signal is $s = [1 \quad -1]'$.

9.12.   Repeat Problem 9.10 if the signal is $s = [1 \quad 2 \quad -1]'$.

9.13.   Repeat Problem 9.10 if the signal is $s = [1 \quad 1 \quad -1]'$.

9.14.   Let $V = \{v_1, v_2, \ldots\}$ be the set of three-dimensional column vectors, where

$$v_1 = \begin{bmatrix} a_1 \\ a_2 \\ a_3 \end{bmatrix}, \qquad v_2 = \begin{bmatrix} b_1 \\ b_2 \\ b_3 \end{bmatrix}, \qquad \ldots$$

Which of the following are valid inner products? That is, which satisfy the properties in Definition 8.2.2?
(a) $\langle v_1 | v_2 \rangle = \min\{a_i b_i\}$
(b) $\langle v_1 | v_2 \rangle = \max\{a_i b_i\}$

(c) $\langle v_1 | v_2 \rangle = v_1' \bar{v}_2$, where the bar denotes reversal, i.e., the vector v_2 upside down

(d) $\langle v_1 | v_2 \rangle = \text{trace}(v_1 v_2')$, where trace (A) is the sum of elements on the main diagonal of square matrix A.

9.15. Given the following vectors:

$$v_1 = \begin{bmatrix} 1 \\ -2 \\ 1 \end{bmatrix} \quad v_2 = \begin{bmatrix} -3 \\ 2 \\ 1 \end{bmatrix} \quad v_3 = \begin{bmatrix} -1 \\ 0 \\ 1 \end{bmatrix} \quad v_4 = \begin{bmatrix} 1 \\ 2 \\ 3 \end{bmatrix}$$

test the following sets for dependence.

(a) $\{v_1, v_2, v_3\}$
(b) $\{v_1, v_2, v_4\}$
(c) $\{v_1, v_3, v_4\}$
(d) $\{v_2, v_3, v_4\}$

9.16. Let V be the set of all polynomials of degree 2 or less, and define the inner product for this space as

$$\langle v_1 | v_2 \rangle = \int_0^1 v_1(t) \, v_2(t) \, dt$$

Let $v_1(t) = 1$, $v_2(t) = t$, $v_3(t) = t^2$.

(a) Show that $\{v_1, v_2, v_3\}$ is a basis for V.
(b) Construct an orthonormal basis $\{e_1, e_2, e_3\}$ for V.

9.17. Let V be the set of all discrete-time signals of length 3, and define the inner product for this space as

$$\langle v_1 | v_2 \rangle = \sum_{n=1}^{3} v_1(n) \, v_2(n)$$

Let $v_1(n) = \{1, 1, 0\}$, $v_2(n) = \{1, 0, 1\}$, $v_3(n) = \{0, 1, 1\}$.

(a) Show that $\{v_1, v_2, v_3\}$ is a basis for V.
(b) Construct an orthonormal basis $\{e_1, e_2, e_3\}$ for V.

9.18. Let $s_1 \in$ class 1, and $s_2 \in$ class 2. Design a pattern classifier using (a) type A templates, and (b) type B templates for the following patterns:

$$s_1 = \begin{bmatrix} 1 & 1 & 1 \\ 1 & 1 & 0 \\ 1 & 0 & 0 \end{bmatrix} \quad s_2 = \begin{bmatrix} 0 & 0 & 1 \\ 0 & 1 & 1 \\ 1 & 1 & 1 \end{bmatrix}$$

9.19. Suppose that there are three classes in Problem 9.18, with $s_3 \in$ class 3 given by

$$s_3 = \begin{bmatrix} 1 & 1 & 1 \\ 0 & 1 & 1 \\ 0 & 0 & 1 \end{bmatrix}$$

Design three type B templates for the pattern classifier.

9.20.   Suppose that in Problem 9.18 there are two classes, but now we have four samples with two from each class. In addition to the two given in Problem 9.18, we also have $s_3 \in$ class 1, and $s_4 \in$ class 2, where

$$s_3 = \begin{bmatrix} 1 & 1 & 0 \\ 1 & 0 & -1 \\ 0 & -1 & -1 \end{bmatrix} \qquad s_4 = \begin{bmatrix} -1 & -1 & 0 \\ -1 & 0 & 1 \\ 0 & 1 & 1 \end{bmatrix}$$

Design a pattern classifier using (a) type A templates, and (b) type B templates.

Further Reading

1. JAMES A. CADZO, *Foundations of Digital Signal Processing and Data Analysis,* Macmillan, New York, 1987.
2. CHARLES W. THERRIEN, *Discrete Random Signals and Statistical Signal Processing,* Prentice Hall, Englewood Cliffs, NJ, 1992.
3. MAURICE M. TATSUOKA, *Multivariate Analysis,* John Wiley, New York, 1971.
4. RICHARD O. DUDA and PETER E. HART, *Pattern Classification and Scene Analysis,* John Wiley, New York, 1973.

APPENDIX A _____

Table of Normal Curve Areas

The area under the standard normal curve from 0 to z (the shaded area) is $A(z)$.

EXAMPLES. If Z is the standard normal random variable and $z = 1.54$, then

$$A(z) = P\{0 < Z < z\} = 0.4382$$
$$P\{Z > z\} = 0.0618$$
$$P\{Z < z\} = 0.9382$$
$$P\{|Z| < z\} = 0.8764$$

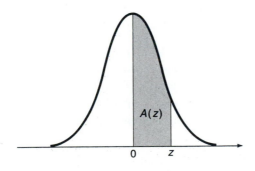

z	.00	.01	.02	.03	.04	.05	.06	.07	.08	.09
0.0	.0000	.0040	.0080	.0120	.0160	.0199	.0239	.0279	.0319	.0359
0.1	.0398	.0438	.0478	.0517	.0557	.0596	.0636	.0675	.0714	.0753
0.2	.0793	.0832	.0871	.0910	.0948	.0987	.1026	.1064	.1103	.1141
0.3	.1179	.1217	.1255	.1293	.1331	.1368	.1406	.1443	.1480	.1517
0.4	.1554	.1591	.1628	.1664	.1700	.1736	.1772	.1808	.1844	.1879
0.5	.1915	.1950	.1985	.2019	.2054	.2088	.2123	.2157	.2190	.2224
0.6	.2257	.2291	.2324	.2357	.2389	.2422	.2454	.2486	.2517	.2549
0.7	.2580	.2611	.2642	.2673	.2704	.2734	.2764	.2794	.2823	.2852
0.8	.2881	.2910	.2939	.2967	.2995	.3023	.3051	.3078	.3106	.3133
0.9	.3159	.3186	.3212	.3238	.3264	.3289	.3315	.3340	.3365	.3389
1.0	.3413	.3438	.3461	.3485	.3508	.3531	.3554	.3577	.3599	.3621
1.1	.3643	.3665	.3686	.3708	.3729	.3749	.3770	.3790	.3810	.3830
1.2	.3849	.3869	.3888	.3907	.3925	.3944	.3962	.3980	.3997	.4015
1.3	.4032	.4049	.4066	.4082	.4099	.4115	.4131	.4147	.4162	.4177
1.4	.4192	.4207	.4222	.4236	.4251	.4265	.4279	.4292	.4306	.4319
1.5	.4332	.4345	.4357	.4370	.4382	.4394	.4406	.4418	.4429	.4441
1.6	.4452	.4463	.4474	.4484	.4495	.4505	.4515	.4525	.4535	.4545
1.7	.4554	.4564	.4573	.4582	.4591	.4599	.4608	.4616	.4625	.4633
1.8	.4641	.4649	.4656	.4664	.4671	.4678	.4686	.4693	.4699	.4706
1.9	.4713	.4719	.4726	.4732	.4738	.4744	.4750	.4756	.4761	.4767
2.0	.4772	.4778	.4783	.4788	.4793	.4798	.4803	.4808	.4812	.4817
2.1	.4821	.4826	.4830	.4834	.4838	.4842	.4846	.4850	.4854	.4857
2.2	.4861	.4864	.4868	.4871	.4875	.4878	.4881	.4884	.4887	.4890
2.3	.4893	.4896	.4898	.4901	.4904	.4906	.4909	.4911	.4913	.4916
2.4	.4918	.4920	.4922	.4925	.4927	.4929	.4931	.4932	.4934	.4936
2.5	.4938	.4940	.4941	.4943	.4945	.4946	.4948	.4949	.4951	.4952
2.6	.4953	.4955	.4956	.4957	.4959	.4960	.4961	.4962	.4963	.4964
2.7	.4965	.4966	.4967	.4968	.4969	.4970	.4971	.4972	.4973	.4974
2.8	.4974	.4975	.4976	.4977	.4977	.4978	.4979	.4979	.4980	.4981
2.9	.4981	.4982	.4982	.4983	.4984	.4984	.4985	.4985	.4986	.4986
3.0	.4987	.4987	.4987	.4988	.4988	.4989	.4989	.4989	.4990	.4990

_____

Gauss–Jordan Matrix Inversion

A square matrix $B = [b_{ij}]$ of dimension $N \times N$ has an inverse if its determinant is not 0. The Gauss–Jordan algorithm for constructing an inverse of such a matrix proceeds as follows.

First construct an augmented matrix $A = [B|I]$, which is a new matrix of N rows and $2N$ columns, where

$A_{ij} = B_{ij}$ for $i = 1, 2, \ldots, N$, and $j = 1, 2, \ldots, N$
$A_{ij} = I$, the identity matrix for $i = 1, 2, \ldots, N$, and $j = N + 1, \ldots, 2N$.

Thus if

$$B = \begin{bmatrix} 1 & 2 & 1 \\ 3 & -1 & 1 \\ 1 & -2 & -3 \end{bmatrix} \quad \text{then} \quad A = \begin{bmatrix} 1 & 2 & 1 & 1 & 0 & 0 \\ 3 & -1 & 1 & 0 & 1 & 0 \\ 1 & -2 & -3 & 0 & 0 & 1 \end{bmatrix}$$

The algorithm now proceeds to operate on A using elementary row, column operations to change the first part of A into the identity matrix while changing the second part into the inverse. Here is the step-by-step algorithm.

Step 1: Let $i = 1$.
Step 2: Set $p = a_{ii}$. Divide each element in the ith row of A by p to make $a_{ii} = 1$.
Step 3: Change the off-diagonal elements in the ith column to 0. To do this, multiply the ith row, which now has $a_{ii} = 1$, by $-a_{ki}$ and add to the kth row ($k \neq i$). This changes the value of each element in the kth row, and in particular changes the a_{ki} element to 0. Repeat this for each row with $k \neq i$.
Step 4: Stop when $i = N$.
Step 5: Set $i = i + 1$ and go to Step 2.

As an example, let us apply this algorithm to the matrix above.

Step 1: $i = 1$.
Step 2: $p = a_{11} = 1$. Division by 1 leaves the top row unchanged.
Step 3: $a_{21} = 3$. Multiply row 1 by $-a_{21}$ to get

$$\begin{bmatrix} -3 & -6 & -3 & -3 & 0 & 0 \\ 3 & -1 & 1 & 0 & 1 & 0 \\ 1 & -2 & -3 & 0 & 0 & 1 \end{bmatrix}$$

Add row 1 to row 2 to get

$$\begin{bmatrix} 1 & 2 & 1 & 1 & 0 & 0 \\ 0 & -7 & -2 & -3 & 1 & 0 \\ 1 & -2 & -3 & 0 & 0 & 1 \end{bmatrix}$$

(We have replaced row 1 by its original values.) Next, $a_{31} = 1$. Multiply row 1 by -1 and add to row 3 to get

$$\begin{bmatrix} 1 & 2 & 1 & 1 & 0 & 0 \\ 0 & -7 & -2 & -3 & 1 & 0 \\ 0 & -4 & -4 & -1 & 0 & 1 \end{bmatrix}$$

Step 4: $i \neq 3$.
Step 5: Set $i = 2$ and go to step 2.
Step 2. $p = a_{22} = -7$. Divide row 2 by -7 to get

$$\begin{bmatrix} 1 & 2 & 1 & 1 & 0 & 0 \\ 0 & 1 & \frac{2}{7} & \frac{3}{7} & -\frac{1}{7} & 0 \\ 0 & -4 & -4 & -1 & 0 & 1 \end{bmatrix}$$

Step 3. Multiply row 2 by -2 and add the result to row 1. This places a 0 in the 1,2 position, and changes other entries in the top row to give

$$\begin{bmatrix} 1 & 0 & \frac{3}{7} & \frac{1}{7} & \frac{2}{7} & 0 \\ 0 & 1 & \frac{2}{7} & \frac{3}{7} & -\frac{1}{7} & 0 \\ 0 & -4 & -4 & -1 & 0 & 1 \end{bmatrix}$$

This should be enough for you to see the general trend. This algorithm is easy to implement in any high-level language. Here is the pseudocode to accomplish Gauss–Jordan matrix inversion. We assume that the *A* matrix has been constructed and is available to the program.

```
procedure Gauss-Jordan
define t(2N)    /* temporary storage for one row of A */
i = 1
while i ≤ N
begin
    p = a(i,i)
    j = 1
    while j ≤ 2N
    begin
        if(p ≠ 0) let a(i,j) = a(i,j)/p
        else print "a diagonal element is zero" and exit
        j = j + 1
    end for j
    k = 1
    while k ≤ N
    begin
        if(k ≠ i)
        {
            p = -a(k,i)
            j = 1
            while j ≤ 2N
            begin
                t(i) = p * a(i,j)
                j = j + 1
            end for j
            j = 1
            while j ≤ 2N
            begin
                a(k,j) = a(k,j) + t(j)
                j = j + 1
            end for j
        }
    end for k
end for i
```

APPENDIX C _____

Symbolic Differentiation

Several times in this book we differentiate with respect to a vector or matrix. This is called symbolic differentiation because it involves finding the partial derivative of a function with respect to each element of the vector or matrix. The resulting derivatives are then arranged to form a vector or matrix of the same type. If $f(x)$ is a function of the elements of vector x, the symbolic derivative $\partial f(x)/\partial x$ is the column vector whose elements are

$$\frac{\partial f}{\partial x_1}, \quad \frac{\partial f}{\partial x_2}, \quad \ldots, \quad \frac{\partial f}{\partial x_n}$$

The most common forms for the function f are the inner products $x'y$, $x'Ay$, and the quadratic form $x'Ax$. It is therefore useful to have a general formula for writing out the symbolic derivatives of these forms. We show how it works for the simplest cases when x is a two-dimensional vector and A is a 2×2 matrix. This then generalizes to any dimension. Knowing how these forms can be derived for these specific functions will allow you to easily derive the appropriate form for different functions.

We begin with $x'Ax$. Since

$$x'Ax = [x_1 \quad x_2] \begin{bmatrix} a_{11} & a_{12} \\ a_{21} & a_{22} \end{bmatrix} \begin{bmatrix} x_1 \\ x_2 \end{bmatrix} = a_{11}x_1^2 + (a_{12} + a_{21})x_1x_2 + a_{22}x_2^2$$

the derivatives are given by

$$\frac{\partial}{\partial x_1}(x'Ax) = 2a_{11}x_1 + (a_{12} + a_{21})x_2$$

$$\frac{\partial}{\partial x_2}(x'Ax) = (a_{12} + a_{21})x_1 + 2a_{22}x_2$$

Collecting these into a column vector gives

$$\frac{\partial}{\partial x}(x'Ax) = \begin{bmatrix} 2a_{11}x_1 + (a_{12} + a_{21})x_2 \\ (a_{12} + a_{21})x_1 + 2a_{22}x_2 \end{bmatrix} = \begin{bmatrix} 2a_{11} & (a_{12} + a_{21}) \\ (a_{12} + a_{21}) & 2a_{22} \end{bmatrix}\begin{bmatrix} x_1 \\ x_2 \end{bmatrix}$$

$$= \left(\begin{bmatrix} a_{11} & a_{12} \\ a_{21} & a_{22} \end{bmatrix} + \begin{bmatrix} a_{11} & a_{21} \\ a_{12} & a_{22} \end{bmatrix}\right)\begin{bmatrix} x_1 \\ x_2 \end{bmatrix} = (A + A')x$$

Therefore our first result in symbolic differentiation says

$$\frac{\partial}{\partial x}(x'Ax) = (A + A')x \tag{C.1}$$

Although we derived this for two-dimensional vectors, the result extends without modification to any finite dimension.

Similar derivations lead to the two results

$$\frac{\partial}{\partial x}(x'Ay) = Ay \tag{C.2}$$

$$\frac{\partial}{\partial x}(y'Ax) = A'y \tag{C.3}$$

Notice that we are following the dictates of our original intent, namely to arrange the final result in a matrix of the same type as x. That is the reason this last result is of the form $A'y$, which is a column vector, rather than the form $y'A$, which would be a row vector. Likewise, we have

$$\frac{\partial}{\partial x}(x'y) = y \tag{C.4}$$

$$\frac{\partial}{\partial x}(y'x) = y \tag{C.5}$$

These are not the only formulas you may need, but you should be able to derive any needed derivatives with the method outlined for the quadratic form. Let us now consider symbolic differentiation with respect to a matrix.

Symbolic differentiation with respect to a matrix follows the same concept as differentiation with respect to a column vector. If $f(A)$ is a function of the elements of an $n \times m$ matrix A, the symbolic derivative is the $n \times m$ matrix whose elements are

$$\frac{\partial f}{\partial a_{11}}, \qquad \frac{\partial f}{\partial a_{12}}, \qquad \cdots, \qquad \frac{\partial f}{\partial a_{nm}}$$

These are arranged in the same order as the elements of matrix A, with $\partial f/\partial a_{ij}$ replacing element a_{ij}. We will again derive the result for the two-dimensional case, for the result applies to any finite dimension.

Let

$$Q = x'Ax = a_{11}x_1^2 + (a_{12} + a_{21})x_1x_2 + a_{22}x_2^2$$

Then

$$\frac{\partial Q}{\partial a_{11}} = x_1^2 \qquad \frac{\partial Q}{\partial a_{12}} = x_1x_2 \qquad \frac{\partial Q}{\partial a_{21}} = x_1x_2 \qquad \frac{\partial Q}{\partial a_{22}} = x_2^2$$

Collecting these in a 2 × 2 matrix gives

$$\frac{\partial Q}{\partial A} = \begin{bmatrix} x_1^2 & x_1x_2 \\ x_1x_2 & x_2^2 \end{bmatrix} = \begin{bmatrix} x_1 \\ x_2 \end{bmatrix} [x_1 \quad x_2] = xx'$$

Therefore the general result is

$$\frac{\partial}{\partial A}(x'Ax) = xx' \qquad \text{if } A \text{ is nonsymmetrical} \qquad (C.6)$$

Hidden is the assumption that A is nonsymmetrical. If A is symmetrical (i.e., $A = A'$), we get a different result, because $a_{12} = a_{21}$. To see how this makes a difference, let $b = a_{12} = a_{21}$ and write

$$Q = a_{11}x_1^2 + 2bx_1x_2 + a_{22}x_2^2$$

Now

$$\frac{\partial Q}{\partial a_{11}} = x_1^2 \qquad \frac{\partial Q}{\partial a_{12}} = \frac{\partial Q}{\partial a_{21}} = \frac{\partial Q}{\partial b} = 2x_1x_2 \qquad \frac{\partial Q}{\partial a_{22}} = x_2^2$$

Placing these results into a matrix gives

$$\frac{\partial Q}{\partial A} = \begin{bmatrix} x_1^2 & 2x_1x_2 \\ 2x_1x_2 & x_2^2 \end{bmatrix} = \begin{bmatrix} x_1 \\ x_2 \end{bmatrix} [x_1 \quad x_2] + \begin{bmatrix} 0 & x_1x_2 \\ x_1x_2 & 0 \end{bmatrix}$$

This somewhat messy form can be summarized by doubling xx' and subtracting the diagonal elements x_i^2. This gives the general form

$$\frac{\partial}{\partial A}(x'Ax) = 2xx' - D(xx') \qquad (C.7)$$

where $D(xx')$ denotes the diagonal matrix with $x_1^2, x_2^2, \ldots, x_n^2$ as its diagonal elements (all off-diagonal elements are 0). Therefore, for symmetrical matrices we should use Eq. C.7, while C.6 is appropriate for nonsymmetrical matrices.

Another common form is $x'Ay$. The symbolic derivative with respect to A is given by

$$\frac{\partial}{\partial A} x'Ay = xy' \qquad\qquad \text{for nonsymmetrical } A \qquad (C.8)$$

$$\frac{\partial}{\partial A} x'Ay = xy' + yx' - D(xy') \qquad \text{for symmetrical } A \qquad (C.9)$$

From these examples we see that it makes a difference whether or not A is symmetrical. You should now be able to successfully derive any symbolic derivative.

Index